Principles of Plant Breeding for PGR Management

Principles of Plant Breeding for PGR Management

Deepak Kumar Mishra
Plant Breeding, KVK, Dariyapur, Raebareli. Uttar Pradesh

Chanchal Bhargava
Krishi Vigyan Kendra, JNKVV, Chhindwara, M.P.

Pramod Kumar Pandey
College of Agriculture, Central Agricultural University (Imphal)
Kyrdemkulai, Meghalaya

Y. Vinod Kumar Reddy
School of Agriculture, Loyola Academy Degree and P.G. College
Osmania State University, Hyderabad

Neha Rohila
Crop Improvement Division, ICAR - CSSRI, Karnal, Haryana

Sheetanshu Gupta
Department of Biochemistry, G.B. Pant University of
Agriculture & Technology, Pantnagar, Uttarakhand

-EPH-
Elite Publishing House

CRC Press
Taylor & Francis Group
Boca Raton London New York

CRC Press is an imprint of the
Taylor & Francis Group, an informa business

First published 2026
by CRC Press
4 Park Square, Milton Park, Abingdon, Oxon, OX14 4RN

and by CRC Press
2385 NW Executive Center Drive, Suite 320, Boca Raton FL 33431

CRC Press is an imprint of Informa UK Limited

British Library Cataloguing-in-Publication Data
A catalogue record for this book is available from the British Library

ISBN: 9781041247340 (hbk)
ISBN: 9781041280293 (pbk)
ISBN: 9781003757412 (ebk)

DOI: 10.1201/9781003757412

Typeset in A Caslon Pro
by Elite Publishing House

-EPH-

Contents

Preface

The field of plant breeding has continually evolved, integrating both classical principles and modern technological advancements to improve crop productivity, resilience, and quality. This book, Principles of Plant Breeding for PGR Management, is designed to provide a comprehensive exploration of the foundational concepts and emerging trends in plant breeding, with a focus on the management of plant genetic resources (PGR). It aims to bridge the gap between traditional plant breeding techniques and the contemporary methods influenced by molecular biology and biotechnological innovations. The goal of this book is to offer students, researchers, and professionals in the field a thorough understanding of plant breeding principles, practices, and their applications in sustainable crop improvement.

Plant breeding is the art and science of improving the genetic potential of plants to meet human needs. With increasing challenges such as climate change, biotic and abiotic stresses, and the need for higher yields, efficient plant breeding strategies are crucial for food security and agricultural sustainability. This book is structured to guide readers through a series of well-organized chapters, each addressing critical aspects of plant breeding, from traditional methods to cutting-edge molecular approaches. The inclusion of plant genetic resources management ensures that readers understand the importance of preserving and utilizing genetic diversity, which is essential for long-term crop improvement and adaptability.

The contents of this book are organized into three units, covering a wide range of topics from basic breeding principles to advanced molecular breeding techniques. Each chapter is designed to provide in-depth insights into the specific aspects of plant breeding, making it a valuable resource for students studying plant sciences, genetics, and crop improvement.

Authors

About the Authors

Dr. Deepak Kumar Mishra was born in 1976 and permanent resident of Village Tajopur, Post Tajopu, District Mau (U.P.). He received his B.Sc. Ag & A.H from CSAUA&T, Kanpur in 1998. He was awarded with chancellor gold medal and four other gold medals during his graduation program. He did his M.Sc. Ag and Ph.D in Genetics and Plant Breeding from GBPUA&T, Pantnagar in 2000 & 2003 respectively. He has an experience of over 20 years in Research and Extension. He has good experience in research breeding work of major pulses and sunflowers. He has good experience in communication skills in the upscaling of improved agricultural technologies among the Farming community of Bundelkhand and Central Plain Zone of U.P. He has to his credit more than 20 research papers published in National and International scientific journals. He is a member of many academic and professional societies in the country. He is presently working as a Scientist in Plant Breeding in KVK, Dariyapur, Raebareli.

Chanchal Bhargava is a renowned extension worker and a young researcher, with nearly 13 years of illustrious extension, teaching, and research experience in the field of plant breeding and genetics. She has contributed vastly to the research and extension activities related to the multiplication and distribution of several cultivated field crop species. Especially in the field of seed research, she has several achievements regarding seed production technologies. She has 6 years of administrative experience too. She is pursuing an In-Service doctorate from Jawaharlal Nehru Krishi Vishwa Vidhyalaya (JNKVV), Jabalpur, MP. At present, she is working as a Senior Technical Assistant at Krishi Vigyan Kendra, JNKVV, Chhindwara, MP. She has published more than 15 research papers during her career present in Reynold Journals worldwide. She is a lifetime member of several agricultural-based societies. She has been a recipient of several awards from different societies like Young Plant Breeder Award in 4thNational Annual Conference on Advances tools and techniques in agriculture, horticulture, environmental, biotechnological Science in current era (ATTAHEBS -2021), Best Extension Worker Award From Home Minister of Govt. of MP Hon'ble Shri Narottam Mishra on the occasion of Republic Day on dated 26 January 2022, Young Women Scientist Award on the occasion of International conference on "One Health-one World" organized by RVSKVV Gwalior MP and Agri-Meet Foundation UP in the year 2023, Excellence in Research Award on the occasion of 5th international Conference in hybrid mode on Innovation and Current Advances in Agriculture & Allied Sciences (ICAAAS-2023) held at Dubai, Krishak Seva Ratn Alankaran Award from Bharat Krishak Samaj District Jabalpur by Cabinet Minister in the year 2024, Young Scientist Award on the occasion of 4 th national conference on "Agriculture in 2050: Technology and Dissemination" organized Society of Krishi Vigyan held at JNKVV, Jabalpur in the year 2024. She has published more than 55 Extension Books/Bulletins/Booklets. With the multi-disciplinary approach, she is Project In-charge of the Tribal Sub Plan funded by ICAR, New Delhi, entitled "Fresh Water Pearl Cultivation. She is the one and only master trainer for pearl cultivation in the JNKVV. She has successfully recorded many video lectures on the Agri-Diksha portal of ICAR.

Dr. Pramod Kumar Pandey is working at the College of Agriculture, Kyrdemkulai, Meghalaya, which is part of the Central Agricultural University headquartered in Imphal, Manipur. He was born in the village of Kondari, which is in the Ghazipur District of Uttar Pradesh. He did his undergraduate studies at PG College, Ghazipur, U.P., and obtained his postgraduate degree in Biotechnology from KIIT University, Bhubaneswar, Odisha. He obtained his Ph.D. in Biotechnology from Dibrugarh University, Dibrugarh, Assam. He has authored numerous scholarly research papers, and articles and contributed to various book chapters. He possesses about 10 years of professional and academic expertise in the field of Endophytes. He is currently involved in maize breeding activities and has more than 6 years of experience in the field of maize breeding. He serves as the In-Charge of AICRP on Maize, CAU Brapanai Centre.

Dr. Y. Vinod Kumar Reddy completed his Ph.D in Agriculture (Genetics & Plant Breeding) from SHUATS (Formerly Allahabad Agricultural Institute), Prayagraj (Allahabad). He has carried out his Doctoral Research from ICRISAT, (Sorghum Breeding), Hyderabad, under a DBT Fellowship. He has worked on Introgression study & Gene Pyramiding in sweet and high biomass sorghum lines by transferring bmr traits through Marker-assisted back cross-breeding, which was carried out at ICRISAT during his doctoral study. He has completed his Master's degree in Agriculture (Genetics and Plant Breeding) from SHUATS, (Formerly Allahabad Agricultural Institute), Prayagraj (Allahabad). He has completed his Graduation, BSc. (Hons). Agriculture, from the School of Agriculture, Loyola Academy Degree and P.G. College, Osmania State University, Hyderabad. He has 10 Research paper publications; he has written a book chapter on Wheat breeding (Genomic Selection-Driven Wheat Breeding for Superior Genetic Gains: Status Quo and Future Steps) Springer publication. He has attended 3 International conferences and 6 National Conferences. Dr. Y. Vinod Kumar Reddy has 3 years of Teaching Experience and External Paper setter and evaluator for U.G & P.G, Agriculture Students, he is expertized in the subjects of Agriculture Genetics, Plant Breeding, Agriculture Biotechnology, Seed Science and Technology, and Crop Improvement, Commercial Plant Breeding, Genomics in Crop Improvement, At present, Dr. Y. Vinod Kumar Reddy working as an Assistant Professor at the School of Agriculture, Loyola Academy Degree and P.G. College, Osmania State University, Hyderabad.

Dr. Neha Rohila has two years of research experience in the area of Genetics and Plant Breeding. She has contributed vastly to the research activities related to the multiplication of cultivated as well as wild forage species, sustainable production of forage crops, and biological studies of cultivated as well as wild breeding crops. Dr. Neha Rohila has a Doctorate in Genetics and Plant Breeding and M.Sc. in Genetics and Plant Breeding and a B.Sc. (Hons.) Agriculture from Chaudhary Charan Singh Haryana Agricultural University, Hisar. She is working as a Senior Research Fellow in the Crop Improvement Division of ICAR - CSSRI, Karnal. She has contributed to research through working on different development and research projects bringing out more than 6 publications, 17 articles, and 4 book chapters. She has participated in several National and International conferences.

Dr. Sheetanshu Gupta working in Biochemistry, Molecular Biology, and Biotechnology for the last 12 years, an alumnus of G.B. Pant University of Agriculture & Technology, Pantnagar, Uttarakhand, India, ICAR ARS NET (Plant Biochemistry and Plant Physiology) and CSIR NET. Dr. Gupta gained valuable experience in the areas of environmental problems, aeroponics and hydroponics, and protein biochemistry. His work showcased his expertise in applying innovative farming techniques and exploring novel approaches to plant protection. Dr. Gupta researched on tissue culture hydroponics and mushroom farming, contributing to the advancement of these fields. Dr. Gupta's academic qualifications, research experience, and expertise in biochemistry and biotechnology make him a valuable asset in the field. With his passion for teaching and research, dedication to innovative farming techniques, and student advocacy, Dr. Gupta continues to contribute significantly to the advancement of agricultural and biochemical sciences with more than 15 published books and nearly 20 Research/ Review papers.

Objectives of Plant Breeding

Plant breeding involves several specific objectives aimed at improving the economic and agronomic value of crops. These objectives can vary based on the crop species and the particular needs of agriculture in different regions.

1. **Higher Yield**: One of the primary objectives of plant breeding is to increase crop yield, whether it be grain, fiber, or oil production. Higher yields help ensure food security and improve farmers' income. Breeding efforts focus on developing high-yielding varieties or hybrids that make efficient use of available resources like water, nutrients, and sunlight. For instance, high-yielding varieties of wheat, rice, and maize have been key contributors to global food supply.
2. **Improved Quality**: Quality traits are crop-specific, and improving these traits is often a critical objective in plant breeding. For example, in wheat, quality characteristics like grain size, milling, and baking qualities are important, while in fruits and vegetables, traits like size, color, flavor, and nutritional value take precedence. Similarly, breeders focus on improving oil content in oilseeds and fiber length and strength in cotton. Enhancing these traits makes the crop more valuable in the marketplace and contributes to consumer satisfaction.
3. **Abiotic Resistance**: Crops face a range of abiotic stresses, such as drought, salinity, and extreme temperatures, which can severely impact productivity. One of the major goals of plant breeding is to develop varieties that are resistant to these environmental factors. For example, drought-resistant varieties of crops like maize and rice have been developed to ensure higher productivity in water-scarce regions. Salt-tolerant varieties are also critical for maintaining productivity in areas with saline soils.
4. **Biotic Resistance**: Pests and diseases are constant threats to agricultural productivity. By developing genetic resistance to these biotic stresses, plant breeders can reduce the need for chemical pesticides and mitigate yield losses. For example, resistance to common diseases such as rust in wheat or bacterial blight in rice is crucial for maintaining yield stability. Breeding for biotic resistance is one of the most effective and sustainable ways to protect crops from pathogens and insect pests.
5. **Change in Maturity Duration / Earliness**: Earliness is an advantageous trait in many crops, as it allows farmers to reduce the growing season and lower production costs. Early-maturing varieties of crops such as wheat and pigeon pea allow farmers to plant additional crops in a single season or avoid the risks associated with late-season environmental conditions. Breeding for reduced maturity duration has resulted in crops that can be harvested sooner, helping farmers manage their land more efficiently.
6. **Determinate Growth**: In crops like pigeon pea, mung bean, and cotton, determinate growth is a desirable trait. Determinate plants stop growing after reaching a certain size, which allows for more synchronized flowering and fruiting. This characteristic simplifies management practices and makes harvesting more efficient. Breeding for determinate growth has proven beneficial in crops where uniformity in growth and maturity is desired.
7. **Dormancy**: In some crops, premature seed germination can occur before the harvest, leading

Dr. Neha Rohila has two years of research experience in the area of Genetics and Plant Breeding. She has contributed vastly to the research activities related to the multiplication of cultivated as well as wild forage species, sustainable production of forage crops, and biological studies of cultivated as well as wild breeding crops. Dr. Neha Rohila has a Doctorate in Genetics and Plant Breeding and M.Sc. in Genetics and Plant Breeding and a B.Sc. (Hons.) Agriculture from Chaudhary Charan Singh Haryana Agricultural University, Hisar. She is working as a Senior Research Fellow in the Crop Improvement Division of ICAR - CSSRI, Karnal. She has contributed to research through working on different development and research projects bringing out more than 6 publications, 17 articles, and 4 book chapters. She has participated in several National and International conferences.

Dr. Sheetanshu Gupta working in Biochemistry, Molecular Biology, and Biotechnology for the last 12 years, an alumnus of G.B. Pant University of Agriculture & Technology, Pantnagar, Uttarakhand, India, ICAR ARS NET (Plant Biochemistry and Plant Physiology) and CSIR NET. Dr. Gupta gained valuable experience in the areas of environmental problems, aeroponics and hydroponics, and protein biochemistry. His work showcased his expertise in applying innovative farming techniques and exploring novel approaches to plant protection. Dr. Gupta researched on tissue culture hydroponics and mushroom farming, contributing to the advancement of these fields. Dr. Gupta's academic qualifications, research experience, and expertise in biochemistry and biotechnology make him a valuable asset in the field. With his passion for teaching and research, dedication to innovative farming techniques, and student advocacy, Dr. Gupta continues to contribute significantly to the advancement of agricultural and biochemical sciences with more than 15 published books and nearly 20 Research/ Review papers.

Chapter - 1

Objectives and Genetic basis of Plant breeding

A. Definition of Plant Breeding

Plant breeding can be defined as the art, science, and technology of improving the genetic makeup of plants to enhance their usefulness and value for humankind. It involves manipulating the heredity of plants through various methods to develop varieties that exhibit desirable traits, such as increased productivity, improved quality, and greater resistance to diseases or environmental stresses. Historically, plant breeding was primarily an art, relying on the breeder's skill in selecting the best plants based on observation. Over time, with advancements in genetics and biotechnology, it has evolved into a highly scientific discipline. Today, plant breeding is a combination of both art and science, utilizing both traditional techniques and modern molecular tools to bring about the desired changes in crops.

The essence of plant breeding is to enhance the genetic potential of plants, whether by selecting superior varieties, hybridizing different species, or employing advanced techniques such as genetic engineering. This makes plant breeding a critical part of global agricultural progress, ensuring that crops can meet the food and economic needs of growing populations while addressing challenges posed by changing environments.

Aim of Plant Breeding

The overarching aim of plant breeding is to improve the characteristics of plants to make them more desirable from an agronomic and economic standpoint. Whether it's increasing yields, enhancing quality, or boosting resilience to environmental stresses, plant breeders work towards making crops better suited to human needs. The specific aims vary depending on the crop species being targeted and the desired traits. For example, while the goal for grain crops may be higher yields, for fiber crops like cotton, breeders might focus on improving fiber length and strength. Another key objective of plant breeding is adaptability, enabling crops to thrive in diverse climates and conditions, and addressing challenges like climate change, water scarcity, or poor soil quality. By enhancing these traits, plant breeders can ensure that crops remain viable and productive in different agricultural settings, making plant breeding essential for the future of sustainable agriculture.

DOI: 10.1201/9781003757412-1

Objectives of Plant Breeding

Plant breeding involves several specific objectives aimed at improving the economic and agronomic value of crops. These objectives can vary based on the crop species and the particular needs of agriculture in different regions.

1. **Higher Yield**: One of the primary objectives of plant breeding is to increase crop yield, whether it be grain, fiber, or oil production. Higher yields help ensure food security and improve farmers' income. Breeding efforts focus on developing high-yielding varieties or hybrids that make efficient use of available resources like water, nutrients, and sunlight. For instance, high-yielding varieties of wheat, rice, and maize have been key contributors to global food supply.
2. **Improved Quality**: Quality traits are crop-specific, and improving these traits is often a critical objective in plant breeding. For example, in wheat, quality characteristics like grain size, milling, and baking qualities are important, while in fruits and vegetables, traits like size, color, flavor, and nutritional value take precedence. Similarly, breeders focus on improving oil content in oilseeds and fiber length and strength in cotton. Enhancing these traits makes the crop more valuable in the marketplace and contributes to consumer satisfaction.
3. **Abiotic Resistance**: Crops face a range of abiotic stresses, such as drought, salinity, and extreme temperatures, which can severely impact productivity. One of the major goals of plant breeding is to develop varieties that are resistant to these environmental factors. For example, drought-resistant varieties of crops like maize and rice have been developed to ensure higher productivity in water-scarce regions. Salt-tolerant varieties are also critical for maintaining productivity in areas with saline soils.
4. **Biotic Resistance**: Pests and diseases are constant threats to agricultural productivity. By developing genetic resistance to these biotic stresses, plant breeders can reduce the need for chemical pesticides and mitigate yield losses. For example, resistance to common diseases such as rust in wheat or bacterial blight in rice is crucial for maintaining yield stability. Breeding for biotic resistance is one of the most effective and sustainable ways to protect crops from pathogens and insect pests.
5. **Change in Maturity Duration / Earliness**: Earliness is an advantageous trait in many crops, as it allows farmers to reduce the growing season and lower production costs. Early-maturing varieties of crops such as wheat and pigeon pea allow farmers to plant additional crops in a single season or avoid the risks associated with late-season environmental conditions. Breeding for reduced maturity duration has resulted in crops that can be harvested sooner, helping farmers manage their land more efficiently.
6. **Determinate Growth**: In crops like pigeon pea, mung bean, and cotton, determinate growth is a desirable trait. Determinate plants stop growing after reaching a certain size, which allows for more synchronized flowering and fruiting. This characteristic simplifies management practices and makes harvesting more efficient. Breeding for determinate growth has proven beneficial in crops where uniformity in growth and maturity is desired.
7. **Dormancy**: In some crops, premature seed germination can occur before the harvest, leading

to losses. Breeding for dormancy helps prevent this issue. For example, introducing dormancy traits in crops like barley or mung bean can help avoid pre-harvest sprouting, which can negatively affect yield and quality. Conversely, in some cases, it may be desirable to eliminate dormancy to allow for immediate germination, depending on the cropping system.

8. **Desirable Agronomic Characteristics**: Agronomic traits such as plant height, tillering capacity, and growth habit are important for improving crop performance. For instance, dwarf varieties of rice and wheat are preferred because they are less prone to lodging, allowing for higher fertilizer inputs and improved yields. In fodder crops, traits like tallness and profuse branching are desirable to increase biomass production.

9. **Elimination of Toxic Substances**: Some crops naturally produce toxic compounds that are harmful to humans or animals. For example, khesari (Lathyrus sativus) contains a neurotoxin that can cause paralysis, and gossypol in cotton seeds is toxic to humans. Breeding efforts aim to eliminate these substances to make the crops safer for consumption and improve their nutritional value.

10. **Non-shattering Characteristics**: Shattering, where seeds or pods are lost before harvest, can be a significant issue in crops like green gram. Breeding for non-shattering traits ensures that more of the harvestable produce is retained, reducing losses and improving yield.

11. **Synchronous Maturity**: Synchronous maturity refers to the simultaneous maturation of all plants in a field, allowing for a single harvest rather than multiple pickings. This trait is particularly useful in crops like cotton and green gram, where it can reduce labor costs and simplify harvesting operations.

12. **Photo and Thermo Insensitivity**: Developing varieties that are insensitive to changes in day length (photoperiod) and temperature (thermo-insensitivity) allows for more flexible cropping systems. For example, photo- and thermo-insensitive varieties of wheat and rice can be cultivated in regions with varying climatic conditions, expanding the areas where these crops can be grown.

13. **Wider Adaptability**: Crops that are widely adaptable can perform well under a variety of environmental conditions. Breeding for adaptability helps ensure stable yields across different regions and seasons, making crops more resilient to fluctuations in climate and other environmental factors.

14. **Varieties for New Seasons**: Traditionally, certain crops are grown in specific seasons. However, plant breeders have developed varieties that allow for cultivation outside these traditional seasons. For instance, maize, which is typically a kharif crop, can now be grown in the rabi and zaid seasons as well. Similarly, mung bean can be cultivated in the summer in addition to the kharif season, providing farmers with more flexibility in crop production.

Future Prospects of Plant Breeding: Scope and Opportunities

Plant breeding has been an integral part of human progress in agriculture, contributing to significant advancements in crop production and food security throughout history. The application of plant breeding is evolving rapidly with the integration of modern technologies, especially in the wake of global challenges like population growth, climate change, and resource scarcity. As the demand for sustainable agricultural practices increases, plant breeding continues to play a pivotal role in shaping the future of crop improvement.

Genetic Engineering and Biotechnology in Plant Breeding

One of the most transformative developments in modern plant breeding is the application of genetic engineering and biotechnology. Traditional plant breeding relies on crossing selected parents with desirable traits, followed by selection in subsequent generations. However, the emergence of molecular tools has revolutionized this process, allowing breeders to directly manipulate genes responsible for key traits. This shift has enabled the development of genetically modified (GM) crops, which offer enhanced traits such as pest resistance, improved nutritional content, and higher yields.

GM crops like genetically engineered rice, maize, soybean, cotton, and oilseed rape have been developed and are undergoing field testing. These crops are designed to resist biotic stresses like pests and diseases, as well as abiotic stresses such as drought and salinity. In the future, we can expect the commercialization of GM crops to become even more widespread, as ongoing research continues to improve the precision and efficiency of gene editing technologies like CRISPR-Cas9. For instance, genes from various organisms are being introduced into crop plants to bolster their resilience to environmental stressors. Such efforts are particularly important as climate change intensifies the challenges faced by global agriculture. Moreover, crops may also be genetically engineered to produce valuable compounds such as pharmaceuticals. The production of hirudin, an anti-thrombin protein, in transgenic *Brassica napus* is a prime example of how plant breeding can intersect with medicine. This convergence of biotechnology and plant breeding is expected to open new avenues for the cultivation of crops for non-food purposes, such as biofuels and biopharmaceuticals.

Enhancing Resistance to Biotic and Abiotic Stresses

One of the most promising future applications of plant breeding lies in developing crop varieties that are resilient to both biotic and abiotic stresses. This is particularly important in the face of growing concerns about food security due to increasing environmental instability. Crops engineered to resist pests, diseases, drought, salinity, and temperature fluctuations will be critical for sustaining agricultural productivity in regions most affected by climate change.

Researchers are already making strides in this area by incorporating stress-resistance genes from various organisms into crops like maize, wheat, and rice. In the future, more targeted approaches using molecular breeding and precision gene editing will allow for the development of crops that can adapt to extreme environmental conditions. This will not only help stabilize food production but also ensure that crops can be grown in previously unsuitable areas, thus expanding the geographic range of agriculture.

Plant Breeding for Nutritional Enhancement

In addition to improving yield and stress resistance, plant breeding is expected to make significant contributions to enhancing the nutritional quality of crops. Biofortification, or increasing the nutritional value of food crops through breeding, is a key area of focus for future plant breeding efforts. By introducing genes responsible for higher levels of vitamins, minerals, and essential nutrients into staple crops like rice, wheat, and maize, breeders aim to combat malnutrition in developing countries. For instance, golden rice, a genetically engineered variety that produces beta-carotene, is designed to address vitamin A deficiency, a serious public health issue in many parts of the world. Future plant breeding efforts will likely focus on developing more biofortified crops to improve food quality and public health outcomes, particularly in regions with limited access to diverse diets.

Sustainable Agriculture and Climate-Resilient Crops

Sustainability is a central theme in the future of plant breeding. With the growing awareness of the environmental impact of agriculture, breeders are focusing on developing crop varieties that require fewer inputs, such as water, fertilizers, and pesticides. This not only reduces the cost of production but also minimizes the environmental footprint of farming.

One promising area is the development of drought-tolerant and water-efficient crop varieties. As water scarcity becomes a major challenge in many agricultural regions, breeding for crops that use water more efficiently will be crucial for ensuring sustainable food production. Additionally, breeding for nitrogen-use efficiency is gaining attention as a way to reduce the reliance on synthetic fertilizers, which contribute to environmental pollution and greenhouse gas emissions.

Plant Breeding for Novel Applications: Beyond Food Production

Another emerging aspect of plant breeding is its application beyond food production. Future plant breeding efforts will likely focus on developing crops for bio-based industries, such as biofuels and biodegradable materials. Crops engineered for higher biomass production or specific chemical compounds can serve as raw materials for these industries, reducing dependence on fossil fuels and contributing to a more circular economy. In the pharmaceutical industry, transgenic plants are already being explored as factories for producing vaccines, therapeutic proteins, and other biologically active compounds. This approach, known as "pharming," has the potential to revolutionize drug production by providing a cost-effective and scalable method for producing biologics. With continued advancements in plant biotechnology, crops may be specifically bred to serve as biofactories for a wide range of industrial applications.

History and Development of Plant Breeding

Plant breeding is a field with deep historical roots that has evolved significantly over the centuries. From its origins in simple seed selection to the complex, biotechnology-driven approaches of today, plant breeding has been essential in shaping the agricultural systems that sustain human populations. Over the years, the discipline has moved from an art practiced by early farmers to a science grounded in genetics and molecular biology. The development of plant breeding has mirrored advances in scientific knowledge, particularly in genetics, cytogenetics, and biotechnology, making it a continually evolving and dynamic field. In this section, we will explore the history and development of plant breeding, focusing on key milestones, significant figures, and technological breakthroughs.

Early Domestication: The Roots of Plant Breeding

Plant breeding began with the domestication of wild plant species. The earliest form of human intervention in the genetic composition of plants involved selecting individuals with desirable traits such as higher yields, better taste, or increased resistance to environmental stressors. This process, known as domestication, transformed wild plants into the crops we depend on today. Domestication was the first step toward plant breeding, marking a transition from natural selection to human-directed selection.

In ancient civilizations like Mesopotamia and Egypt, farmers observed that certain plants produced better results and began to selectively cultivate those plants. The domestication of wheat, barley, and legumes in the Fertile Crescent around 10,000 years ago was a significant moment in agricultural history, laying the foundation for human civilization. Over millennia, domestication continued, and as human societies migrated, they carried their cultivated crops with them, further influencing plant evolution through selective breeding. For example, as people moved into different ecological zones, their crops adapted to varying climates and soils, leading to the development of region-specific varieties. One of the earliest recorded examples of deliberate plant breeding comes from ancient Babylon, where farmers as early as 700 BC artificially pollinated date palms to improve fruit production. The Babylonians, Assyrians, and Egyptians were among the first to employ techniques resembling modern plant breeding. This process of moving plants to new environments and selecting for desirable traits was a rudimentary but effective form of plant breeding that persisted for thousands of years.

Classical Plant Breeding: The 17th and 18th Centuries

The practice of plant breeding gained momentum in the 17th and 18th centuries, as agricultural societies in Europe began to understand more about plant genetics and inheritance. In 1717, British horticulturist Thomas Fairchild produced the first artificial hybrid, known as Fairchild's mule, by crossing a carnation with sweet William. This experiment demonstrated that new plant varieties could be created through controlled pollination, setting the stage for more sophisticated breeding techniques.

The first plant breeding company, Vilmorin, was established in France in 1727. The Vilmorin family made significant contributions to agriculture, particularly with their work on sugar beets. They developed the progeny test method, which involves selecting plants based on the performance of their offspring, a technique still in use today. Around the same time, Joseph Koelreuter, a German botanist, conducted extensive hybridization experiments on tobacco plants between 1760 and 1766, confirming that hybrids inherit traits from both parent plants. This work laid the foundation for understanding inheritance patterns and hybrid vigor.

Other notable contributions came from Thomas Andrew Knight, who in the late 18th and early 19th centuries used artificial hybridization to develop new varieties of fruit crops. Knight's work focused on improving traits like fruit size and disease resistance, and he was one of the first to apply scientific principles to plant breeding. His efforts were followed by Patrick Shireff, whose research on cereal crops in the mid-19th century contributed to the understanding of heritable variation and natural hybridization.

The Birth of Modern Plant Breeding: Mendelian Genetics

The modern era of plant breeding began with the rediscovery of Gregor Mendel's laws of inheritance in 1900 by three botanists: Hugo de Vries, Carl Correns, and Erich von Tschermak. Mendel's work, originally published in 1866, had gone largely unnoticed until the turn of the century, when it became the foundation for understanding how traits are passed from one generation to the next. Mendel's experiments with pea plants revealed the basic principles of heredity, including the segregation of traits and independent assortment. This discovery revolutionized plant breeding by providing a scientific framework for predicting the outcomes of crosses.

Following the rediscovery of Mendel's work, scientists like Nilsson-Ehle and Johannsen built upon his principles. Nilsson-Ehle, working in Sweden, developed the individual plant selection method, which allowed breeders to isolate and propagate plants with desirable traits. Johannsen's pure line theory, proposed in 1903, further refined the practice of plant selection by emphasizing that true breeding lines could be developed through self-pollination and selection over generations. These advancements were complemented by discoveries in cytogenetics, the study of chromosomes, which revealed that chromosomes carry the genetic material responsible for inheritance. The discovery of chromosomes as the carriers of genes paved the way for chromosome engineering, a technique that allows breeders to manipulate plant genetics more precisely.

Molecular Biology and Biotechnology: Revolutionizing Plant Breeding

The 20th century saw rapid advancements in molecular biology that transformed plant breeding into a highly sophisticated science. The discovery of the structure of DNA by James Watson and Francis Crick in 1953 marked the beginning of a new era in genetics. With the development of molecular markers and gene mapping technologies, breeders could identify specific genes responsible for important traits, such as disease resistance or drought tolerance, and incorporate them into new varieties more efficiently.

The advent of genetic engineering in the latter half of the 20th century further revolutionized plant breeding. Genetically modified (GM) crops, such as herbicide-resistant soybeans and insect-resistant cotton, were developed using recombinant DNA technology. This allowed for the direct transfer of genes from one organism to another, bypassing the need for traditional cross-breeding methods. By the 1990s, GM crops were being cultivated on a large scale, particularly in the United States, China, and Brazil. The ability to engineer crops with specific traits opened up new possibilities for addressing global challenges like food security, climate change, and pest management. Biotechnology also enabled the production of crops with enhanced nutritional profiles, such as Golden Rice, which is genetically modified to produce beta-carotene, a precursor of vitamin A. This development highlights the potential of plant breeding to not only increase agricultural productivity but also improve human health and nutrition.

Contributions of Eminent Scientists

The field of plant breeding has been shaped by the contributions of numerous scientists over the years. Some of the most notable figures include:

» **Gregor Johann Mendel**: The father of genetics, whose work on inheritance patterns in pea plants laid the foundation for modern plant breeding.

» **Hugo de Vries, Carl Correns, and Erich von Tschermak**: Rediscovered Mendel's work in 1900, bringing it to the attention of the scientific community.

» **Thomas Andrew Knight**: A pioneer in artificial hybridization who developed new fruit varieties and applied scientific principles to plant breeding.

» **Patrick Shireff**: Contributed to the understanding of heritable variation and natural hybridization in cereal crops.

» **M.S. Swaminathan**: Led India's Green Revolution by developing high-yielding varieties of wheat and rice, which helped India achieve self-sufficiency in food production.

S. No.	Year	Contribution/Development	Scientist/Contributor
1	700 BC	Artificial pollination of date palms	Babylonians and Assyrians
2	1717	First artificial hybrid ("Fairchild's mule") produced by crossing carnation with sweet William	Thomas Fairchild
3	1727	Establishment of the first plant breeding company in France	Vilmorin family
4	1760-1766	Extensive hybridization experiments in tobacco	Joseph Koelreuter
5	1799-1835	Developed new fruit varieties through artificial hybridization	Thomas Andrew Knight
6	1856	Introduction of progeny testing for sugar beet improvement	Vilmorin family
7	1871	Creation of the Department of Agriculture in India	Government of India
8	1873	Work published on cereal variety selection; concluded that heritable variation responds to selection	Patrick Shireff
9	1900	Rediscovery of Mendel's laws of inheritance by three botanists	Hugo de Vries, Carl Correns, Erich von Tschermak
10	1901-1905	Establishment of Agricultural Colleges in India (Kanpur, Pune, Sabour, Coimbatore, Llyalpur)	Indian government
11	1903	Proposal of the pure line theory, providing a genetic basis for plant selection	Johannsen
12	1905	Establishment of Imperial Agricultural Research Institute (IARI) in Pusa, Bihar	Government of India
13	1921	Establishment of the Indian Central Cotton Committee, leading to the development of improved cotton varieties	Indian government
14	1929	Establishment of the Imperial Council of Agricultural Research (ICAR)	Government of India
15	1936	Shift of the IARI from Pusa, Bihar to New Delhi	Government of India
16	1946	Name change from Imperial Council of Agricultural Research to Indian Council of Agricultural Research (ICAR)	Indian government
17	1956	Initiation of the Project for Intensification of Regional Research on Cotton, Oilseeds, and Millets (PIRRCOM)	Government of India

18	1957	Launch of the All India Coordinated Maize Improvement Project	ICAR
19	1960	Establishment of India's first agricultural university at Pantnagar, Nainital, Uttar Pradesh	Indian government
20	1961	Release of the first hybrid maize varieties	ICAR
21	1970	Development of the world's first hybrid cotton variety	C.T. Patel
22	1990s	Commercialization of genetically modified crops (e.g., insect-resistant cotton, herbicide-resistant soybean)	Various
23	1996-1997	Introduction of Bt cotton in the United States and later in other countries	Various
24	Ongoing	Application of CRISPR-Cas9 for precise gene editing in plant breeding	Various

Key Contributions of Eminent Scientists in Plant Breeding

S. No.	Scientist	Contribution/Discovery
1	Gregor **Johann** Mendel	Father of genetics, discovered laws of inheritance through experiments with pea plants
2	Hugo de Vries, Carl Correns, Erich von Tschermak	Rediscovered Mendel's work, leading to the foundation of modern genetics
3	Thomas Andrew Knight	Pioneer in artificial hybridization, developed new fruit varieties
4	Patrick Shireff	Work on heritable variation and natural hybridization in cereal crops
5	Johannsen	Proposed the pure line theory, establishing a genetic basis for plant selection
6	Nilsson-Ehle	Developed the individual plant selection method in Sweden
7	M.S. Swaminathan	Leader of the Green Revolution in India, developed high-yielding varieties of wheat and rice
8	C.T. Patel	Developed the world's first hybrid cotton variety in 1970
9	T.S. Venkatraman	Developed sugarcane varieties by transferring traits from tropical cane to North Indian canes
10	B.P. Pal	Developed superior disease-resistant wheat varieties (N.P. varieties)
11	N.G.P. Rao	Eminent sorghum breeder
12	Pushkarnath	Famous potato breeder
13	Yuan Longping	Father of hybrid rice, developed the first high-yielding hybrid rice varieties

14	N.I.Vavilov	Developed the theory of centers of diversity, contributing to the understanding of genetic resources
15	T.T. Chang	Played a key role in developing semi-dwarf rice varieties at the International Rice Research Institute (IRRI)
16	G.W. Burton	Notable forage breeder
17	E.E. Hartwig	Renowned soybean breeder
18	D.S. Athwal	Famous pearl millet breeder
19	K. Ramaiah	Renowned rice breeder
20	Dharampal Singh	Eminent oil-seed breeder

Plant Breeding-Phases

1. Domestication

Domestication is the foundational phase of plant breeding, representing the first instance in which humans intentionally altered the genetic makeup of wild plant species to meet their needs. This process, which began over 9,000 years ago, involves the transformation of wild species into cultivated forms, making them more suitable for human use. Domestication is regarded as the most basic method of plant breeding and remains relevant today as new crops are continuously brought under cultivation. In its earliest form, domestication involved selecting plants with favorable traits such as larger seeds, better flavor, or increased yield. Through this selection, humans inadvertently began the process of altering the genetic makeup of these plants, which over generations led to the development of cultivated varieties. As humans migrated, they carried their crops with them, leading to the spread of domesticated species to new regions. Over time, these cultivated species became dependent on human intervention and artificial environments for their continued survival, marking a critical shift in the relationship between humans and plants.

Natural selection also played a role in the domestication process. While humans were selected for specific traits, environmental pressures such as climate and soil conditions influenced the evolution of domesticated species. These dual forces of human selection and natural adaptation resulted in the creation of plants that were uniquely suited to particular environments, giving rise to what are known as "centers of origin" for various crops. These regions, which include parts of the Middle East, South Asia, and Central America, are where many of the world's staple crops—such as wheat, rice, and maize—were first domesticated. One key concept in domestication is the development of landraces. Landraces are locally adapted plant varieties that evolved through a combination of human selection and natural pressures. Unlike modern varieties, which are often genetically uniform, landraces exhibit a high degree of genetic diversity, making them particularly resilient to environmental fluctuations. For example, rice landraces such as *Oryza sativa* subspecies *indica* (developed in South Asia) and *Oryza sativa* subspecies *japonica* (developed in China) are highly adapted to their respective environments. Another outcome of domestication is the creation of cultigens. A cultigen is a plant that has been deliberately altered by humans through selection or breeding. Cultigens are distinct from their wild relatives in that they often exhibit traits that would be disadvantageous but are beneficial under human cultivation. These traits include larger fruits, reduced seed dispersal mechanisms (to prevent seed loss), and increased yield.

Domestication is not a one-time event but an ongoing process. Even today, new crops are being domesticated, and existing crops are continuously improved through breeding programs. For example, timber trees and medicinal plants are currently undergoing domestication as researchers seek to enhance their value for industrial and medicinal purposes. Additionally, domestication is likely to continue in the future as the need for new crops to adapt to changing climates and growing population demands becomes more urgent. Domestication is the process by which wild species are transformed into cultivated crops through human intervention. It is the first phase of plant breeding and forms the basis for all subsequent breeding efforts. This process, which began thousands of years ago, continues to play a vital role in modern agriculture as breeders work to adapt plants to new environmental and societal challenges. Domestication has resulted in the creation of landraces and cultigens, both of which have contributed to the genetic diversity and adaptability of the world's food supply.

2. Selection

Selection is a critical phase in plant breeding, where breeders choose individuals with desirable traits to propagate and improve crop varieties. Selection can occur naturally, but in plant breeding, it is predominantly an artificial process driven by human intervention. By selecting plants with favorable characteristics—such as higher yields, disease resistance, or improved quality—breeders can shape the genetic makeup of a population over generations. Selection has been practiced for millennia and remains one of the most important tools in plant breeding. There are different methods of selection, each suited to specific breeding goals.

i) Mass Selection

Mass selection is one of the oldest and simplest methods of plant breeding. In this process, a large number of plants are grown, and the best individuals are selected based on visible traits. These selected plants are then allowed to reproduce, and their offspring become the next generation of the population. Mass selection is particularly useful in maintaining genetic diversity because it allows for the selection of multiple individuals, each contributing their genes to the next generation. This method is commonly used in self-pollinated crops such as wheat and rice. The primary advantage of mass selection is its simplicity. It does not require detailed knowledge of the genetic makeup of the plants, and selection is based purely on phenotype (observable traits). However, the drawback is that it may not always result in uniformity or high genetic gain, as undesirable recessive traits may persist in the population.

ii) Pureline Selection

Pureline selection is a more advanced form of selection, primarily used in self-pollinated crops. In this method, individual plants are selected, and their progeny are grown in isolation to create pure lines. A pure line is a genetically uniform population derived from a single plant. Over successive generations, breeders can refine these lines by selecting the best-performing individuals and eliminating those with undesirable traits. This process results in highly uniform and stable varieties that exhibit consistent performance across different environments.

Pureline selection is particularly effective in crops where uniformity is desirable, such as grains and legumes. However, it requires more time and resources than mass selection because the process of isolating and testing pure lines can take several years.

iii) Recurrent Selection

Recurrent selection is a method used primarily in cross-pollinated crops, such as maize and sugarcane. In this process, selected individuals are crossed with each other over multiple generations to accumulate favorable traits. Unlike mass or pure line selection, which focuses on individual plants, recurrent selection aims to improve the overall population by gradually increasing the frequency of desirable alleles. There are different types of recurrent selection, including selection for general combining ability (GCA) and specific combining ability (SCA). GCA refers to the ability of a plant to combine well with a wide range of other plants, while SCA focuses on the performance of specific crosses. Recurrent selection is particularly useful in hybrid breeding programs, where the goal is to produce hybrids with superior performance. Selection is a fundamental phase of plant breeding that involves choosing the best individuals to propagate the next generation. Whether through mass selection, pure line selection, or recurrent selection, breeders use this process to improve the genetic makeup of crops and achieve desired traits. Each method has its advantages and limitations, and the choice of method depends on the crop species and breeding objectives.

3. Hybridization

Hybridization is a powerful phase in plant breeding that involves the crossing of genetically different plants to create hybrids with improved traits. The primary goal of hybridization is to combine the desirable characteristics of two or more parent plants into a single offspring. This method is widely used in both self-pollinated and cross-pollinated crops to produce superior varieties that exhibit hybrid vigor or heterosis—a phenomenon where the hybrid offspring outperforms its parents in terms of yield, growth rate, or resistance to environmental stresses.

i) Intervarietal/Intraspecific Hybridization

Intervarietal hybridization, also known as intraspecific hybridization, occurs when two different varieties or strains of the same species are crossed. This method is commonly used in crops like rice, wheat, and maize, where breeders aim to combine specific traits from each parent variety. For example, one parent may have high yield potential while the other is resistant to a particular disease. By crossing these varieties, breeders can develop hybrids that possess both traits. Interspecific hybridization is relatively straightforward because the parents belong to the same species, and there are fewer barriers to successful fertilization and seed development. The resulting hybrids are often fertile and can be used for further breeding. However, it is essential to conduct extensive field trials to evaluate the performance of the hybrids across different environments, as not all crosses result in superior offspring.

ii) Distant Hybridization

Distant hybridization refers to the crossing of plants from different species (interspecific) or even different genera (intergeneric). This type of hybridization is more challenging than intervarietal hybridization because genetic incompatibilities between the parents may result in sterility or other developmental issues. However, distant hybridization offers the potential to introduce entirely new traits into a crop species, such as resistance to pests or tolerance to environmental stresses.

a) Interspecific/Intrageneric Hybridization

Interspecific hybridization occurs between different species within the same genus. For example, in the genus *Brassica*, breeders have successfully crossed species like *Brassica rapa* (turnip) and *Brassica oleracea* (cabbage) to develop hybrids with improved traits. Intrageneric hybridization can lead to the creation of new crop varieties that are more adaptable to different environments or possess unique characteristics not found in either parent species.

b) Intergeneric Hybridization

Intergeneric hybridization is the crossing of plants from different genera. This type of hybridization is rare and often requires advanced techniques such as embryo rescue or tissue culture to overcome genetic barriers. An example of intergeneric hybridization is the creation of *Raphanobrassica*, a hybrid between radish (*Raphanus*) and cabbage (*Brassica*). While intergeneric hybrids are usually sterile, they can serve as a bridge for transferring beneficial traits between unrelated species.

c) Somatic Hybridization

Somatic hybridization is a form of hybridization that does not rely on sexual reproduction. Instead, it involves the fusion of somatic cells (non-reproductive cells) from different plants to create a hybrid with a mix of genetic material from both parents. This technique is typically carried out in a laboratory using protoplasts—plant cells with the cell wall removed. The fused protoplasts regenerate into whole plants, which can then be evaluated for desirable traits. Somatic hybridization is particularly useful for overcoming sexual incompatibility between distant species or genera. Hybridization is a crucial phase of plant breeding that allows breeders to combine the best traits of different plants into a single hybrid. Whether through intervarietal, interspecific, intergeneric, or somatic hybridization, this process has led to the development of many of the world's most successful crop varieties. Hybridization remains one of the most effective ways to improve yield, quality, and resilience in crops.

4. Creation of Novel Genetic Variations

The creation of novel genetic variations is an essential phase in plant breeding that involves generating new traits or combinations of traits that do not naturally occur in the plant population. Novel genetic variations can arise through several mechanisms, including mutation, polyploidy, somaclonal variation, and gametoclonal variation. These methods allow breeders to introduce new traits, such as disease resistance or drought tolerance, into a crop species, thereby enhancing its adaptability and productivity.

i) Mutation

Mutation breeding involves the deliberate induction of mutations in a plant's DNA to create genetic variation. Mutations can occur spontaneously in nature, but breeders often use chemical mutagens (such as ethyl methanesulfonate) or physical agents (such as radiation) to accelerate the process. Mutagenesis can result in a wide range of changes, from point mutations that alter a single gene to larger chromosomal rearrangements. The resulting plants are screened for desirable traits, such as improved yield, resistance to pests, or tolerance to environmental stresses. Mutation breeding has been used to develop several successful crop varieties, including disease-resistant wheat, high-yielding rice, and improved oilseed crops. The main advantage of mutation breeding is that it allows breeders to introduce new traits without the need for crossbreeding, which can be time-consuming and may result in undesirable combinations of traits.

ii) Polyploidy

Polyploidy refers to the condition of having multiple sets of chromosomes. While most plants are diploid (with two sets of chromosomes), polyploid plants have three or more sets. Polyploidy can occur naturally or be induced artificially through the use of chemicals like colchicine, which disrupts normal cell division and results in the duplication of chromosomes. Polyploid plants often exhibit increased size, improved vigor, and enhanced resistance to environmental stresses. Polyploidy is particularly common in certain crops, such as wheat, cotton, and potatoes. For example, modern bread wheat (*Triticum aestivum*) is a hexaploid species with six sets of chromosomes. The increased genetic diversity associated with polyploidy allows breeders to exploit a broader range of traits, making it a valuable tool in plant breeding.

iii) Somaclonal Variation

Somaclonal variation refers to the genetic changes that occur in plants regenerated from tissue culture. This process can result in novel traits that are not present in the parent plant. Somaclonal variation is a form of spontaneous mutation that arises during the process of cell division and differentiation in tissue culture. Breeders can exploit this variation by selecting plants with desirable traits, such as improved disease resistance or increased yield. Somaclonal variation has been used to develop several commercially successful crop varieties, including disease-resistant bananas and improved sugarcane. It is a valuable tool in plant breeding because it allows breeders to generate genetic diversity without the need for crossbreeding or mutagenesis.

iv) Gametoclonal Variation

Gametoclonal variation is similar to somaclonal variation but occurs in plants regenerated from gametic cells (such as pollen or egg cells) rather than somatic cells. This process can result in unique genetic combinations that are not present in the parent plant. Gametoclonal variation is particularly useful for introducing new traits into a breeding population, as it allows for the creation of novel genetic combinations that would not be possible through conventional breeding methods. The creation of novel genetic variations is a key phase of plant breeding that allows breeders to introduce new traits and enhance the genetic diversity of crop species. Whether through mutation, polyploidy, somaclonal variation, or gametoclonal variation, these methods have been instrumental in the development of improved crop varieties that can meet the challenges of modern agriculture.

5. Genetic Engineering/Recombinant DNA Technology

Genetic engineering, also known as recombinant DNA technology, is one of the most advanced phases of plant breeding. It involves the direct manipulation of an organism's DNA to introduce new traits or enhance existing ones. Unlike traditional breeding methods, which rely on crossing and selection, genetic engineering allows breeders to precisely insert specific genes from one organism into another, regardless of species barriers. This technology has revolutionized plant breeding by enabling the development of genetically modified (GM) crops with enhanced traits such as pest resistance, herbicide tolerance, and improved nutritional content.

Recombinant DNA technology involves several key steps. First, the gene of interest is identified and isolated from the donor organism. This gene is then inserted into a plasmid (a circular piece of DNA) or a vector, which is used to transfer the gene into the target plant. Once inside the plant, the gene is incorporated into the plant's genome, where it can be expressed to produce the desired trait. The genetically modified plant is then propagated and tested to ensure that the new trait is stable and effective. One of the most well-known applications of genetic engineering in plant breeding is the development of Bt crops. Bt crops, such as Bt cotton and Bt maize, contain a gene from the bacterium *Bacillus thuringiensis*, which produces a protein toxic to certain insect pests. By incorporating this gene into the plant's genome, breeders have created crops resistant to pests such as the cotton bollworm and the European corn borer, reducing the need for chemical insecticides. Another important application of genetic engineering is the development of herbicide-tolerant crops. For example, genetically modified soybean varieties that are resistant to glyphosate (a common herbicide) allow farmers to control weeds without damaging the crop. This technology has significantly reduced the labor and cost associated with weed management in agriculture.

Genetic engineering has also been used to enhance the nutritional content of crops. One of the most famous examples is Golden Rice, which has been genetically modified to produce beta-carotene, a precursor of vitamin A. Golden Rice was developed to address vitamin A deficiency, a major public health issue in many developing countries. By increasing the nutritional value of rice, this technology can potentially improve the health of millions of people who rely on rice as a staple food. These applications, genetic engineering has opened up new possibilities for crop improvement that were previously unimaginable. For example, researchers are exploring the use of genetic engineering to develop crops that can produce pharmaceuticals, biofuels, and industrial chemicals. These "bioreactor" crops could serve as sustainable alternatives to traditional manufacturing processes, reducing the environmental impact of industrial production. However, genetic engineering is not without its challenges. There are concerns about the potential environmental and health risks associated with GM crops, such as the development of resistance in pests and weeds, and the unintended effects on non-target organisms. Regulatory agencies around the world have established strict guidelines for the testing and approval of GM crops to ensure their safety for human consumption and the environment.

Mode of Reproduction in Plants

Reproduction in plants is a critical biological process that determines the propagation and continuation of plant species. For plant breeders, the mode of reproduction is essential as it significantly influences breeding strategies and the success of genetic improvement efforts. The two primary modes of reproduction in plants are **asexual reproduction** and **sexual reproduction**, each with its mechanisms, advantages, and limitations. These modes play a vital role in how plants are selected, propagated, and improved for agricultural purposes.

1. Asexual Reproduction

i) Vegetative Reproduction

Vegetative reproduction is a mode of asexual reproduction that involves the growth of new plants from vegetative structures such as roots, stems, or leaves. This process can occur naturally, through structures like tubers, rhizomes, and bulbs, or artificially, through techniques like grafting and cuttings.

This method is particularly advantageous for breeders because it ensures the preservation of desirable traits, which are passed on to the offspring without genetic variation.

a) Natural Vegetative Reproduction

Natural vegetative reproduction occurs without human intervention. It involves various plant structures that have evolved to propagate new plants. These include:

» **Rhizomes**: These are horizontal underground stems that grow laterally and produce new shoots. Each segment of the rhizome can develop into a new plant. For example, **ginger** (*Zingiber officinale*) and **turmeric** (*Curcuma longa*) propagate through rhizomes. These plants spread naturally in their environment through this process, which ensures the rapid colonization of available space.

» **Tubers**: Tubers are swollen underground stems that store nutrients and can give rise to new plants. The most well-known example of a tuber is the **potato** (*Solanum tuberosum*), which develops underground tubers that can sprout new plants when conditions are favorable. Tubers contain buds, known as "eyes," from which new stems and roots grow, providing a natural method of reproduction.

» **Corms**: Corms are short, vertical, swollen underground stems used by some plants to store nutrients. New shoots grow from buds located at the top of the corm, and roots form from its base. Examples of plants that reproduce through corms include **taro** (*Colocasia esculenta*) and **crocus** species.

» **Bulbs**: Bulbs are storage organs made of a short stem surrounded by fleshy leaves that contain nutrients. Bulbs serve as a method of reproduction in plants like **onions** (*Allium cepa*) and **garlic** (*Allium sativum*). The fleshy leaves provide energy for the development of new shoots when conditions are right, and each bulb can grow into a new plant.

» **Runners and Stolons**: Some plants produce horizontal stems, called runners or stolons, which grow along the surface of the soil and form new plants at their nodes. **Strawberries** (*Fragaria* species) and **mint** (*Mentha* species) propagate naturally in this way. The new plants are clones of the parent and are connected by the runner until they establish their own root systems.

» **Bulbils**: Bulbils are specialized structures that form in place of flowers. When they fall to the ground, they can develop into new plants. This form of reproduction is common in **garlic** and some species of **lilies**. Bulbils allow the plant to spread without relying on seeds, which may be less reliable or take longer to germinate.

Natural vegetative reproduction offers several advantages. It ensures the rapid propagation of plants, allowing them to occupy space quickly and efficiently. It also guarantees genetic uniformity, which is beneficial for maintaining the desirable traits of crops like size, shape, and flavor. However, this uniformity can also be a disadvantage, as it reduces genetic diversity, making the population more vulnerable to diseases and environmental stresses.

b) Artificial Vegetative Reproduction

Artificial vegetative reproduction is a method that involves human intervention to propagate plants. This method is widely used in agriculture, horticulture, and forestry to maintain uniformity in crops and rapidly produce large numbers of plants with desirable traits. There are several key techniques involved in artificial vegetative reproduction:

» **Stem Cuttings**: Stem cuttings involve cutting a piece of the stem from a parent plant and planting it to produce a new plant. This method is used extensively in crops like **sugarcane** (*Saccharum officinarum*), **roses**, and **grapes** (*Vitis vinifera*). The cuttings develop roots and grow into new plants that are genetically identical to the parent. Stem cutting is a quick and effective method for mass propagation, especially in ornamental plants and fruit trees.

» **Root Cuttings**: Root cuttings involve the use of sections of roots to propagate new plants. This method is common in crops like **sweet potatoes** and some citrus varieties. When the cut section is planted, it produces new shoots and roots, developing into a full plant. Root cuttings are often used for plants that are difficult to propagate from seeds or require rapid multiplication.

» **Layering**: In layering, a stem is bent to the ground, and a portion is covered with soil while still attached to the parent plant. Roots form at the buried section, and once established, the stem is cut from the parent plant to grow independently. This method is particularly useful for woody plants like **roses** and **blackberries**. Layering ensures a high success rate, as the new plant remains attached to the parent plant during root formation.

» **Grafting**: Grafting involves joining the tissues of two plants so that they grow as one. The root system of one plant, called the rootstock, is combined with the shoot system (scion) of another plant. This method is used in fruit trees like **apple**, **pear**, and **citrus** species to combine the desirable traits of both plants, such as disease resistance from the rootstock and high fruit yield from the scion. Grafting is also used to repair damaged plants or trees and in horticulture to produce ornamental plants with unique characteristics.

» **Micropropagation**: Micropropagation is a modern technique that involves growing plants in a laboratory from small tissue samples. This method uses sterile conditions and specific growth media to produce large numbers of genetically identical plants from a small piece of plant tissue. Micropropagation is widely used in commercial agriculture for crops like **banana**, **orchids**, and **pineapple**. It is also used in conservation efforts to propagate endangered plant species. The primary advantage of micropropagation is its ability to produce large numbers of plants in a short time, all of which are clones of the parent plant.

Vegetative reproduction, whether natural or artificial, is essential in modern agriculture. It provides a reliable method for mass production of crops while maintaining consistency in desirable traits. However, the lack of genetic diversity can make crops more vulnerable to pests, diseases, and environmental changes. As a result, breeders often use a combination of vegetative and sexual reproduction methods to balance uniformity and diversity in crop populations.

ii) Apomixis

Apomixis is an asexual form of reproduction where seeds are produced without fertilization, resulting in offspring that are genetically identical to the parent plant. This process bypasses the typical sexual reproduction mechanism and allows for the preservation of hybrid vigor and desirable traits over multiple generations. Apomixis is observed in several plant species, and its application in plant breeding holds significant potential for improving crop production.

There are two primary types of apomixis:

» Adventive Embryony

» Gametophytic Apomixis

a) Adventive Embryony

Adventive embryony is a form of apomixis where the embryo develops directly from somatic cells of the ovule, such as the nucellus or integuments, without the involvement of the embryo sac or fertilization. This process allows for the direct formation of embryos from diploid cells, resulting in offspring that are clones of the parent plant. Adventive embryony is common in some species of **citrus**, where seeds may contain multiple embryos, one of which is derived from the adventive process.

The significance of adventive embryony lies in its ability to maintain the genetic integrity of the parent plant across generations. In agriculture, this form of apomixis can be beneficial for the propagation of high-quality fruit crops, where the consistency of traits like size, sweetness, and disease resistance is critical. However, adventive embryony can also be a challenge in breeding programs aimed at introducing genetic variation, as it limits the ability to create new hybrid combinations.

b) Gametophytic Apomixis

Gametophytic apomixis involves the development of an embryo from the egg cell or other cells within the embryo sac, without fertilization. This form of apomixis is further classified into two types:

» **Apospory**: In apospory, certain vegetative cells of the ovule develop into unreduced embryo sacs through mitotic divisions without undergoing meiosis. This process bypasses the typical reduction in chromosome number, resulting in an embryo sac that retains the diploid chromosome number. Apospory is observed in species like **grasses** and **orchids**, where the embryo develops without fertilization, maintaining the genetic identity of the parent plant.

» **Diplospory**: Diplospory involves the development of an embryo sac directly from the megaspore mother cell, without meiosis. In this process, the embryo sac retains the diploid chromosome number, and the embryo forms without the need for fertilization. Diplospory is observed in species like dandelions and some members of the **Asteraceae** family. Diplospory allows for the maintenance of hybrid vigor in crops, as the offspring are genetically identical to the parent plant.

Gametophytic apomixis is particularly valuable in breeding programs because it allows for the propagation of hybrids without the need for repeated crossbreeding. This is especially useful in crops where hybrid vigor results in higher yields, improved disease resistance, and better adaptability to environmental stresses. By preserving the genetic integrity of the hybrid, breeders can produce multiple generations of high-performing crops without the risk of losing desirable traits through genetic segregation.

- **Significance of Apomixis in Plant Breeding**: Apomixis holds immense potential for revolutionizing plant breeding. It allows breeders to bypass the challenges of maintaining hybrid vigor through repeated cross-pollination, providing a reliable method for producing uniform, high-quality crops. In crops like **grasses**, **citrus**, and **forage plants**, apomixis can ensure the consistent production of high-yielding, disease-resistant varieties. However, the challenge remains in harnessing apomixis for a wider range of crops, as not all species exhibit this form of reproduction. Future research into the genetic mechanisms of apomixis may unlock its potential for broader application in crop improvement programs.

2. Sexual Reproduction

Sexual reproduction is the process through which plants produce offspring by the fusion of male and female gametes, resulting in seeds that grow into new individuals. This mode of reproduction introduces genetic variation, which is crucial for the adaptability and survival of species. In agriculture, sexual reproduction is fundamental for plant breeding programs aimed at developing new varieties with improved traits such as disease resistance, higher yield, and better quality. The process of sexual reproduction involves several stages, including pollination, fertilization, and seed formation.

Sexual reproduction can be classified into two main types:

- Self-Pollination (Autogamy)
- Cross-Pollination (Allogamy)

i) Self-Pollination (Autogamy)

Self-pollination occurs when pollen from the male part (anther) of a flower fertilizes the female part (stigma) of the same flower or another flower on the same plant. This mode of reproduction is common in hermaphroditic plants, where both male and female reproductive organs are present in the same flower. Self-pollination leads to genetic uniformity, as the offspring inherit traits from a single parent, ensuring consistency in the crop's characteristics.

a) Mechanism of Self-Pollination

The process of self-pollination involves the transfer of pollen from the anther to the stigma within the same flower or plant. Several mechanisms promote self-pollination in plants:

- **Bisexuality**: In plants with bisexual or hermaphroditic flowers, the presence of both male and female organs allows self-pollination. For example, crops like **rice** (*Oryza sativa*), **wheat** (*Triticum aestivum*), and **barley** (*Hordeum vulgare*) produce flowers that contain both stamens and pistils, making self-pollination possible.
- **Homogamy**: Homogamy refers to the simultaneous maturation of the anther and stigma within the same flower. This synchrony ensures that pollen is available when the stigma is receptive, promoting self-pollination. For example, **tomato** plants exhibit homogamy, where the flowers open, and both male and female organs mature at the same time.

- **Cleistogamy**: In cleistogamy, the flower does not open at all, and pollination occurs within the closed bud. This mechanism guarantees self-pollination, as pollen is deposited on the stigma before the flower even opens. Cleistogamy is common in crops like **wheat**, **barley**, and **oats** (*Avena sativa*).
- **Chasmogamy**: Chasmogamy refers to the opening of flowers after pollination has already occurred. This prevents the entry of foreign pollen and ensures that self-pollination takes place before the flower opens to the outside environment. This is seen in crops like **rice** and **wheat**, where the flowers remain closed during pollination.

b) Advantages and Disadvantages of Self-Pollination

Self-pollination offers several advantages in plant breeding and crop production:

- **Genetic Uniformity**: Since self-pollination involves a single parent, the offspring are genetically uniform. This uniformity ensures consistency in crop traits, which is especially important in commercial agriculture, where uniformity in size, shape, and quality is desired.
- **Reproductive Assurance**: Self-pollination provides a reliable method of reproduction, especially in environments where pollinators are scarce or unreliable. In isolated areas or under harsh environmental conditions, self-pollination ensures that the plant can reproduce without relying on external agents like wind or insects.
- **Inbreeding Depression**: One of the main disadvantages of self-pollination is the risk of inbreeding depression. Over successive generations, self-pollination can lead to the accumulation of deleterious alleles, reducing the overall fitness, vigor, and yield of the plant population. This can result in lower productivity and increased susceptibility to diseases and pests.

c) Examples of Self-Pollinated Crops

Self-pollination is common in many staple food crops, which have been selectively bred for consistent traits:

- **Wheat** (*Triticum aestivum*): Wheat is a self-pollinating crop, and its uniformity in traits such as grain size, milling quality, and yield is essential for its commercial production. The self-pollinating nature of wheat also makes it easier to breed varieties with specific traits, such as disease resistance or drought tolerance.
- **Rice** (*Oryza sativa*): Like wheat, rice is a self-pollinating crop. Its ability to self-pollinate ensures that desirable traits such as grain quality, cooking characteristics, and yield are consistently passed on to the next generation. Self-pollination has allowed breeders to develop high-yielding rice varieties that are adapted to different environmental conditions.
- **Barley** (*Hordeum vulgare*): Barley is another self-pollinating crop that benefits from genetic uniformity. This uniformity is crucial for producing consistent malt quality, which is important for the brewing industry. Self-pollination also makes barley easier to breed for specific traits such as disease resistance and drought tolerance.

ii) Cross-Pollination (Allogamy)

Cross-pollination occurs when pollen from the anther of one plant is transferred to the stigma of another plant of the same species. This mode of reproduction introduces genetic diversity, as the offspring inherit traits from two different parent plants. Cross-pollination is essential for increasing the genetic variation within a population, which is crucial for adapting to changing environmental conditions and resisting diseases and pests.

a) Mechanism of Cross-Pollination

Cross-pollination is facilitated by various external agents, including wind, insects, water, and animals. The mechanisms that promote cross-pollination in plants include:

- **Dichogamy**: Dichogamy refers to the differential timing of maturation between the anther and stigma within the same flower. This prevents self-pollination and encourages cross-pollination. Dichogamy can be further divided into two types:
  - **Protandry**: In protandry, the anther matures before the stigma, preventing self-pollination and promoting cross-pollination. This is common in crops like **maize** (*Zea mays*) and **carrots** (*Daucus carota*).
  - **Protogyny**: In protogyny, the stigma matures before the anther, ensuring that pollen from another plant fertilizes the flower. **Pearl millet** (*Pennisetum glaucum*) is an example of a crop that exhibits protogyny.
- **Heterostyly**: Heterostyly refers to the presence of different lengths of styles and filaments in flowers of the same species. This physical difference promotes cross-pollination by ensuring that pollen from one flower is more likely to reach the stigma of a flower with a different style length. **Linseed** (*Linum usitatissimum*) is an example of a crop that exhibits heterostyly
- **Self Incompatibility**: Some plants have mechanisms that prevent self-pollination, even if pollen from the same plant reaches the stigma. Self-incompatibility is a genetic mechanism that prevents fertilization by the plant's own pollen, ensuring cross-pollination. This is common in crops like **brassicas** (e.g., **cabbage**, **broccoli**) and **radish** (*Raphanus sativus*).
- **Monoecy and Dioecy**: In **monoecious** plants, male and female flowers are separate but occur on the same plant, as in **maize** and **cucumber** (*Cucumis sativus*). In **dioecious** plants, male and female flowers occur on separate plants, requiring cross-pollination for reproduction. Examples of dioecious plants include **papaya** (*Carica papaya*) and **kiwifruit** (*Actinidia deliciosa*).

b) Advantages and Disadvantages of Cross-Pollination

Cross-pollination offers several advantages in plant breeding:

- **Genetic Diversity**: Cross-pollination increases genetic diversity within a population, allowing plants to adapt to changing environmental conditions. This genetic variation is also crucial for plant breeders, as it provides a larger pool of traits to select from, such as disease resistance, drought tolerance, and improved yield.
- **Hybrid Vigor (Heterosis)**: Cross-pollination often leads to hybrid vigor, where the offspring exhibit superior traits compared to the parent plants. Hybrid vigor can result in increased yield,

faster growth, and greater resilience to environmental stresses. This phenomenon is widely exploited in crop breeding programs, particularly for maize and other cross-pollinated crops.

» **Dependency on Pollinators**: One disadvantage of cross-pollination is the reliance on external agents, such as wind, insects, or animals, for pollen transfer. Environmental factors that affect pollinator populations, such as habitat loss, pesticide use, or climate change, can reduce the success of cross-pollination and negatively impact crop yields.

c) Examples of Cross-Pollinated Crops

Cross-pollination is common in many crops, especially those that benefit from genetic diversity and hybrid vigor:

» **Maize** (*Zea mays*): Maize is a wind-pollinated crop that relies on cross-pollination for genetic diversity. Hybrid maize varieties have been developed through selective cross-pollination, resulting in improved yield, disease resistance, and adaptability to different environmental conditions.

» **Sunflowers** (*Helianthus annuus*): Sunflowers are primarily pollinated by insects, particularly bees. Cross-pollination in sunflowers enhances genetic diversity and contributes to improved seed production, oil content, and resistance to pests and diseases.

» **Sorghum** (*Sorghum bicolor*): Sorghum is another crop that relies on cross-pollination to maintain genetic diversity. Breeding programs for sorghum often use cross-pollination to develop hybrids with improved traits such as drought tolerance and increased biomass production.

3. Mixed Reproduction Systems

Some plants exhibit both sexual and asexual reproduction, allowing them to adapt to a variety of environmental conditions. These plants can reproduce asexually through vegetative structures or apomixis, while also producing seeds through sexual reproduction. Mixed reproductive systems offer the flexibility to propagate desirable traits while also introducing genetic variation when necessary. Some plant species are capable of both **sexual and asexual reproduction**, allowing them to use different reproductive strategies based on environmental conditions. These systems provide plants with a unique advantage, offering flexibility in how they propagate. When environmental conditions are favorable, plants may use sexual reproduction to generate genetic diversity, while in more challenging environments, they might switch to asexual reproduction to ensure survival and consistent propagation of traits. Mixed reproductive systems are particularly important for plant breeders because they offer opportunities for maintaining desirable characteristics while also allowing for the introduction of new traits through genetic diversity. Let's further explore the significance, advantages, and limitations of mixed reproductive systems.

i) Importance of Mixed Reproduction Systems

Mixed reproductive systems offer several benefits in plant breeding and natural ecosystems. These plants can balance the need for genetic diversity with the ability to reproduce rapidly and maintain consistent traits through asexual means.

» **Adaptability**: Plants with mixed reproduction systems can adapt to fluctuating environments by alternating between sexual and asexual reproduction. During periods of environmental stress or when pollinators are scarce, asexual reproduction can ensure the survival and propagation of the species. Conversely, in favorable conditions, sexual reproduction introduces genetic diversity, allowing the population to evolve and adapt to new challenges.

» **Stability and Flexibility**: In agricultural systems, crops with mixed reproduction methods offer both stability and flexibility. For example, when a farmer needs to rapidly propagate a high-performing crop, asexual reproduction ensures that desirable traits are maintained. However, by allowing sexual reproduction, breeders can introduce new traits such as disease resistance or higher yield.

ii) Mechanisms of Mixed Reproduction

Several mechanisms allow plants to use both sexual and asexual reproduction. These mechanisms include:

» **Vegetative Reproduction with Sexual Capacity**: Plants that reproduce vegetatively through structures like tubers, stolons, or rhizomes may also produce seeds through sexual reproduction. For example, **potatoes** propagate asexually through tubers but can also flower and produce seeds sexually. This dual capacity allows potatoes to spread rapidly through asexual means while also introducing genetic variation through seed production.

» **Facultative Apomixis**: In facultative apomixis, plants can reproduce both sexually and asexually through apomixis. In species like **dandelions** (*Taraxacum officinale*), apomixis allows for clonal reproduction, ensuring the preservation of traits, while occasional sexual reproduction introduces genetic variation. This system provides a balance between maintaining desirable traits and adapting to environmental changes.

iii) Advantages and Disadvantages of Mixed Reproduction Systems

» **Advantages**:

- **Genetic Diversity and Uniformity**: Mixed reproduction systems allow plants to maintain genetic diversity through sexual reproduction while ensuring uniformity through asexual reproduction. This is particularly useful in breeding programs that require both stable, high-performing crops and opportunities for genetic improvement.
- **Survival and Resilience**: Plants with mixed reproductive systems are more resilient to environmental changes. If sexual reproduction is compromised due to a lack of pollinators or unfavorable conditions, the plant can still reproduce asexually. This ensures the continuation of the species even in challenging environments.
- **Increased Breeding Opportunities**: For breeders, mixed reproductive systems offer increased opportunities to develop new varieties. By selecting plants that can reproduce both sexually and asexually, breeders can work with a larger gene pool while maintaining desirable traits across generations.

» **Disadvantages**:

- **Reduced Genetic Diversity**: While mixed systems provide the option of asexual reproduction, they can also lead to reduced genetic diversity if sexual reproduction is not frequently used. Over-reliance on asexual reproduction can result in a population that is more susceptible to diseases and pests due to a lack of genetic variability.

- **Complex Breeding Programs**: Managing plants with mixed reproductive systems can be more complex for breeders. While asexual reproduction ensures uniformity, the introduction of genetic diversity through sexual reproduction requires careful planning to avoid losing desirable traits.

iv) Examples of Mixed Reproduction Systems

Several plant species exhibit mixed reproductive strategies:

» **Strawberries** (*Fragaria* spp.): Strawberries reproduce asexually through runners, which spread along the ground and form new plants. However, strawberries also produce flowers and seeds, allowing for sexual reproduction and genetic diversity.

» **Grasses**: Many grasses, such as **Kentucky bluegrass** (*Poa pratensis*), can reproduce both sexually through seeds and asexually through rhizomes. This dual strategy helps grasses to spread rapidly while also allowing for genetic recombination through sexual reproduction.

4. Mode of Pollination

Pollination is the process by which pollen is transferred from the anther (male organ) to the stigma (female organ) of flowers, facilitating fertilization and the formation of seeds. The mode of pollination is a critical factor in determining the reproductive success of plants and plays a significant role in plant breeding. There are two main types of pollination: **self-pollination** and **cross-pollination**. Each type has distinct genetic consequences and affects how breeders approach crop improvement.

i) Self-Pollination (Autogamy)

Self-pollination, or autogamy, occurs when pollen from the anther of a flower fertilizes the stigma of the same flower or another flower on the same plant. As a result, the offspring are genetically similar to the parent plant, leading to uniformity in traits. Self-pollination is common in many crops, particularly those that have been selectively bred for consistent quality and yield.

» **Mechanisms Promoting Self-Pollination**: Several mechanisms ensure self-pollination:

  - **Bisexuality**: In plants with hermaphroditic flowers, the presence of both male and female organs promotes self-pollination.
  - **Homogamy**: The simultaneous maturation of anthers and stigmas ensures that pollen is transferred to the stigma without external agents.
  - **Cleistogamy**: Some plants produce closed flowers where pollination occurs within the bud, ensuring self-pollination.

» **Examples**: Common self-pollinating crops include **rice**, **wheat**, **barley**, **peas**, and **beans**. These crops are highly uniform and have been bred to maintain specific desirable traits.

ii) Cross-Pollination (Allogamy)

Cross-pollination, or allogamy, occurs when pollen is transferred from the anther of one plant to the stigma of a different plant of the same species. This process introduces genetic diversity, as the offspring inherit a combination of traits from two different parent plants. Cross-pollination is vital for increasing the adaptability and resilience of plants to environmental changes, diseases, and pests.

» **Mechanisms Promoting Cross-Pollination**: Several mechanisms promote cross-pollination in plants:
  - **Dichogamy**: The anther and stigma mature at different times, preventing self-pollination and promoting cross-pollination.
  - **Self-Incompatibility**: Some plants have genetic mechanisms that prevent pollen from fertilizing the same flower, ensuring cross-pollination.
  - **Monoecy and Dioecy**: In monoecious plants, male and female flowers are separate but occur on the same plant. In dioecious plants, male and female flowers occur on different plants, requiring cross-pollination.

» **Examples**: Cross-pollinating crops include **maize, sunflowers, sorghum, apples**, and **grapes**. These crops exhibit high genetic diversity and are often bred for hybrid vigor, where cross-pollination leads to offspring that outperform their parent plants.

5. Significance of Reproduction and Pollination in Plant Breeding

The mode of reproduction and pollination plays a pivotal role in plant breeding strategies. Breeders must carefully select the appropriate breeding methods based on the reproductive biology of the crop species.

» **In Self-Pollinated Crops**: Breeding programs focus on creating **homozygous lines** that consistently express desirable traits. Methods like **pure line selection** and **pedigree breeding** are commonly used to develop high-yielding, disease-resistant varieties.

» **In Cross-Pollinated Crops**: Breeders aim to maintain **genetic diversity** and exploit hybrid vigor through **hybrid breeding** and **synthetic variety development**. Cross-pollinated crops offer greater opportunities for genetic improvement, but breeding them requires careful management of pollination and genetic variability.

B. Genetic Basis of Breeding Self-Pollinated and Cross-Pollinated Crops

Breeding self-pollinated and cross-pollinated crops requires a deep understanding of the genetic principles that govern inheritance, variability, and selection. The genetic makeup of these crops differs significantly due to the mode of reproduction, and as a result, the breeding strategies also vary. Self-pollinated crops tend to be more homozygous and uniform, while cross-pollinated crops exhibit more genetic diversity and heterozygosity.

1. Genetic Basis of Breeding Self-Pollinated Crops

Breeding self-pollinated crops, also known as autogamous crops, is a unique and crucial component of modern plant breeding. These crops reproduce through the transfer of pollen from the anther to the stigma of the same flower or plant, resulting in the offspring receiving identical alleles from both parents. Over time, this mode of reproduction leads to high levels of homozygosity, genetic uniformity, and stabilization of traits. Self-pollination is particularly advantageous for the creation of pure lines, but it also presents challenges, such as inbreeding depression and limited genetic diversity.

i) Homozygosity and Genetic Uniformity

The defining genetic feature of self-pollinated crops is their homozygosity, a condition where the plant's alleles at particular gene loci are identical due to self-pollination. Since the offspring inherit genes from only one parent, they maintain the same genetic composition as the parent plant, leading to high uniformity within the population. This genetic structure has several key implications for the breeding process:

a) Pure Line Development

One of the primary goals in breeding self-pollinated crops is the development of **pure lines**. A pure line is a population of plants that are genetically identical and homozygous for the traits of interest. Pure line development is essential for ensuring that the desirable traits are consistently expressed across generations. To achieve this, breeders subject the plant population to repeated cycles of self-pollination and selection, gradually eliminating heterozygous individuals and fixing homozygous alleles.

For instance, in **wheat** (*Triticum aestivum*) and **rice** (*Oryza sativa*), pure line selection is extensively used to ensure uniformity in traits such as grain size, milling quality, and disease resistance. In rice breeding programs, pure lines are crucial for the development of high-yielding varieties that can withstand diseases like bacterial blight (*Xanthomonas oryzae*). In **wheat**, the creation of pure lines allows breeders to stabilize traits related to yield, straw strength, and resistance to fungal diseases such as **rusts** (*Puccinia* spp.). The importance of pure line development is not only limited to maintaining uniformity but also extends to facilitating the efficient production of **breeder seed**, which forms the foundation for commercial seed production. Breeders must ensure that these pure lines remain genetically stable to meet the requirements of large-scale cultivation.

b) Genetic Stability

The genetic stability of self-pollinated crops is another critical advantage for breeders. In self-pollinated species, once a desirable trait is fixed through homozygosity, it remains stable across generations because the same alleles are passed down without introducing genetic variability. This stability is highly beneficial for breeding programs that require consistent expression of specific traits, such as yield, quality, and resistance to biotic (e.g., pests and diseases) and abiotic (e.g., drought, salinity, temperature) stresses. For example, in **barley** (*Hordeum vulgare*), pure line development ensures consistent performance in traits like lodging resistance (the ability of the plant to remain upright) and malt quality for the brewing industry. Similarly, in **soybeans** (*Glycine max*), genetic stability in pure lines ensures that seed quality, oil content, and protein levels remain consistent from one generation to the next, which is crucial for food processing and animal feed industries. However, the genetic stability achieved through self-pollination also presents a **trade-off**: while it ensures uniformity, it limits genetic diversity within the population. This limited variability can make the population vulnerable to changing environmental conditions or emerging diseases, as there is little genetic buffer to allow for adaptation. For example, during the mid-20th century, the **Irish potato famine** was exacerbated by the lack of genetic diversity in potato crops, which were clonally propagated and highly uniform, making them susceptible to **Phytophthora infestans**, the pathogen responsible for late blight.

c) Inbreeding Depression

Although self-pollination leads to homozygosity and genetic uniformity, it can also result in **inbreeding depression** in certain cases. Inbreeding depression occurs when the accumulation of deleterious recessive alleles in a homozygous state reduces the overall fitness, vigor, and productivity of the plant population. This phenomenon is particularly evident in crops where heterozygosity is important for maintaining vigor, such as cross-pollinated species. However, most self-pollinated crops are less susceptible to severe inbreeding depression because they have evolved mechanisms to tolerate high levels of homozygosity. For instance, in crops like **beans** (*Phaseolus vulgaris*) and **peas** (*Pisum sativum*), self-pollination does not lead to significant inbreeding depression because these species have been domesticated over centuries to perform well under self-pollinated conditions. However, in some cases, self-pollination can lead to reduced seed set, smaller plant sizes, or decreased resistance to environmental stresses, which can negatively impact crop performance. Breeders can mitigate the effects of inbreeding depression by carefully selecting for **vigorous lines** that maintain productivity and by introducing **genetic variation** through methods like **cross-breeding** or **mutation breeding**. For example, in **cotton** (*Gossypium hirsutum*), where hybrid varieties are used to combat the effects of inbreeding depression, breeders often introduce genetic diversity from wild relatives to improve yield and fiber quality.

ii) Selection Methods in Self-Pollinated Crops

Selection is the cornerstone of plant breeding, especially in self-pollinated crops. The process of selection involves identifying and propagating individuals that possess favorable traits, while eliminating those with undesirable characteristics. Since self-pollinated crops tend to become homozygous over time, the primary goal of selection in these crops is to fix favorable alleles in the population and develop uniform, high-performing lines. Several key selection methods are used in the breeding of self-pollinated crops, including **mass selection**, **pure line selection**, and **pedigree selection**. Each method has specific advantages and challenges, and plant breeders choose the most suitable method based on the crop species, breeding goals, and available genetic resources.

1. Mass Selection

Mass selection is one of the simplest and oldest methods used in plant breeding, especially in self-pollinated crops. It involves the selection of a large number of individuals based on their phenotypic traits. The selected individuals are allowed to self-pollinate, and their seeds are harvested to form the next generation. This process is repeated over several generations, gradually increasing the frequency of favorable alleles in the population. While mass selection is an effective method for improving certain traits, it does not eliminate genetic variability, as some heterozygous loci may persist within the population. Despite its simplicity, mass selection remains a valuable tool in the breeding of self-pollinated crops such as **beans** (*Phaseolus vulgaris*) and **lentils** (*Lens culinaris*).

a) Principles and Advantages of Mass Selection

The primary advantage of mass selection lies in its simplicity and ease of implementation. Breeders select plants that exhibit desirable traits, such as high yield, disease resistance, or drought tolerance, based on visual assessment or phenotypic markers. Once these individuals are selected, they are allowed to self-pollinate, and their seeds are collected to form the basis of the next generation. This process is repeated for several generations, gradually improving the population by increasing the frequency of favorable traits.

Mass selection is particularly useful in crops where quantitative traits—traits controlled by multiple genes—are important. For example, in **beans** (*Phaseolus vulgaris*), breeders use mass selection to improve yield, disease resistance, and seed size. The wide genetic variation in bean populations allows breeders to select individuals with the best performance in these traits, leading to gradual genetic improvement over time. Mass selection also provides breeders with the flexibility to respond to changing environmental conditions. Since it maintains some genetic variability within the population, breeders can select new traits or adapt to evolving biotic and abiotic challenges. For example, in **lentils** (*Lens culinaris*), breeders can use mass selection to develop varieties that are more tolerant to drought or resistant to specific diseases like **Ascochyta blight** (*Ascochyta lentis*).

b) Limitations of Mass Selection

While mass selection is an effective breeding method, it has certain limitations. One of the key challenges is that it does not eliminate all genetic variability within the population. As a result, some heterozygous loci may persist, and the population may not become as genetically uniform as it would through other methods such as pure line selection. This lack of uniformity can lead to variability in crop performance, particularly in traits that are controlled by single genes or simple inheritance patterns. Additionally, mass selection is often less efficient for improving highly complex traits, especially when the heritability of these traits is low. For example, improving traits like **seed protein content** or **nutrient use efficiency** may require more refined selection methods, as these traits are influenced by multiple genes and environmental factors.

2. Pure Line Selection

Pure line selection is a more precise and refined breeding method compared to mass selection. The goal of pure line selection is to identify a single superior plant within a genetically diverse population and propagate it over several generations through self-pollination. This process leads to the development of a **pure line**, which is a genetically uniform population of homozygous plants that consistently express the desired traits. Pure line selection is widely used in crops where uniformity in specific traits is essential for commercial production, such as **wheat** (*Triticum aestivum*), **barley** (*Hordeum vulgare*), and **rice** (*Oryza sativa*).

a) Principles of Pure Line Selection

The process of pure line selection begins by selecting a single plant that exhibits superior performance in traits of interest, such as yield, disease resistance, or grain quality. This plant is then allowed to self-pollinate, and its seeds are collected and sown in the following generation. Over successive generations of self-pollination, the population becomes increasingly homozygous, leading to a genetically uniform pure line. Each plant within the pure line is genetically identical and expresses the same traits, ensuring uniformity in crop performance. Pure line selection is particularly effective for improving traits that are controlled by single genes or have high heritability. For example, in **wheat** breeding, pure line selection is used to stabilize traits such as grain size, milling quality, and disease resistance. By developing pure lines, breeders ensure that these traits are consistently expressed across generations, allowing for reliable crop performance in different environments. Improving yield and quality traits, pure line selection is also used to develop varieties with **resistance to biotic stresses** such as diseases and pests. For instance, in **barley** (*Hordeum vulgare*), pure line selection has been used to develop varieties resistant to fungal diseases like **powdery mildew** (*Blumeria graminis*) and **leaf rust** (*Puccinia hordei*). These disease-resistant pure lines are critical for ensuring stable crop production, particularly in regions where disease pressure is high.

b) Applications and Advantages of Pure Line Selection

The primary advantage of pure line selection is the genetic uniformity it provides. Once a pure line is developed, all plants within that line are genetically identical and exhibit the same traits. This uniformity is essential for commercial crop production, as it ensures consistency in yield, quality, and other important traits. For example, in **rice** (*Oryza sativa*), pure lines are used to develop varieties with uniform grain size, cooking quality, and resistance to diseases such as **rice blast** (*Magnaporthe oryzae*). These uniform traits are important for meeting market demands and ensuring high-quality produce.

Another advantage of pure line selection is that it simplifies the production of **certified seed**. Since pure lines are genetically stable and uniform, the seed produced from these lines can be reliably propagated without losing the desired traits. This is particularly important in crops like **wheat** and **barley**, where farmers rely on certified seeds to ensure consistent crop performance and meet quality standards. However, one limitation of pure line selection is that it can take several generations to develop a fully homozygous population. This process requires careful management and evaluation to ensure that the selected line remains genetically stable over time. Additionally, pure line selection is most effective for traits with high heritability, and it may be less efficient for improving complex traits that are influenced by multiple genes and environmental factors.

3. Pedigree Selection

Pedigree selection is a selection method that involves tracking the ancestry, or pedigree, of selected plants over several generations. This method is particularly useful for improving complex traits that are controlled by multiple genes, as it allows breeders to monitor the inheritance of desirable traits and make informed decisions about which individuals to propagate. Pedigree selection is widely used in crops like **rice** (*Oryza sativa*) and **soybeans** (*Glycine max*), where breeders aim to improve traits such as yield, disease resistance, and cooking quality.

a) Principles of Pedigree Selection

In pedigree selection, breeders start by crossing two genetically diverse parents that possess complementary traits. The resulting offspring, or **F1 generation**, are evaluated for desirable traits, and individuals with the best performance are selected. These selected individuals are allowed to self-pollinate, and their offspring, or **F2 generation**, are further evaluated. The process continues over several generations, with breeders carefully recording the pedigree of each selected individual and monitoring the inheritance of traits. One of the key advantages of pedigree selection is that it allows breeders to track the performance of specific lines over time. By maintaining detailed records of each plant's ancestry, breeders can identify which crosses produce the best combinations of traits and select individuals that consistently express these traits. For example, in **rice** breeding, pedigree selection is used to develop varieties with high yield potential, resistance to diseases like **bacterial blight** (*Xanthomonas oryzae*), and improved grain quality for different market preferences.

b) Advantages and Challenges of Pedigree Selection

The main advantage of pedigree selection is its ability to improve complex traits that are influenced by multiple genes. By tracking the inheritance of traits over multiple generations, breeders can identify which combinations of alleles produce the best performance and select individuals with superior genetic potential. This makes pedigree selection particularly valuable in crops where breeders aim to improve multiple traits simultaneously, such as **yield**, **disease resistance**, and **quality traits**.

Another advantage of pedigree selection is that it allows breeders to evaluate the performance of lines in different environments. Since pedigree selection involves multiple generations of self-pollination, breeders can test how well a particular line performs under various environmental conditions, such as drought, heat, or pest pressure. This is especially important in crops like **soybeans**, where breeders aim to develop varieties that are adaptable to different climates and growing conditions. However, one challenge of pedigree selection is that it requires careful record-keeping and management. Breeders must maintain detailed records of each plant's ancestry and traits, which can be time-consuming and labor-intensive. Additionally, pedigree selection may be less effective for traits with low heritability, as these traits are more strongly influenced by environmental factors and may not be consistently inherited.

iii) Breeding Objectives in Self-Pollinated Crops

The breeding of self-pollinated crops is driven by a range of objectives that aim to improve crop productivity, quality, and resilience under diverse environmental conditions. Self-pollinated crops have unique genetic characteristics, such as high levels of homozygosity, which influence how breeders approach the development of new varieties. The breeding objectives focus on critical factors such as **yield improvement, disease and pest resistance**, and **tolerance to abiotic stresses**. These objectives are central to addressing food security, enhancing crop adaptability, and reducing dependency on chemical inputs.

1. Yield Improvement

Improving yield is the most fundamental objective in the breeding of self-pollinated crops. Yield is a complex trait influenced by multiple genetic and environmental factors, making it a critical target for breeders. Self-pollinated crops like **wheat, rice**, and **barley** are major staples worldwide, and their productivity directly impacts global food security. Over the years, significant progress has been made in developing high-yielding varieties through breeding strategies that combine desirable genetic traits.

a) Yield as a Primary Breeding Objective

Yield is a quantitative trait, meaning that it is controlled by many genes, each contributing a small effect. This polygenic control of yield makes breeding for higher yields challenging, as it requires careful selection of multiple traits that contribute to overall productivity. Breeders focus on improving both **biological yield** (total plant biomass) and **economic yield** (harvestable product, such as grain or seed).

For example, in **wheat breeding**, the development of high-yielding varieties was a key component of the **Green Revolution** in the 1960s. **Norman Borlaug**, often referred to as the father of the Green Revolution, developed semi-dwarf wheat varieties that were highly responsive to fertilizers and irrigation, leading to dramatic increases in wheat production. These varieties had shorter stems, which reduced lodging (falling over of the crop due to wind or rain) and allowed for higher planting densities, ultimately improving grain yield. Similarly, in **rice breeding**, the development of high-yielding varieties has been a priority for increasing global rice production. Breeders have focused on improving traits such as grain size, tillering capacity (the ability to produce side shoots), and resistance to lodging. High-yielding varieties like **IR8** (often referred to as "Miracle Rice") were instrumental in boosting rice production in Asia during the Green Revolution. These varieties combined high yield potential with resistance to pests and diseases, making them suitable for cultivation in diverse environments.

b) Strategies for Yield Improvement

Breeding for yield improvement involves multiple strategies, including:

- **Selection for Yield-Related Traits**: Breeders select for traits that contribute directly to yield, such as grain number, grain size, biomass production, and harvest index (the ratio of harvestable product to total plant biomass). For example, in **rice**, breeders select plants with higher numbers of grains per panicle (flower cluster) and greater grain filling capacity to improve yield.
- **Hybridization and Selection**: Cross-breeding high-yielding varieties with other lines that possess desirable traits, such as disease resistance or drought tolerance, is a common strategy. In self-pollinated crops like **wheat**, breeders develop high-yielding hybrid lines by selecting parents that exhibit complementary traits. These hybrids often display **heterosis** (hybrid vigor), resulting in increased yield and overall performance.
- **Use of Genetic Markers**: Marker-assisted selection (MAS) allows breeders to identify specific genes or genetic markers associated with high yield potential. By using molecular markers, breeders can accelerate the selection process and ensure that desirable alleles related to yield are incorporated into new varieties. This approach is increasingly used in crops like **rice** and **barley**, where identifying yield-related genes can speed up the development of high-performing varieties.

2. Disease and Pest Resistance

Breeding for disease and pest resistance is critical for reducing yield losses and minimizing the need for chemical inputs such as pesticides. Many self-pollinated crops are susceptible to a variety of diseases caused by fungi, bacteria, viruses, and pests. The incorporation of genetic resistance into crop varieties is one of the most effective and sustainable ways to combat these biotic stresses.

a) Importance of Disease Resistance in Self-Pollinated Crops

Self-pollinated crops, due to their uniform genetic structure, can be highly susceptible to epidemics if a pathogen overcomes their genetic resistance. Therefore, breeding for resistance to major diseases is a priority for maintaining crop productivity. Resistance breeding involves identifying and incorporating resistance genes from wild relatives, landraces, or other crop varieties into elite lines. For example, in **rice breeding**, the two most significant diseases that affect global rice production are **blast disease** (*Magnaporthe oryzae*) and **bacterial blight** (*Xanthomonas oryzae pv. oryzae*). Blast disease can cause severe yield losses in tropical and subtropical regions, while bacterial blight affects rice in both tropical and temperate climates. Breeders have developed rice varieties that incorporate resistance genes such as **Pi54** for blast resistance and **Xa21** for bacterial blight resistance. These genes provide a durable defense against pathogen attack, reducing the reliance on chemical control measures.

b) Strategies for Disease and Pest Resistance Breeding

There are several strategies for breeding disease- and pest-resistant self-pollinated crops:

» **Incorporation of Resistance Genes**: Breeders identify resistance genes in wild relatives or landraces and incorporate them into elite crop varieties through hybridization and backcrossing. For example, in **wheat**, resistance to **stem rust** (*Puccinia graminis*) has been achieved by introducing the **Sr2** gene from wild wheat species into modern wheat varieties. This gene provides durable resistance to the stem rust pathogen, which poses a major threat to wheat production worldwide.

» **Pyramiding of Resistance Genes**: Pyramiding involves stacking multiple resistance genes into a single variety to provide broad-spectrum and long-lasting resistance. This strategy helps prevent the pathogen from overcoming resistance, as it would need to overcome several different resistance mechanisms. In **rice**, breeders have successfully pyramided resistance genes for blast, bacterial blight, and sheath blight to create varieties with multi-disease resistance.

» **Breeding for Pest Resistance**: Breeding for resistance to insect pests is also crucial for minimizing yield losses. For instance, in **rice**, resistance to the **brown planthopper** (*Nilaparvata lugens*), a major pest, has been achieved by incorporating genes like **Bph3** and **Bph4** from wild rice species into cultivated rice varieties. These resistance genes reduce the feeding success of the planthopper and protect the plant from damage.

3. Abiotic Stress Tolerance

Abiotic stresses, such as drought, salinity, extreme temperatures, and soil nutrient deficiencies, are major constraints to crop production. Breeding for tolerance to these stresses is essential for improving the resilience of self-pollinated crops, especially in regions that experience harsh environmental conditions. The goal of breeding for abiotic stress tolerance is to develop varieties that can maintain productivity under suboptimal growing conditions.

a) Importance of Abiotic Stress Tolerance

Self-pollinated crops, like all crops, are affected by abiotic stresses, which can lead to significant yield losses. However, developing varieties that can withstand these stresses is particularly important for ensuring food security in regions prone to **drought**, **saline soils**, or **temperature extremes**. Improving abiotic stress tolerance through breeding helps farmers cope with climate change and reduce the risk of crop failure. For example, in **barley breeding**, drought tolerance is a key breeding objective, as barley is often grown in arid and semi-arid regions. Breeders have focused on developing varieties with improved water-use efficiency, deeper root systems, and tolerance to low soil moisture. These varieties can maintain high yields under water-limited conditions, making them suitable for cultivation in regions with unpredictable rainfall patterns.

b) Strategies for Breeding Abiotic Stress Tolerance

Breeding for abiotic stress tolerance involves several strategies, including:

» **Selection for Physiological Traits**: Breeders select for traits that enhance the plant's ability to withstand abiotic stresses. For example, **drought-tolerant varieties** of **wheat** have been developed by selecting for traits such as deeper root systems, delayed leaf senescence, and osmotic adjustment (the plant's ability to retain water under drought conditions).

» **Incorporating Stress-Tolerance Genes**: Molecular breeding techniques, such as marker-assisted selection (MAS), are used to identify and incorporate genes that confer tolerance to abiotic stresses. For instance, in **rice**, the **Saltol** gene has been introduced into high-yielding varieties to improve tolerance to soil salinity, making these varieties suitable for cultivation in coastal and saline-affected areas.

» **Breeding for Temperature Tolerance**: High temperatures during critical growth stages, such as flowering, can significantly reduce yield in crops like **rice** and **barley**. Breeders select for traits such as **heat tolerance** and **early flowering** to ensure that the crop can complete its life cycle before extreme heat sets in. For example, heat-tolerant wheat varieties have been developed for cultivation in regions with rising temperatures due to climate change.

» **Development of Climate-Resilient Varieties**: The challenge of climate change has prompted breeders to develop climate-resilient varieties that can withstand multiple abiotic stresses. For instance, in **barley**, breeders are developing varieties that combine drought tolerance, heat tolerance, and resistance to nutrient deficiencies. These varieties are essential for ensuring stable yields in the face of climate variability.

2. Genetic Basis of Breeding Cross-Pollinated Crops

Breeding cross-pollinated crops, also known as allogamous crops, presents unique challenges and opportunities due to the fundamental reproductive mechanisms that promote genetic diversity. In cross-pollinated crops, pollen from one plant fertilizes the ovule of another plant, ensuring a continuous exchange of genetic material. This mode of reproduction results in offspring that are genetically different from both parents, leading to high levels of heterozygosity and genetic variation within the population. The genetic diversity inherent in cross-pollinated crops is a crucial factor in their adaptability, resilience, and overall performance under various environmental conditions. The breeding of cross-pollinated crops focuses on managing genetic variation, maintaining heterozygosity, exploiting heterosis, and developing hybrid varieties that leverage the advantages of cross-pollination.

i) Heterozygosity and Genetic Diversity

The most distinctive genetic feature of cross-pollinated crops is their high degree of heterozygosity, which is the presence of different alleles at a particular locus. This genetic heterogeneity is the result of cross-pollination, where pollen from one plant fertilizes the ovule of another, producing offspring that carry different combinations of alleles. The resulting genetic diversity is a key factor in the adaptability and resilience of cross-pollinated crops, providing a broad pool of genetic material for breeders to work with. This diversity has several important implications for breeding:

a) Genetic Variation in Cross-Pollinated Crops

Genetic variation is the cornerstone of breeding programs for cross-pollinated crops. This variation arises from the recombination of alleles during sexual reproduction, which introduces new combinations of traits in each generation. The high level of genetic diversity in cross-pollinated crops provides breeders with a rich resource for selecting and improving traits such as yield, quality, resistance to diseases and pests, and tolerance to abiotic stresses like drought and salinity. For example, in **maize** (*Zea mays*), one of the most widely grown cross-pollinated crops, the genetic diversity within populations has been extensively exploited to develop high-yielding, drought-tolerant, and disease-resistant varieties. Maize exhibits a wide range of phenotypic traits due to the large pool of genetic variation available for selection. Breeders have utilized this variation to improve critical traits such as kernel size, plant height, ear placement, and resistance to fungal diseases like **corn leaf blight** (*Bipolaris maydis*) and **corn rust** (*Puccinia sorghi*).

The importance of genetic diversity is also evident in crops like **alfalfa** (*Medicago sativa*), where breeders aim to maintain a diverse gene pool to ensure high forage yield and quality. Alfalfa is a perennial forage crop that is grown in a variety of climates, and its ability to thrive under different environmental conditions is largely due to the genetic diversity present within the population. This diversity allows breeders to select for traits like drought tolerance, winter hardiness, and resistance to pests such as **alfalfa weevil** (*Hypera postica*). In contrast to self-pollinated crops, where homozygosity is the goal, the genetic variation in cross-pollinated crops is intentionally preserved to maintain adaptability and resilience. This variation enables breeders to respond to new challenges, such as emerging diseases or changing climatic conditions, by selecting individuals that carry the most favorable combinations of alleles.

b) Heterozygous Balance and Population Vigor

In cross-pollinated crops, a **heterozygous balance** is maintained between homozygous and heterozygous loci within the population. This balance contributes to the overall vigor, resilience, and adaptability of the crop. Heterozygosity, which results from the combination of different alleles at a gene locus, often confers greater adaptability and fitness to the plant, as heterozygous individuals tend to be more vigorous and better able to withstand environmental stresses than homozygous individuals.

For example, in **alfalfa**, breeders aim to maintain a heterozygous balance to ensure that the population retains its high forage yield and quality. Alfalfa plants that are heterozygous at key loci exhibit greater vigor, faster growth, and better resistance to diseases like **anthracnose** (*Colletotrichum trifolii*) and **verticillium wilt** (*Verticillium albo-atrum*). This heterozygous balance is also important for maintaining high seed set and fertility in the population, as heterozygous plants are often more fertile and produce more viable seeds than homozygous individuals.

In crops like **sunflower** (*Helianthus annuus*), maintaining heterozygosity is essential for ensuring high yields and good oil content. Sunflower is an important oilseed crop, and heterozygous plants tend to produce larger seeds with higher oil content compared to homozygous plants. Breeders maintain heterozygous populations through cross-pollination, ensuring that the plants retain their vigor and productivity. The **heterozygous advantage**, also known as **heterosis** or **hybrid vigor, which** is a key concept in the breeding of cross-pollinated crops. Heterosis occurs when the offspring of two genetically diverse parents exhibit superior traits compared to either parent. This phenomenon is widely exploited in the development of hybrid varieties, where breeders cross two inbred lines to produce offspring that display enhanced vigor, yield, and overall performance.

c) Inbreeding Depression in Cross-Pollinated Crops

Unlike self-pollinated crops, which tolerate high levels of homozygosity, cross-pollinated crops are highly susceptible to **inbreeding depression.** Inbreeding depression occurs when a cross-pollinated crop is subjected to self-pollination, leading to the accumulation of deleterious recessive alleles. These alleles, when expressed in a homozygous state, reduce the plant's vigor, fertility, and yield. This reduction in fitness is because many harmful mutations are recessive and are only expressed when both alleles at a locus are identical.

In cross-pollinated crops like **sunflowers** (*Helianthus annuus*), inbreeding depression can result in reduced seed set, smaller plant size, and lower oil content. Sunflowers are highly heterozygous, and when they are forced to self-pollinate, the loss of heterozygosity leads to a decline in overall plant fitness. To avoid inbreeding depression, breeders maintain cross-pollination in sunflower populations, ensuring that the plants remain heterozygous and retain their vigor. Similarly, in **maize**, inbreeding depression is a major concern. Maize is a highly cross-pollinated crop, and when it is self-pollinated for several generations, the accumulation of recessive alleles leads to a decline in yield, plant height, and resistance to diseases. To combat this, maize breeders develop **hybrid varieties** by crossing two inbred lines, which restores heterozygosity and maximizes yield potential through heterosis.

ii) Exploitation of Heterosis (Hybrid Vigor)

One of the most important breeding strategies in cross-pollinated crops is the exploitation of **heterosis**, or hybrid vigor. Heterosis refers to the phenomenon where the offspring of two genetically distinct parents exhibit superior traits compared to either parent. This improvement in traits, such as yield, growth rate, or stress tolerance, is the result of increased heterozygosity, which allows for better complementation of alleles and enhanced performance.

a) Development of Hybrid Varieties

In cross-pollinated crops, hybrid varieties are developed to maximize the benefits of heterosis. Hybrid breeding involves the crossing of two genetically distinct inbred lines, each of which carries desirable traits. The resulting **F1 hybrid** exhibits improved performance due to the combination of favorable alleles from both parents. Hybrid varieties are widely used in crops like **maize, sunflower, sorghum** (*Sorghum bicolor*), and **canola** (*Brassica napus*), where heterosis leads to significant improvements in yield, disease resistance, and adaptability.

In **maize,** hybrid varieties have revolutionized global maize production. By crossing two inbred lines, breeders create F1 hybrids that display greater vigor, faster growth, and higher yield compared to either parent. The success of hybrid maize has been a key factor in increasing maize production worldwide, particularly in regions like the United States, where hybrid maize dominates the market.

Similarly, in **sorghum**, hybrid breeding has been used to develop varieties with improved drought tolerance, resistance to pests like the **sorghum midge** (*Contarinia sorghicola*), and higher grain yield. Sorghum hybrids are widely cultivated in arid and semi-arid regions, where their enhanced performance under water-limited conditions makes them a valuable crop for food, fodder, and biofuel production.

b) Heterosis for Disease Resistance

In addition to improving yield, heterosis is also exploited for developing disease-resistant hybrids in cross-pollinated crops. For example, in **sunflowers**, breeders develop hybrid varieties that are resistant to diseases such as **downy mildew** (*Plasmopara halstedii*) and **verticillium wilt** (*Verticillium dahliae*). These hybrids not only produce higher yields but also have improved resistance to pathogens, reducing the need for chemical treatments and contributing to more sustainable agricultural practices. Hybrid breeding for disease resistance is also common in **canola**, where F1 hybrids are developed to resist fungal diseases like **blackleg** (*Leptosphaeria maculans*) and **sclerotinia stem rot** (*Sclerotinia sclerotiorum*). The heterozygosity in hybrid canola varieties provides a genetic buffer against disease outbreaks, ensuring that the crop remains productive even in disease-prone regions.

c) Challenges of Hybrid Breeding

Despite the advantages of heterosis, hybrid breeding presents several challenges. The production of hybrid seeds requires careful control of pollination to prevent self-pollination or unintended cross-pollination. In some crops, breeders use **male sterility systems** to facilitate hybrid seed production, as male sterile plants cannot produce viable pollen and must be fertilized by another plant. However, maintaining male sterility and ensuring high-quality hybrid seed production can be labor-intensive and costly. Additionally, hybrid seeds must be produced anew each generation, as the heterosis observed in the F1 generation is lost in subsequent generations due to segregation and recombination. Farmers must purchase new hybrid seeds each year to maintain the benefits of heterosis, which can increase production costs.

ii) Breeding Methods in Cross-Pollinated Crops

The breeding of cross-pollinated crops requires specialized methods to maintain or enhance genetic diversity while improving desirable traits such as yield, disease resistance, and stress tolerance. Cross-pollinated crops, being allogamous, rely on genetic recombination through the exchange of pollen between different plants. This results in high heterozygosity and genetic diversity, which can be leveraged for crop improvement. The challenge in breeding cross-pollinated crops is balancing the need for genetic diversity with the selection of specific, desirable traits. Several key breeding methods are employed in cross-pollinated crops, including **recurrent selection, hybrid breeding**, and **synthetic variety development**. Each method is tailored to harness the genetic characteristics of cross-pollinated crops and optimize their performance in various environments.

1. Recurrent Selection

Recurrent selection is one of the most widely used breeding methods in cross-pollinated crops. This method involves selecting individuals with favorable traits and intercrossing them over multiple generations to improve the frequency of desirable alleles in the population. The goal of recurrent selection is to accumulate favorable alleles while maintaining genetic diversity, ensuring that the population remains adaptable and resilient to environmental changes. Recurrent selection is particularly useful for improving complex traits that are controlled by multiple genes, such as yield, disease resistance, and drought tolerance.

a) Principles of Recurrent Selection

The basic principle of recurrent selection is to repeatedly select and intermate individuals with the best performance in a specific trait or set of traits. This process is repeated over several generations, with each cycle of selection and intercrossing gradually increasing the frequency of favorable alleles in the population. Recurrent selection allows breeders to maintain genetic variation within the population, which is important for long-term adaptability and the ability to respond to new challenges, such as emerging diseases or changing climate conditions. There are different types of recurrent selection, including **phenotypic recurrent selection** and **genotypic recurrent selection.** In phenotypic recurrent selection, individuals are selected based on their observable traits (phenotype), while in genotypic recurrent selection, individuals are selected based on their genetic makeup (genotype). Both methods aim to enhance the population's genetic potential by increasing the frequency of desirable alleles.

b) Applications of Recurrent Selection

Recurrent selection is widely used in cross-pollinated crops like **maize** (*Zea mays*), **alfalfa** (*Medicago sativa*), and **rye** (*Secale cereale*). In **maize**, recurrent selection has been instrumental in improving traits such as grain yield, drought tolerance, and resistance to diseases like **northern corn leaf blight** (*Exserohilum turcicum*) and **southern corn leaf blight** (*Bipolaris maydis*). Breeders select high-performing plants within a population, intercross them, and repeat the process over several generations, gradually improving the population's overall performance.

In **alfalfa**, recurrent selection is used to develop varieties with improved forage yield, quality, and resistance to pests such as the **alfalfa weevil** (*Hypera postica*). Since alfalfa is a perennial crop, recurrent selection allows breeders to improve traits over multiple growing seasons, ensuring that the population remains productive and adaptable to different environmental conditions.

c) Advantages and Challenges of Recurrent Selection

One of the main advantages of recurrent selection is that it allows breeders to improve complex traits while maintaining genetic diversity. This is particularly important in cross-pollinated crops, where genetic variation is essential for the population's long-term adaptability. Recurrent selection is also flexible, as it can be applied to a wide range of traits and breeding goals. However, recurrent selection has its challenges. The process is time-consuming, as it requires several cycles of selection and intercrossing to achieve significant improvements. Additionally, phenotypic recurrent selection can be influenced by environmental factors, making it difficult to distinguish between genetic and environmental effects on trait performance. To overcome this, breeders often use **genotypic recurrent selection** or combine recurrent selection with **marker-assisted selection** (MAS) to ensure that the best genetic material is selected for breeding.

2. Hybrid Breeding

Hybrid breeding is one of the most powerful and widely used methods in the breeding of cross-pollinated crops. The development of hybrid varieties involves crossing two genetically distinct parents to produce offspring (hybrids) that exhibit **heterosis** or **hybrid vigor**. Hybrids often outperform their parents in terms of yield, vigor, and stress tolerance due to the combination of favorable alleles from both parents. Hybrid breeding is particularly important in crops like **maize**, **sunflowers** (*Helianthus annuus*), and **sorghum** (*Sorghum bicolor*), where the benefits of heterosis can lead to significant improvements in productivity.

a) Principles of Hybrid Breeding

The principle of hybrid breeding is based on the concept of **heterosis**, which occurs when the offspring of two genetically diverse parents exhibit superior traits compared to either parent. This performance improvement is due to the increased heterozygosity of the hybrid, which allows for better complementation of alleles and enhanced physiological efficiency. To develop hybrid varieties, breeders first create **inbred lines** by self-pollinating individuals over several generations to achieve homozygosity at key loci. Once the inbred lines are stable, they are crossed to produce **F1 hybrids**. The F1 hybrids typically show greater vigor, higher yield, and improved stress tolerance compared to their inbred parents. Hybrid breeding is especially effective for traits like yield, disease resistance, and drought tolerance, where heterosis can lead to significant improvements in crop performance.

b) Applications of Hybrid Breeding

Hybrid breeding has revolutionized the production of several cross-pollinated crops, particularly **maize** (*Zea mays*), where hybrid varieties have played a crucial role in increasing global maize production. **Hybrid maize** varieties exhibit superior yield, disease resistance, and adaptability compared to open-pollinated varieties. The development of hybrid maize during the 20th century was one of the key drivers of the **Green Revolution**, which dramatically increased crop productivity in many parts of the world.

In **sunflowers** (*Helianthus annuus*), hybrid breeding has been used to develop varieties with improved oil content, disease resistance, and drought tolerance. Sunflower hybrids are widely cultivated for their high oil yield, making them a valuable crop for both food and industrial uses. Hybrid breeding has also been used to develop sunflower varieties that are resistant to diseases like **downy mildew** (*Plasmopara halstedii*) and **Sclerotinia** stem rot (*Sclerotinia sclerotiorum*).

In **sorghum** (*Sorghum bicolor*), hybrid breeding has been used to develop varieties that are tolerant to drought, resistant to pests like the **sorghum midge** (*Contarinia sorghicola*), and have improved grain yield. Sorghum hybrids are widely grown in arid and semi-arid regions, where their ability to withstand water-limited conditions makes them a valuable crop for food, fodder, and biofuel production.

c) Advantages and Challenges of Hybrid Breeding

The main advantage of hybrid breeding is the exploitation of heterosis, which can lead to substantial improvements in yield, vigor, and stress tolerance. Hybrid varieties often outperform their parents and provide farmers with a more productive and reliable crop. Additionally, hybrid breeding allows for the rapid incorporation of desirable traits from both parents, making it an effective method for developing disease-resistant and climate-resilient varieties. However, hybrid breeding has several challenges. The production of hybrid seeds requires careful control of pollination to prevent self-pollination or unintended cross-pollination. In some crops, breeders use **male sterility systems** to facilitate hybrid seed production, as male-sterile plants cannot produce viable pollen and must be fertilized by another plant. However, maintaining male sterility and ensuring high-quality hybrid seed production can be labor-intensive and costly. Additionally, hybrid seeds must be produced anew each generation, as the heterosis observed in the F1 generation is lost in subsequent generations due to segregation and recombination. Farmers must purchase new hybrid seeds each year to maintain the benefits of heterosis, which can increase production costs.

3. Synthetic Variety Development

Synthetic varieties are developed by intercrossing multiple parents with desirable traits and allowing their progeny to cross-pollinate freely. This method maintains genetic diversity while selecting for favorable traits, making it suitable for crops that benefit from high genetic variability. Synthetic varieties are particularly useful in forage crops like **alfalfa** (*Medicago sativa*) and **rye** (*Secale cereale*), where maintaining genetic diversity is important for adaptability and resilience.

a) Principles of Synthetic Variety Development

The development of synthetic varieties involves selecting several genetically diverse parents that possess complementary traits, such as yield, quality, or disease resistance. These parents are intercrossed, and their offspring are allowed to cross-pollinate freely. The resulting population, known as a **synthetic variety**, maintains a high level of genetic diversity, which contributes to the overall resilience and adaptability of the crop. Unlike hybrid varieties, which are developed by crossing two inbred lines, synthetic varieties retain a higher degree of heterozygosity, allowing them to adapt to changing environmental conditions. This makes synthetic varieties particularly valuable in crops where genetic diversity is important for maintaining productivity over time.

b) Applications of Synthetic Variety Development

Synthetic variety development is widely used in forage crops like **alfalfa** and **rye**, where genetic diversity is important for traits like forage yield, quality, and resistance to diseases and pests. In **alfalfa**, synthetic varieties are developed by intercrossing multiple parents with high forage yield and resistance to pests like the **alfalfa weevil** (*Hypera postica*). These synthetic varieties maintain genetic diversity within the population, allowing them to perform well under a range of environmental conditions. In **rye** (*Secale cereale*), synthetic varieties are used to develop crops with improved grain yield, winter hardiness, and resistance to diseases like **powdery mildew** (*Blumeria graminis*). Rye is often grown in regions with harsh winters, and maintaining genetic diversity within synthetic varieties helps ensure that the crop remains productive under cold and challenging conditions.

c) Advantages and Challenges of Synthetic Variety Development

The main advantage of synthetic varieties is that they maintain genetic diversity, which contributes to the population's long-term adaptability and resilience. This makes synthetic varieties particularly suitable for crops that are grown in variable environments or where genetic diversity is important for traits like disease resistance and stress tolerance. However, the performance of synthetic varieties is generally lower than that of hybrid varieties, as synthetic populations do not exhibit the same level of heterosis. Additionally, the development of synthetic varieties requires the careful selection of parents to ensure that the population retains a high level of genetic diversity while still expressing desirable traits.

iii) Exploiting Heterosis in Cross-Pollinated Crops

Heterosis, also known as **hybrid vigor**, refers to the phenomenon where the offspring (F1 generation) of two genetically distinct parents exhibit superior performance compared to their parents. This concept is one of the most significant advantages of cross-pollination and is widely utilized in the breeding of cross-pollinated crops. In cross-pollinated species, the combination of diverse genetic material results in a high degree of heterozygosity, which can enhance traits such as yield, disease resistance, stress tolerance, and overall plant vigor. Exploiting heterosis through hybrid breeding has revolutionized agriculture and has led to the development of high-performing hybrid varieties in several important crops. The phenomenon of heterosis is particularly prominent in cross-pollinated crops because of their natural tendency to maintain a high level of genetic diversity. When two genetically distinct inbred lines are crossed, the resulting hybrid offspring often display increased growth, yield, and adaptability, making hybrid breeding an essential tool in crop improvement programs. Below are key examples of how heterosis is exploited in the breeding of cross-pollinated crops.

1. Heterosis in Maize

Maize (*Zea mays*) is one of the most successful examples of heterosis in crop breeding, and hybrid maize varieties have played a critical role in transforming global agriculture. The development and commercialization of hybrid maize in the early 20th century marked a turning point in maize production, as hybrid varieties consistently outperformed traditional open-pollinated varieties in terms of yield, disease resistance, and adaptability. The success of hybrid maize is largely attributed to the exploitation of heterosis, which has led to significant increases in global maize yields and contributed to the **Green Revolution**.

a) Hybrid Maize Development

The process of developing hybrid maize involves creating two inbred lines through repeated self-pollination. These inbred lines are genetically uniform and homozygous at key loci, but they often exhibit lower vigor and yield due to inbreeding depression. However, when two genetically distinct inbred lines are crossed, the resulting **F1 hybrid** shows remarkable improvement in traits such as yield, plant height, and disease resistance. This improvement is the result of heterosis, which allows the hybrid plants to capitalize on the complementary alleles inherited from both parents. Hybrid maize varieties are known for their **superior grain yield** and **adaptability to various growing conditions**. For example, hybrid maize varieties are developed to perform well in a range of environments, from tropical regions to temperate climates. Breeders select parents based on their complementary traits, such as disease resistance, drought tolerance, and nutrient efficiency. By combining the strengths of each parent, breeders create hybrids that are not only high-yielding but also resilient to environmental stresses.

b) Impact of Hybrid Maize on Agriculture

The impact of hybrid maize on global agriculture has been profound. In the United States, for instance, the adoption of hybrid maize in the 1930s and 1940s led to a dramatic increase in maize yields, contributing to the country's position as a leading maize producer. Hybrid maize varieties are now widely grown in many parts of the world, including **Africa**, **Asia**, and **Latin America**, where they contribute to food security and economic growth. In addition to yield improvement, hybrid maize varieties offer better resistance to **pests and diseases**. For example, hybrid maize varieties have been bred to resist **corn borers** (*Ostrinia nubilalis*) and **corn rootworms** (*Diabrotica spp.*), which are major pests in maize production. Hybrid maize varieties also show improved resistance to diseases such as **northern corn leaf blight** (*Exserohilum turcicum*) and **gray leaf spot** (*Cercospora zeae-maydis*). By incorporating resistance genes from diverse parental lines, hybrid maize offers farmers a more reliable and productive crop.

c) Future of Hybrid Maize Breeding

The future of hybrid maize breeding lies in the integration of modern technologies such as **marker-assisted selection** (MAS), **genomic selection**, and **biotechnology**. These tools allow breeders to more efficiently select parental lines with the most favorable alleles for hybridization. Additionally, the use of **genome editing** technologies, such as **CRISPR-Cas9**, may enable breeders to introduce specific genetic improvements directly into maize genomes, further enhancing hybrid performance.

2. Heterosis in Sunflowers

Sunflowers (*Helianthus annuus*) are another cross-pollinated crop where heterosis is exploited to improve key traits such as **seed yield, oil content**, and **disease resistance**. Sunflower hybrids have become the preferred choice for both oilseed production and ornamental cultivation due to their superior performance compared to open-pollinated varieties.

a) Hybrid Sunflower Breeding

The development of hybrid sunflower varieties follows a similar process to that of maize. Inbred lines are created through self-pollination, and these lines are then crossed to produce F1 hybrids. The F1 hybrids exhibit heterosis, resulting in increased seed production, better oil quality, and improved resistance to diseases such as **downy mildew** (*Plasmopara halstedii*) and **Sclerotinia stem rot** (*Sclerotinia sclerotiorum*). Hybrid sunflowers are widely grown for their high oil yield, which is used for both cooking and industrial purposes. Breeders select parental lines with complementary traits to ensure that the hybrid offspring inherit the best characteristics from both parents. For example, one parent may be selected for its high oil content, while the other parent may be chosen for its resistance to diseases or environmental stresses. The resulting hybrid displays the desired combination of traits, making it highly valuable for commercial production.

b) Exploiting Heterosis for Oil Content

One of the main objectives in sunflower breeding is to improve **oil content and quality**. Sunflower oil is prized for its high levels of **unsaturated fatty acids**, particularly **linoleic acid,** and **oleic acid**, which are important for human nutrition and industrial applications. Hybrid sunflowers consistently produce seeds with higher oil content and better oil quality compared to open-pollinated varieties. This is due to the combination of favorable alleles from both parents, which enhances the efficiency of oil synthesis in the hybrid plants. Hybrid sunflowers also exhibit greater resistance to **drought** and other abiotic stresses, making them suitable for cultivation in regions with challenging environmental conditions. The ability to withstand water-limited environments is particularly important in sunflower-growing regions such as **Eastern Europe, Argentina**, and the **United States**.

c) Disease Resistance in Hybrid Sunflowers

Disease resistance is another important target in sunflower hybrid breeding. **Downy mildew** (*Plasmopara halstedii*) and **Sclerotinia stem rot** (*Sclerotinia sclerotiorum*) are two major diseases that can significantly reduce sunflower yields. Breeders exploit heterosis to develop hybrids with enhanced resistance to these diseases, reducing the need for chemical treatments and improving the sustainability of sunflower production. For example, hybrid sunflowers bred for resistance to downy mildew are developed by incorporating resistance genes from diverse parental lines. These hybrids can resist infection and maintain high yields, even in regions where the disease is prevalent. The exploitation of heterosis for disease resistance is particularly valuable in organic farming systems, where chemical control options are limited.

3. Heterosis in Sorghum

Sorghum (*Sorghum bicolor*) is a staple crop in many parts of the world, particularly in **Africa** and **Asia**, where it is grown for food, fodder, and biofuel production. Sorghum is known for its ability to withstand drought and other environmental stresses, making it a valuable crop in regions with limited water resources. Hybrid sorghum varieties are developed to exploit heterosis for traits such as **drought tolerance**, **biomass production**, and **grain yield**.

a) Hybrid Sorghum Development

Hybrid sorghum varieties are created by crossing two genetically diverse parental lines, resulting in F1 hybrids that exhibit improved performance compared to either parent. The development of hybrid sorghum involves selecting parents that possess complementary traits, such as high grain yield and drought tolerance. The resulting hybrids show **heterosis**, leading to superior performance under both normal and stress conditions. In regions with erratic rainfall and poor soil conditions, hybrid sorghum varieties are a valuable tool for farmers. These hybrids can produce high yields even in the face of drought and nutrient-poor soils, making them an important crop for food security and income generation in resource-limited areas.

b) Drought Tolerance in Hybrid Sorghum

One of the key advantages of hybrid sorghum is its ability to tolerate **drought**. Sorghum is naturally adapted to dry environments, but hybrid varieties have been developed to further enhance this trait. By crossing parental lines with different drought tolerance mechanisms, breeders create hybrids that are more resilient to water-limited conditions. These hybrids can maintain high grain and biomass yields even when rainfall is scarce. Hybrid sorghum is particularly important in **sub-Saharan Africa**, where drought is a major constraint on agricultural production. By improving drought tolerance through heterosis, hybrid sorghum varieties help farmers achieve higher yields and more stable production in challenging environments.

c) Biomass Production in Hybrid Sorghum

In addition to grain production, sorghum is widely grown for **biomass**, which is used for animal fodder and biofuel production. Hybrid sorghum varieties exhibit increased biomass production due to heterosis, making them an important crop for both food and energy production. For example, hybrid sorghum varieties used for biofuel production are selected for traits such as rapid growth, high biomass yield, and efficient water use. Hybrid sorghum is also used as a forage crop for livestock, as it provides high-quality fodder with improved digestibility and nutritional content. The exploitation of heterosis in forage sorghum has led to the development of hybrids that are well-suited to intensive livestock production systems.

3. Genetic Consequences of Breeding Self- and Cross-Pollinated Crops

The genetic consequences of breeding self- and cross-pollinated crops arise due to differences in their modes of reproduction. These differences impact the strategies used by breeders, the level of genetic variation in populations, and the ability of crops to respond to environmental pressures. For breeders, understanding these consequences is critical to designing programs that either maintain or enhance genetic diversity while improving crop productivity. The contrasting nature of self- and cross-pollination results in distinct genetic outcomes in terms of **homozygosity**, **heterozygosity**, **inbreeding depression**, and the expression of desirable traits across generations.

i) Genetic Consequences in Self-Pollinated Crops

Self-pollinated crops reproduce by transferring pollen from the anther to the stigma of the same flower or plant, leading to genetic uniformity over successive generations. This mode of reproduction is also referred to as **autogamy**. Due to the lack of genetic mixing from outside sources, the primary consequence of breeding self-pollinated crops is an increase in homozygosity, which comes with both advantages and disadvantages.

1. Homozygosity

One of the most defining genetic consequences of self-pollination is the rapid and sustained increase in **homozygosity**. Homozygosity refers to the presence of identical alleles at specific gene loci. In self-pollinated crops, the repeated self-pollination of individuals leads to a highly uniform population where nearly all loci become homozygous. This genetic uniformity ensures that traits that are homozygous dominant (or recessive) are consistently expressed across multiple generations, making self-pollinated crops ideal for producing stable varieties with predictable traits.

- **Benefits of Homozygosity**: The primary advantage of increased homozygosity is the **stability** it brings to breeding programs. In self-pollinated crops like **wheat** (*Triticum aestivum*) and **rice** (*Oryza sativa*), this uniformity allows for the reliable production of varieties with desirable characteristics, such as grain size, milling quality, and resistance to specific diseases. Since all individuals in a pure line are genetically similar, breeders can ensure that traits such as yield, quality, and stress tolerance are consistently passed on.
- **Drawbacks of Homozygosity**: Despite its benefits, homozygosity also presents challenges. The lack of **genetic diversity** in self-pollinated crops can limit the plant's ability to adapt to changing environmental conditions or emerging diseases. This is because self-pollinated crops often have fewer alleles available to respond to new selective pressures, such as the emergence of a novel pathogen or extreme weather events.

2. Inbreeding Depression

While self-pollinated crops are generally less susceptible to **inbreeding depression** compared to cross-pollinated crops, excessive selfing can still lead to negative outcomes. **Inbreeding depression** occurs when the accumulation of deleterious recessive alleles reduces the overall fitness, vigor, and productivity of the crop population.

» **Causes of Inbreeding Depression**: Inbreeding depression arises from the expression of harmful recessive alleles that are normally masked in heterozygous individuals. Since self-pollination leads to the increased homozygosity of these recessive alleles, it can result in reduced plant vigor, lower fertility, and diminished yields. In crops like **barley** (*Hordeum vulgare*), inbreeding depression can manifest as reduced resistance to biotic stresses like pests and diseases, making populations more vulnerable to crop failure.

» **Management of Inbreeding**: To manage the potential for inbreeding depression, breeders of self-pollinated crops carefully select for favorable alleles while minimizing the accumulation of harmful recessive traits. The development of **pure lines** through **pedigree selection** or **bulk selection** helps breeders stabilize desirable traits while keeping inbreeding depression at manageable levels.

3. Genetic Uniformity

The high level of genetic uniformity in self-pollinated crops can be both an asset and a liability. On the one hand, it provides **consistency** in crop performance, ensuring that farmers can rely on stable yields and predictable traits. However, the lack of genetic variability means that these crops may not respond well to changing environmental conditions, leading to potential vulnerabilities in the face of climate change, evolving pests, and diseases. For example, the **Uniformity in Wheat**: The extensive cultivation of genetically uniform wheat varieties with limited genetic diversity has raised concerns about the long-term sustainability of wheat production, as uniform populations may lack the resilience to withstand new threats, such as climate-induced stress or novel pathogens.

ii) Genetic Consequences in Cross-Pollinated Crops

In contrast to self-pollinated crops, cross-pollinated crops rely on the exchange of pollen between different plants, leading to high levels of genetic diversity and heterozygosity. This reproductive strategy, also known as **allogamy**, has significant implications for the genetic makeup of cross-pollinated populations, which tend to be more adaptable and resilient than self-pollinated crops.

1. Heterozygosity

The most defining genetic characteristic of cross-pollinated crops is **heterozygosity**. Heterozygosity refers to the presence of different alleles at a gene locus, resulting in a genetically diverse population. In cross-pollinated crops, pollen from one plant fertilizes the ovule of another, leading to offspring that inherit alleles from both parents.

» **Benefits of Heterozygosity**: The high level of heterozygosity in cross-pollinated crops provides them with a broad genetic base, making them more adaptable to changing environmental conditions and more resilient to stresses such as pests, diseases, and drought. This genetic diversity allows cross-pollinated crops to **evolve** more rapidly in response to selective pressures. For instance, in **maize** (*Zea mays*), the genetic diversity within populations has been instrumental in developing varieties with **drought tolerance**, **disease resistance**, and **higher yields**.

» **Maintenance of Genetic Diversity**: Cross-pollinated crops maintain a **heterozygous balance**, which ensures that the population retains its genetic diversity over generations. This balance is particularly important for traits related to **environmental stress resistance** and **yield stability**. Breeders of crops like **alfalfa** (*Medicago sativa*) maintain heterozygosity to ensure high forage yield and adaptability across different growing conditions.

2. Heterosis

Heterosis, or hybrid vigor, is a phenomenon where the offspring of two genetically distinct parents exhibit superior traits compared to their parents. This is a key genetic consequence of cross-pollination and is widely exploited in the breeding of hybrid varieties.

» **Exploitation of Heterosis**: Heterosis is especially valuable in cross-pollinated crops like **maize**, **sunflowers**, and **sorghum**, where hybrid varieties consistently outperform open-pollinated varieties. In maize, for example, hybrid varieties developed through the crossing of inbred lines exhibit **higher yields**, **greater disease resistance**, and **improved stress tolerance**. This has revolutionized maize production, contributing to its global success as a staple crop.

» **Impact on Crop Improvement**: Heterosis enables breeders to achieve significant improvements in traits such as **yield**, **biomass production**, and **resistance to biotic and abiotic stresses**. In **sorghum**, hybrids are developed to withstand **drought conditions**, making them suitable for cultivation in water-scarce regions like sub-Saharan Africa and parts of Asia.

3. Inbreeding Depression

Unlike self-pollinated crops, cross-pollinated crops are highly susceptible to **inbreeding depression**. The accumulation of deleterious recessive alleles in a population due to selfing or close inbreeding can significantly reduce the fitness, vigor, and productivity of cross-pollinated crops.

» **Causes of Inbreeding Depression**: In cross-pollinated crops, inbreeding leads to the expression of harmful recessive alleles, which are normally masked in heterozygous individuals. This can result in reduced fertility, lower yields, and increased susceptibility to diseases. For example, in **sunflowers** (*Helianthus annuus*), inbreeding depression can manifest as reduced **seed set** and **lower oil content**, which are critical traits for sunflower production.

» **Prevention of Inbreeding**: To avoid inbreeding depression, breeders of cross-pollinated crops focus on maintaining **outcrossing** within populations. **Recurrent selection**, **synthetic variety development**, and **hybrid breeding** are some of the methods used to ensure that genetic diversity is preserved and that populations remain heterozygous.

C. Nature of Variability in Plant Breeding

Variability in plant breeding is the cornerstone that enables the development of improved crop varieties and sustains agricultural productivity. Without variability, the process of selection, which forms the backbone of plant breeding, would not be possible. Variability refers to the presence of differences in genetic and phenotypic traits among individuals within a population. These differences allow breeders to select desirable characteristics such as increased yield, improved disease resistance, or enhanced stress tolerance, and to develop new varieties of crops that are better suited to meet the needs of farmers and consumers.

The practice of plant breeding, which has its roots in the early days of agriculture, has been shaped by the variability present within plant populations. Early farmers relied on natural variation to select plants that performed better under their specific environmental conditions. This process, though unintentional at first, laid the foundation for modern plant breeding techniques, where variability is both harnessed and intentionally introduced to improve crops. Today, plant breeders have access to a wealth of knowledge about how variability is generated, maintained, and manipulated. This understanding allows them to create more productive, resilient, and high-quality crops that can meet the increasing demands of a growing global population, while also addressing challenges such as climate change, pest outbreaks, and soil degradation. The manipulation of variability in plant breeding involves understanding both the **genetic** and **phenotypic** aspects of plant populations, as well as the influence of the environment on these traits.

Genetic variability, in particular, is essential for long-term crop improvement. It provides the raw material for the selection of new traits, whether these traits are inherited through traditional breeding methods or introduced through more modern techniques like genetic engineering. Phenotypic variability, on the other hand, represents how these genetic differences are expressed under different environmental conditions. This aspect is equally important, as the performance of a crop is not determined solely by its genetic makeup but also by how it interacts with its environment. Over the past century, advances in plant breeding have significantly expanded our ability to create and maintain variability in crop populations. Techniques such as **hybridization**, **mutation breeding**, and **biotechnology** have provided breeders with new tools to introduce variability into crops, allowing for the development of varieties with enhanced traits. However, managing this variability is critical. Breeders must strike a balance between maintaining genetic diversity to ensure adaptability and creating uniformity in certain traits to meet specific agricultural or market demands.

In the context of plant breeding, variability can arise from multiple sources, including natural processes such as mutations and recombination, as well as artificial interventions like hybridization and genetic modification. Understanding the sources of variability, as well as the different types of variability, is key to developing effective breeding strategies that can lead to long-term improvements in crop productivity and sustainability.

Types of Variability in Plant Breeding

Variability in plant breeding can be broadly classified into **genetic variability** and **phenotypic variability**. These two types of variability are interconnected and play crucial roles in determining the success of a breeding program. While genetic variability is the primary source of heritable traits, phenotypic variability reflects how these traits are expressed in different environments. Both types of variability are essential for plant breeders, as it allows them to select plants that will perform well across a variety of conditions while maintaining the genetic diversity needed for long-term adaptability.

1. Genetic Variability

Genetic variability refers to the differences in the DNA sequences among individuals in a population. These differences are the foundation for the heritable traits that breeders select for improvement. Genetic variability is crucial because it provides the raw material that allows breeders to introduce new traits into crops or enhance existing ones. Without genetic variability, plant populations would become uniform, limiting the potential for further improvement.

Sources of Genetic Variability

Genetic variability can arise from several sources, including **mutations**, **recombination**, **hybridization**, and **introgression from wild relatives**. Each of these sources contributes to the overall diversity of a population, providing breeders with a wide array of traits to select from.

- **Mutations**: Mutations are spontaneous changes in the DNA sequence that can result in new alleles or variants of a gene. These changes can occur naturally or can be induced artificially through chemical or physical mutagens. While many mutations are neutral or deleterious, some can lead to beneficial traits such as increased disease resistance or improved yield. For example, in barley (*Hordeum vulgare*), induced mutations have led to the development of varieties with enhanced drought tolerance and improved nutritional quality.
- **Recombination**: During the process of meiosis, recombination occurs when alleles are shuffled between homologous chromosomes, resulting in new combinations of alleles in the offspring. This process is especially important in sexually reproducing plants, as it increases the genetic diversity within a population. In cross-pollinated crops like maize (*Zea mays*), recombination creates a wide range of genetic variability, allowing breeders to select for traits such as higher yield, better disease resistance, or improved environmental adaptability.
- **Hybridization**: Hybridization is the process of crossing two genetically distinct individuals to produce offspring with a combination of traits from both parents. This process introduces new genetic material into a population, creating variability that can be selected for crop improvement. Hybridization is particularly useful in crops like rice (*Oryza sativa*) and wheat (*Triticum aestivum*), where breeders cross parents with complementary traits to develop new, high-yielding varieties.
- **Introgression from Wild Relatives**: Wild relatives of cultivated crops often carry alleles for traits that have been lost in domesticated varieties, such as resistance to pests, diseases, or environmental stresses. Breeders can use these wild relatives as sources of genetic variability by crossing them with domesticated varieties and selecting for the desired traits. For example, the wild species *Aegilops tauschii* has been used to introduce disease-resistance genes into modern wheat varieties, enhancing their ability to resist rust diseases.

Importance of Genetic Variability

The importance of genetic variability in plant breeding cannot be overstated. It allows breeders to develop new varieties that are better suited to the challenges of modern agriculture, such as climate change, emerging pests and diseases, and the need for higher productivity. In self-pollinated crops, breeders focus on maintaining a certain level of genetic variability to avoid the negative consequences of inbreeding depression, which can occur when genetic diversity is too low. In cross-pollinated crops, maintaining genetic variability is essential for preserving heterozygosity and ensuring that crops can adapt to changing environmental conditions. Genetic variability also plays a key role in the development of **hybrid varieties**, which rely on the combination of alleles from genetically diverse parents to produce offspring with superior traits. Hybrid varieties often exhibit **heterosis**, or hybrid vigor, where the offspring outperform their parents in terms of yield, disease resistance, or stress tolerance. This phenomenon is widely exploited in crops like maize and sunflower (*Helianthus annuus*), where hybrid varieties have revolutionized agricultural productivity.

2. Phenotypic Variability

While genetic variability refers to the underlying genetic differences among individuals, phenotypic variability refers to the observable differences in traits that result from the interaction between an individual's genetic makeup and its environment. Phenotypic traits include characteristics such as plant height, leaf shape, flower color, yield, and resistance to diseases or pests. Phenotypic variability is influenced by both genetic factors and environmental conditions, making it an important consideration in plant breeding.

Sources of Phenotypic Variability

Phenotypic variability arises from several sources, including **environmental influences**, **genetic-environmental interactions**, and **plasticity**.

» **Environmental Influences**: The environment plays a significant role in shaping the phenotype of a plant. Factors such as soil quality, water availability, temperature, light intensity, and nutrient levels can all affect how a plant grows and develops. For example, a maize plant may exhibit different heights, leaf sizes, or ear lengths depending on the environmental conditions in which it is grown. Breeders must consider environmental variability when selecting for traits, as a variety that performs well in one environment may not perform as well in another.

» **Genetic-Environmental Interactions**: The interaction between an individual's genetic makeup and its environment is a key source of phenotypic variability. This interaction, often referred to as **G x E interaction**, occurs when the expression of genetic traits is influenced by environmental conditions. For example, a wheat variety may exhibit higher yields in a dry climate than in a wet one due to its genetic predisposition for drought tolerance. Breeders often evaluate G x E interactions through multi-environment trials to identify varieties that perform consistently across different environments.

» **Phenotypic Plasticity**: **Plasticity** refers to the ability of an organism to adjust its phenotype in response to environmental conditions. In plants, phenotypic plasticity allows individuals to modify their growth and development based on the resources available in their environment. For example, a tomato plant may produce larger leaves and longer stems when grown in low light conditions, as it attempts to capture more sunlight for photosynthesis. Breeders select for appropriate levels of plasticity depending on the crop and the environment in which it will be grown, ensuring that the variety can adapt to variable conditions without sacrificing yield or quality.

Importance of Phenotypic Variability

Phenotypic variability is critical in plant breeding because it provides a visible manifestation of the genetic diversity within a population. By observing how different individuals express their genetic potential in response to environmental conditions, breeders can identify those that are best suited for specific environments or that exhibit desirable traits under a range of conditions. Phenotypic variability also allows breeders to select traits that are important for crop performance, such as **yield**, **disease resistance**, and **stress tolerance**. In crops where environmental conditions can vary significantly from one growing season to the next, selecting for phenotypic plasticity is essential for ensuring consistent performance. As we discussed earlier, **phenotypic variability** is the observable expression of genetic variability in plants, shaped by both genetics and the environment. The nature of this variability is complex, and plant breeders rely heavily on it to achieve their goals of improving crops. To understand how phenotypic variability works, it is crucial to consider several more aspects:

i) Stability and Adaptability of Phenotypes

Stability refers to the ability of a plant to express a particular trait consistently across different environments, while **adaptability** refers to the plant's ability to adjust and perform well under varying environmental conditions. These two aspects are critical in crop breeding programs that target regions with fluctuating climates or resource availability.

Stability in plant breeding is essential because breeders aim to develop varieties that perform consistently well across different growing conditions. A crop that shows consistent high yield in diverse environments is considered stable. The selection of stable genotypes requires careful multi-environment trials, where crops are tested across various locations and environmental conditions to observe their phenotypic response. The stability of a crop is not just limited to yield but also applies to other traits such as disease resistance, flowering time, and quality of produce. For example, breeders developing rice (*Oryza sativa*) varieties for tropical and subtropical regions must ensure that the selected genotypes consistently perform well under different rainfall patterns and soil types.

Adaptability, on the other hand, focuses on the plant's capacity to thrive in challenging or variable environments. A crop with high adaptability can adjust to stresses such as drought, salinity, or nutrient deficiencies and still produce acceptable yields. For example, in **wheat breeding**, adaptability to varying moisture levels is crucial in arid and semi-arid regions, where rainfall is unpredictable. Breeders often look for phenotypes that exhibit plasticity, such as the ability to conserve water through reduced transpiration rates or the capacity to grow in soils with high salinity. The balance between **stability** and **adaptability** is an important consideration for plant breeders, especially when working on crops intended for areas with unstable growing conditions. In these regions, breeders often target traits like drought tolerance or pest resistance that allow the crop to adapt and still maintain a stable yield.

ii) Evaluation of Phenotypic Variability in Plant Breeding Programs

Evaluating phenotypic variability requires robust testing methods. Breeders conduct **field trials** where different genotypes are grown under controlled and variable environments to assess their phenotypic expression. These trials can take place over several growing seasons and locations, allowing breeders to evaluate the stability of traits and identify genotypes that perform well across a broad spectrum of environmental conditions. Multi-environment trials are particularly common in crops like maize and wheat, where performance consistency across different agroecological zones is a breeding priority.

In addition to field trials, **phenotypic characterization** also plays a crucial role in evaluating variability. Traits such as plant height, leaf color, seed size, and disease resistance are observed and measured. Technological advancements like **remote sensing** and **phenotyping platforms** now enable breeders to gather detailed phenotypic data at different stages of plant growth, providing deeper insights into how various genotypes respond to environmental factors.

iii) The Role of Phenotypic Variability in Breeding for Climate Change Resilience

With climate change posing increasing challenges to global food production, phenotypic variability plays a crucial role in breeding crops that can withstand extreme conditions. Climate change is leading to more frequent and severe droughts, floods, and temperature fluctuations, which directly affect crop yields. Breeding for **climate resilience** requires selecting genotypes that exhibit high phenotypic plasticity, allowing them to adapt to a range of stress conditions.

For example, **maize breeding** in regions prone to drought focuses on identifying plants with traits that allow them to survive prolonged periods of water scarcity, such as deeper root systems or efficient water use mechanisms. Similarly, in rice breeding, varieties are being developed to tolerate **submergence** and **saline conditions**, which are becoming more common due to rising sea levels and erratic rainfall patterns in coastal areas. Varieties like the **Sub1A gene rice** are bred specifically to survive flooding for extended periods, a trait that is only possible due to the phenotypic plasticity observed in certain wild rice relatives. By selecting genotypes with a broad range of phenotypic responses to environmental stresses, breeders can develop crops that are better equipped to handle the unpredictable effects of climate change.

Genetic Basis of Variability in Plant Breeding

1. Sources of Genetic Variability

As discussed previously, genetic variability refers to the differences in the DNA sequences of individuals within a population. It is the most fundamental form of variability in plant breeding and provides the raw material for selection. Without genetic variability, there would be no opportunity to select for desirable traits, making crop improvement impossible. The genetic basis of variability is complex and arises from several sources, including **mutations**, **recombination**, **gene flow**, and **polyploidy**. Each of these processes introduces new alleles into a population, which can then be selected by breeders for crop improvement.

i) Mutations

Mutations are one of the primary sources of genetic variability in plant populations. A **mutation** is a change in the DNA sequence of an organism that can occur naturally or be induced artificially. Mutations can lead to the development of new traits, which can be beneficial, neutral, or deleterious to the organism. In plant breeding, induced mutations have been widely used to develop varieties with improved traits. **Mutation breeding** involves exposing plants to mutagens such as radiation or chemicals to increase the mutation rate and generate new genetic variability.

For example, **induced mutations** in rice have led to the development of varieties with improved resistance to diseases like bacterial blight, as well as improved grain quality. Similarly, in barley, mutation breeding has been used to improve drought tolerance and enhance nutritional content. While not all mutations result in useful traits, the genetic variability they introduce can be valuable for breeders seeking to improve specific crop characteristics.

ii) Recombination

Recombination is the process by which genetic material is exchanged between homologous chromosomes during meiosis, resulting in new combinations of alleles. This process is particularly important in sexually reproducing crops, as it introduces genetic variability into the population and allows breeders to select for desirable combinations of traits. In **cross-pollinated crops**, such as maize and sunflower, recombination creates a wide range of variability, which breeders can exploit to improve yield, disease resistance, and environmental adaptability.

In **self-pollinated crops**, recombination still occurs, but the level of genetic variability is lower compared to cross-pollinated crops. However, through repeated self-pollination and selection, breeders can still develop varieties with improved traits by selecting individuals that carry the desired combinations of alleles.

iii) Polyploidy

Polyploidy is another important source of genetic variability in plant breeding. Polyploid plants contain more than two sets of chromosomes, which can lead to increased genetic diversity and novel phenotypes. **Polyploidy** can occur naturally or be induced by breeders using chemicals such as colchicine to double the chromosome number in a plant. Polyploid crops often exhibit increased size, vigor, and tolerance to environmental stresses compared to their diploid counterparts.

For example, **wheat** is a polyploid crop, with most cultivated varieties being either tetraploid (*Triticum durum*) or hexaploid (*Triticum aestivum*). The increased chromosome number in polyploid wheat provides greater genetic diversity, allowing breeders to select for traits such as disease resistance, drought tolerance, and improved yield. Polyploidy is also common in other crops like **potatoes** and **bananas**, where it contributes to the development of improved varieties with higher yield and better stress tolerance.

2. Mechanisms for Creating Genetic Variability

Creating and maintaining genetic variability is a critical aspect of plant breeding. Breeders use a variety of methods to introduce new genetic variation into crop populations, which can then be selected for improved traits. These methods include **hybridization, mutation breeding**, and **genetic engineering**.

i) Hybridization

Hybridization is the process of crossing two genetically distinct parents to produce offspring with a combination of traits from both parents. Hybridization is one of the most effective ways to introduce new genetic variability into a population, as it combines the alleles from two different individuals, resulting in offspring with new combinations of traits. In plant breeding, hybridization is widely used to develop **hybrid varieties** that exhibit improved traits such as yield, disease resistance, and stress tolerance. For example, **hybrid maize varieties** have been instrumental in increasing global maize production, as they consistently outperform open-pollinated varieties in terms of yield and adaptability. Hybridization has also been used in crops like **rice** and **wheat**, where breeders cross parents with complementary traits to develop new varieties that combine high yield potential with resistance to diseases like rust or bacterial blight.

ii) Mutation Breeding

Mutation breeding is a technique used to induce genetic changes in plants by exposing them to mutagens such as chemicals, radiation, or even biological agents. The goal is to generate new genetic variability that breeders can exploit to improve traits such as disease resistance, yield, or stress tolerance. While spontaneous mutations occur naturally in all organisms, mutation breeding accelerates this process, increasing the chances of developing desirable traits in a shorter period of time.

Mutation breeding has been widely used to improve a variety of crops, including rice, wheat, barley, and legumes. For instance, induced mutations in rice have led to the development of varieties with enhanced resistance to diseases like bacterial blight and blast, as well as improved grain quality and yield stability under stress conditions. Similarly, mutation breeding in barley has resulted in varieties with improved drought tolerance, better malting quality, and enhanced resistance to fungal diseases.

One of the key advantages of mutation breeding is its ability to introduce specific, targeted changes in a plant's genome without altering the overall genetic structure of the variety. This allows breeders to retain the desirable traits of a variety while introducing new traits that enhance their performance under particular conditions. However, mutation breeding also has limitations. Many mutations are deleterious or neutral, and the process of screening large numbers of plants for desirable traits can be time-consuming. Additionally, the mutation rate can be difficult to control, and the success of the technique depends on the availability of effective mutagenic agents and efficient screening methods. Despite these challenges, mutation breeding remains a valuable tool for introducing new genetic variability into crop populations, especially in cases where traditional breeding methods have reached their limits.

iii) Genetic Engineering and Gene Editing

In recent decades, advances in biotechnology have revolutionized the ability to introduce genetic variability into crops. **Genetic engineering** allows breeders to directly manipulate a plant's genome by inserting, deleting, or modifying specific genes to introduce new traits or enhance existing ones. This approach bypasses the need for traditional cross-breeding and allows for the precise introduction of traits from different species or even entirely different kingdoms of life.

One of the most well-known examples of genetic engineering in plant breeding is the development of **Bt cotton** and **Bt maize**, which contain a gene from the bacterium *Bacillus thuringiensis*. This gene provides the crops with resistance to specific insect pests, reducing the need for chemical pesticides and improving yield in areas with high pest pressure. Similarly, **genetically modified soybeans** have been developed with resistance to herbicides, allowing farmers to control weeds more effectively and increase productivity.

Another powerful tool in modern plant breeding is **gene editing**, particularly the **CRISPR-Cas9** system, which allows for precise and targeted changes in a plant's DNA. Unlike traditional genetic engineering, which often involves introducing foreign DNA, gene editing allows breeders to make small, targeted modifications to a plant's existing genes. This technology has been used to enhance traits such as disease resistance, drought tolerance, and nutrient content in crops like wheat, rice, and tomatoes. Gene editing has the potential to dramatically accelerate the process of crop improvement by enabling breeders to introduce desirable traits more quickly and accurately than ever before. It also allows for more precise control over the genetic changes being introduced, reducing the risk of unintended consequences. However, the regulatory landscape for genetically engineered and gene-edited crops varies by region, and public acceptance of these technologies remains a challenge in some parts of the world.

D. Components of Variation

In plant breeding and genetics, **variation** is the foundation upon which all breeding programs are built. Variation refers to the differences in traits or characteristics within a population of organisms. These differences may be due to genetic factors, environmental influences, or the interaction between the two. the **components of variation** help breeders manipulate these factors to develop new and improved crop varieties with desirable traits, such as higher yield, disease resistance, and stress tolerance. Variation is often divided into two main categories: **genetic variation** and **environmental variation**. Each of these categories can be further broken down into specific components, all of which play a crucial role in plant breeding. A detailed understanding of these components allows breeders to design effective breeding strategies that maximize genetic gain and minimize the impact of environmental influences on crop performance.

To achieve these goals, it is important to break down the various components of variation that contribute to the observed differences among individuals in a population. These components include **genetic variance**, **environmental variance**, and **genotype-environment interactions**.

1. Genetic Variation

Genetic variation is the portion of the total variation in a population that is attributed to differences in the genetic makeup of individuals. This type of variation arises from the alleles inherited from the parents and is the main focus in plant breeding since it provides the raw material for selection. Genetic variation can be further divided into **additive**, **dominance**, and **epistatic variance**.

i) Additive Genetic Variation

Additive genetic variation refers to the cumulative effect of individual alleles on a trait. Each allele contributes a certain amount to the phenotype, and these contributions are summed together to determine the overall expression of the trait. Additive genetic variance is particularly important in plant breeding because it is responsible for the inheritance of traits from one generation to the next. Traits that are governed by additive genetic variation are easier to select for, as their inheritance follows predictable patterns.

For example, in **wheat** (*Triticum aestivum*), traits such as plant height and grain yield are often controlled by additive genetic variation. Breeders can select plants with favorable alleles and combine these alleles in successive generations to develop varieties with improved performance. Additive genetic variation is the primary target in most selection programs because it allows for gradual improvement of traits over time.

ii) Dominance Genetic Variation

Dominance genetic variation occurs when the effect of one allele masks or overrides the effect of another allele at the same locus. This type of variation contributes to the observed phenotype in heterozygous individuals, where one allele is dominant over the other. Dominance variance can complicate breeding programs because it does not follow the simple additive inheritance pattern and may not be expressed consistently across generations.

In **corn** (*Zea mays*), dominance variance is often exploited in hybrid breeding programs. When two inbred lines are crossed to produce a hybrid, the dominance variance leads to **heterosis** or **hybrid vigor**, where the hybrid offspring outperform both parental lines in terms of yield, vigor, and stress tolerance. Hybrid maize varieties benefit from dominance genetic variation, but the dominance effects are not passed down to subsequent generations in the same manner as additive effects.

iii) Epistatic Genetic Variation

Epistatic variation occurs when the expression of one gene is influenced by one or more other genes at different loci. This type of genetic interaction can result in complex inheritance patterns, making it difficult to predict the phenotype based solely on the alleles at individual loci. Epistasis can play a significant role in traits that are controlled by multiple genes, where the combined effect of these genes may be greater or less than the sum of their individual effects.

For example, in **barley** (*Hordeum vulgare*), resistance to certain fungal diseases is influenced by epistatic interactions between multiple genes. Breeders must consider these interactions when developing disease-resistant varieties, as selecting for one resistance gene may not be sufficient to achieve full resistance unless the interacting genes are also taken into account. Epistasis can either enhance or reduce the effectiveness of selection, depending on the nature of the interactions between genes. Breeding programs that target complex traits often need to account for epistatic variance to achieve the desired genetic improvements.

2. Environmental Variation

Environmental variation refers to the differences in phenotype that are caused by external factors, rather than genetic differences. These factors can include temperature, soil quality, water availability, and other environmental conditions. Unlike genetic variation, environmental variation is not heritable and cannot be passed on to future generations. However, it still plays a significant role in shaping the phenotype and must be considered in breeding programs.

i) Macro-Environmental Variation

Macro-environmental variation refers to large-scale environmental factors that affect entire populations or regions. These factors include climate, geography, and seasonal changes that influence plant growth and development. For example, drought conditions in a particular region may reduce crop yield across all individuals, regardless of their genetic makeup. Breeders working in regions with high macro-environmental variability must select for traits that allow crops to perform well under a wide range of conditions.

In **rice breeding** programs, varieties are often selected for their ability to tolerate **flooding** or

drought, depending on the environmental conditions of the target region. Breeders must account for macro-environmental variation to ensure that the selected varieties can perform consistently across different growing seasons and locations.

ii) Micro-Environmental Variation

Micro-environmental variation occurs on a smaller scale and is influenced by factors such as soil composition, irrigation practices, and pest pressure. These factors can vary from field to field or even within the same field, leading to differences in plant performance. Micro-environmental variation can be difficult to control in breeding trials, but it is essential to account for when evaluating the performance of different genotypes. In **tomato breeding** (*Solanum lycopersicum*), micro-environmental factors such as soil fertility and pest management can significantly impact fruit size and quality. Breeders often conduct trials across multiple locations and environments to ensure that selected varieties can maintain their performance under varying micro-environmental conditions.

iii) Interaction Between Environment and Genotype (GxE)

One of the most challenging aspects of plant breeding is understanding and managing the **genotype-environment interaction** (GxE). This interaction occurs when the performance of a genotype varies depending on the environmental conditions. A variety that performs well in one environment may not necessarily perform well in another, making it difficult for breeders to select genotypes that will consistently excel across diverse growing conditions.

GxE interactions can complicate breeding programs because they reduce the predictability of trait expression. To address this challenge, breeders often conduct **multi-environment trials** (METs) to evaluate the performance of genotypes across different environments. This approach helps identify **genotypes with broad adaptability**, which can perform well under a wide range of conditions, as well as **genotypes with specific adaptability** to particular environments. For instance, in **maize breeding**, METs are used to assess the performance of different hybrids in various agro-climatic zones. By analyzing the GxE interactions, breeders can identify hybrids that exhibit stable performance across diverse environments or hybrids that are specifically adapted to high-stress conditions such as drought or low soil fertility.

3. Components of Phenotypic Variance

The **total phenotypic variance** observed in a population can be broken down into several components. These components include genetic variance, environmental variance, and the interaction between genetic and environmental factors. Understanding these components is critical for designing breeding programs that effectively exploit genetic variation while minimizing the impact of environmental variation.

i) Heritability

Heritability is a measure of the proportion of the total phenotypic variance that is due to genetic factors. It indicates the extent to which a trait is controlled by genetic factors and can be passed on to the next generation. High heritability suggests that genetic factors play a significant role in determining the trait, while low heritability indicates that environmental factors are more influential.

Heritability is a key concept in plant breeding, as it determines the potential for genetic improvement through selection. Traits with high heritability, such as plant height and seed size, are easier to improve through selection, while traits with low heritability, such as yield under stress conditions, require more complex breeding strategies that account for both genetic and environmental factors. For example, **yield in wheat** is influenced by both genetic and environmental factors. In regions with stable growing conditions, heritability for yield may be relatively high, allowing breeders to select for high-yielding genotypes. However, in regions with fluctuating environmental conditions, heritability for yield may be lower, requiring breeders to focus on traits like **drought tolerance** or **disease resistance** that can help stabilize yield across different environments.

4. Broad-Sense and Narrow-Sense Heritability

Heritability, as discussed, is a measure of the extent to which genetic factors contribute to the observed variation in a trait. However, heritability can be subdivided into two key categories: **broad-sense heritability** and **narrow-sense heritability**. These two measures of heritability are essential in determining how effectively traits can be passed from one generation to the next and how breeders can select for desirable traits in crop breeding programs.

i) Broad-Sense Heritability (H^2)

Broad-sense heritability refers to the proportion of the total phenotypic variance that can be attributed to **all genetic factors**, including additive, dominance, and epistatic variance. It is expressed as the ratio of the genetic variance to the total phenotypic variance in a population.

Broad-sense heritability is useful in estimating the overall genetic control of a trait, providing a general idea of the potential for improving a trait through selection. However, since it includes all types of genetic effects, it is not as precise for determining the potential for long-term genetic gains, as dominance and epistatic effects may not be passed on consistently across generations. In crops like **maize** and **rice**, broad-sense heritability is often used to assess the genetic potential of complex traits such as **yield** and **disease resistance**, where both additive and non-additive genetic effects play significant roles. For example, in **hybrid maize breeding**, broad-sense heritability can help breeders determine whether a hybrid will consistently outperform its parental lines due to heterosis.

ii) Narrow-Sense Heritability (h^2)

Narrow-sense heritability is the proportion of the total phenotypic variance that is attributable to **additive genetic variance** alone. This measure is more relevant to plant breeders because it represents the genetic variation that can be consistently passed from parents to offspring. Additive genetic variance is the component of genetic variance that is responsible for the predictable inheritance of traits and is the primary target in breeding programs aimed at long-term genetic improvement. In **wheat breeding**, for example, traits such as **plant height** and **grain quality** often have high narrow-sense heritability, making them easier to improve through traditional selection methods. Breeders can select individuals with favorable alleles, knowing that these traits will be passed on predictably to future generations. High narrow-sense heritability ensures that the selection of parents will result in reliable genetic gains in the offspring.

Narrow-sense heritability is also essential in calculating the **response to selection** (R), which is the expected improvement in a trait due to selection. The response to selection is given by the equation:

$$R = h2 \times S$$

where SSS is the selection differential or the difference between the mean trait value of the selected individuals and the mean trait value of the population. A high narrow-sense heritability ensures that selection for a particular trait will lead to substantial improvements in the offspring.

5. Genotype-Environment Interaction (GxE)

The **genotype-environment interaction** (GxE) is a critical concept in plant breeding, particularly when developing varieties that must perform consistently across diverse environments. GxE occurs when the performance of a genotype varies depending on the environmental conditions in which it is grown. This interaction complicates the process of selection because the same genotype may perform well in one environment but poorly in another.

i) Importance of GxE in Plant Breeding

Understanding GxE is essential for breeding programs that target crops for large, geographically diverse regions with varying environmental conditions. If breeders do not account for GxE, they may mistakenly select genotypes that perform well under specific conditions but fail to generalize across different environments. To avoid this, breeders must conduct **multi-environment trials** (METs) to evaluate the performance of genotypes in various locations and across different growing seasons.

For instance, **wheat varieties** are often bred to perform well across a wide range of environments, from temperate regions in Europe to arid regions in Australia and parts of Africa. GxE interactions play a significant role in determining which varieties can thrive in all environments or need to be targeted for specific regions. For example, a variety that performs well in the cool, wet conditions of northern Europe might not perform well in the hot, dry conditions of the Middle East.

ii) Identifying Genotypes with Broad Adaptability

One approach to dealing with GxE interactions is to identify **genotypes with broad adaptability**. These genotypes are able to maintain stable performance across a wide range of environmental conditions, making them ideal candidates for breeding programs aimed at developing crops for global or regional markets. Breeders can identify such genotypes through multi-environment trials and statistical analyses that partition the total variance into components due to genotype, environment, and GxE interactions.

In crops like **maize** and **rice**, breeders often select genotypes with broad adaptability to ensure that the varieties will perform consistently across different agroecological zones. For example, rice varieties developed for regions prone to both drought and flooding must be adaptable to varying water availability, as these conditions can change rapidly during the growing season.

iii) Specific Adaptability and Targeted Breeding

While some breeding programs focus on developing genotypes with broad adaptability, others may target **specific adaptability** to particular environments. In this case, breeders aim to develop varieties that are tailored to thrive in specific environmental conditions, such as high-altitude regions, drought-prone areas, or saline soils.

For example, in **barley breeding**, specific adaptability is often a key objective when developing varieties for cultivation in arid and semi-arid regions. Barley varieties with enhanced **drought tolerance** are selected for their ability to withstand prolonged periods of water scarcity while maintaining acceptable yields. These varieties may not perform as well in more favorable environments, but their adaptability to harsh conditions makes them valuable for farmers in regions where water availability is limited.

6. Methods to Measure and Analyze Genetic Variability

To effectively exploit genetic variability in plant breeding, it is essential to have reliable methods for measuring and analyzing this variability. Several statistical tools and experimental designs are commonly used in plant breeding to quantify the components of genetic variation and their contributions to phenotypic variability.

i) Analysis of Variance (ANOVA)

Analysis of variance (ANOVA) is a statistical technique used to partition the total phenotypic variance into its parts, including genetic variance, environmental variance, and GxE interactions. ANOVA is used in multi-environment trials to determine how much of the observed variation in a trait is due to differences between genotypes, environments, or the interaction between the two. In a typical plant breeding experiment, ANOVA allows breeders to compare the performance of different genotypes across multiple environments. By analyzing the variance components, breeders can determine whether the observed differences in performance are primarily due to genetic factors or whether environmental conditions are playing a more significant role. This information is crucial for making informed decisions about which genotypes to advance in the breeding program.

ii) Heritability Estimates

Estimating **heritability** is another critical method for analyzing genetic variability. As discussed earlier, heritability measures the proportion of phenotypic variation that is attributable to genetic factors. High heritability indicates that the trait is largely controlled by genetics and can be improved through selection, while low heritability suggests that environmental factors are more influential.

In plant breeding programs, heritability estimates are often used to prioritize traits for selection. Traits with high heritability, such as **plant height** or **grain weight**, are more likely to respond to selection efforts, while traits with low heritability, such as **yield under stress**, may require more complex breeding strategies that account for both genetic and environmental factors.

iii) Genotype × Environment Interaction (GxE) Models

Genotype × environment interaction (GxE) models are used to quantify and analyze the effects of GxE on trait expression. These models allow breeders to assess the stability and adaptability of genotypes across different environments. One commonly used GxE model is the **Additive Main effects and Multiplicative Interaction (AMMI)** model, which combines ANOVA with principal component analysis (PCA) to separate the effects of genotypes, environments, and their interactions. In crops like **maize** and **sorghum**, GxE models are used to identify genotypes that perform consistently well across a range of environments or that exhibit specific adaptability to challenging conditions, such as drought or high temperatures. The results from these models help breeders make informed decisions about which genotypes to advance in multi-environment breeding trials.

E. Introduction to General and Specific Combining Ability (Expanded)

In plant breeding, the development of superior crop varieties relies heavily on understanding the genetic mechanisms that control trait inheritance. One critical aspect of this is **combining ability**, which refers to the potential of parental lines to produce offspring with desirable traits when crossed. Combining ability plays a vital role in hybrid breeding programs, where the goal is to maximize heterosis (hybrid vigor) and achieve superior trait expression in the progeny.

Combining ability is divided into two main categories: **General Combining Ability (GCA)** and **Specific Combining Ability (SCA)**. GCA measures the average performance of a parent when crossed with a variety of other lines, emphasizing the additive genetic effects passed on to offspring. On the other hand, SCA evaluates the performance of specific parental combinations that may deviate from the expected outcome based on GCA. This deviation is often attributed to non-additive genetic effects, such as dominance and epistasis, which result in superior or inferior performance compared to the average performance of the parents. The interplay between GCA and SCA is especially crucial in crops where hybrid vigor is exploited, such as maize, rice, and sunflowers. By understanding both components, breeders can identify the best parental combinations that will produce high-performing hybrids. These concepts form the cornerstone of modern hybrid breeding, helping to enhance traits such as yield, disease resistance, and stress tolerance, while ensuring genetic diversity and stability in crop populations.

1. General Combining Ability (GCA)

General Combining Ability (GCA) refers to the average performance of a parental line when crossed with other lines. It is largely influenced by **additive genetic effects**, where the cumulative contribution of alleles at multiple loci determines the expression of a trait. GCA is a critical factor in plant breeding because it reflects the consistent genetic contribution of a parent to its progeny, irrespective of the other parent in the cross. This makes GCA highly valuable in breeding programs aimed at long-term genetic improvement through predictable inheritance patterns.

i) Additive Genetic Variance in GCA (Expanded)

GCA is primarily determined by **additive genetic variance**, which represents the cumulative effect of alleles that consistently contribute to trait expression. Additive genetic effects are important because they are reliably passed from one generation to the next, allowing breeders to make steady progress in improving key traits. Traits that are influenced by additive genetic variance, such as plant height, yield, and disease resistance, are often the focus of selection in breeding programs because they can be easily inherited and selected over generations. In **wheat breeding**, for example, additive genetic variance is essential for improving traits like grain quality, yield stability, and resistance to rust diseases. Breeders can select parental lines with high GCA for these traits, knowing that the favorable alleles will be passed on to the offspring in a predictable manner. This allows for the development of wheat varieties that consistently perform well across different environments and seasons.

ii) Role of GCA in Hybrid Breeding

In hybrid breeding, GCA is critical for identifying parental lines that can consistently contribute to the genetic improvement of key traits. High GCA indicates that a parent has strong additive genetic effects for the trait of interest, making it a valuable contributor in the development of high-performing hybrids. In crops like **rice** and **maize**, breeders often use GCA estimates to select parental lines with superior performance in traits such as yield, drought tolerance, and pest resistance. Hybrid rice varieties developed through crosses involving parents with high GCA for yield and stress tolerance have been instrumental in boosting rice production in Asia. These hybrids consistently outperform open-pollinated varieties due to the additive effects contributed by their parents. Similarly, in **maize**, high-GCA parents are used to develop hybrids that exhibit increased grain yield and adaptability to diverse growing conditions. The ability of these hybrids to perform consistently across environments is a direct result of the additive genetic effects captured through GCA.

iii) GCA and Breeding for Complex Traits

In addition to simple traits like plant height or disease resistance, GCA is also essential for breeding **complex traits**, which are controlled by multiple genes and influenced by environmental factors. Traits like yield, drought tolerance, and nutrient efficiency fall into this category. Complex traits are often difficult to improve because of the interactions between multiple genes and the environment. However, by selecting parental lines with high GCA for these traits, breeders can make progress in improving them over time. In **barley breeding**, for instance, breeders often target complex traits such as drought tolerance, which is influenced by multiple physiological and developmental factors. By selecting parental lines with high GCA for drought tolerance, breeders can develop barley varieties that perform well in arid environments while maintaining acceptable yield levels. These varieties are critical for regions facing water scarcity due to climate change, ensuring food security and sustainable crop production.

2. Specific Combining Ability (SCA)

Specific Combining Ability (SCA) refers to the performance of a specific parental combination that deviates from the expected performance based on the GCA of the parents. SCA captures **non-additive genetic effects**, including dominance and epistasis, which result in higher or lower performance than predicted by the average of the parents. SCA is particularly important in hybrid breeding because it helps identify parental combinations that exhibit **heterosis** or hybrid vigor.

i) Non-Additive Genetic Variance in SCA

Unlike GCA, which focuses on additive genetic effects, SCA is influenced by **non-additive genetic variance**. This includes dominance effects, where one allele masks the effect of another, and **epistatic interactions**, where the expression of one gene is affected by the presence of another gene at a different locus. These non-additive effects contribute to the unique performance of specific hybrids, making them superior or inferior to their parents in terms of traits like yield, stress tolerance, or disease resistance. In **sunflower breeding**, SCA plays a significant role in the development of hybrids with superior oil content and seed yield. Sunflower hybrids often exhibit strong heterosis due to the favorable interaction between alleles from different parents. These hybrids outperform both parental lines, resulting in higher yields and better oil quality. SCA analysis helps breeders identify the best parental combinations for maximizing these non-additive effects.

ii) Role of SCA in Exploiting Heterosis

SCA is the key to exploiting **heterosis**, which occurs when the offspring of two genetically distinct parents exhibit superior performance compared to their parents. Heterosis is widely used in the breeding of cross-pollinated crops like maize, sorghum, and sunflowers, where hybrids often outperform open-pollinated varieties in terms of yield, vigor, and stress tolerance. The success of hybrid breeding programs depends on identifying parental combinations with high SCA, as these combinations produce the most vigorous and productive offspring. In **maize hybrid breeding**, SCA analysis is used to identify crosses that exhibit strong heterosis for traits such as grain yield, drought tolerance, and disease resistance. Maize hybrids developed through high-SCA crosses have revolutionized maize production, leading to significant increases in yield and resilience to environmental stresses. The ability of these hybrids to perform well in diverse environments is due to the non-additive genetic effects captured through SCA.

iii) Importance of SCA in Hybrid Breeding Programs

In hybrid breeding programs, SCA is essential for identifying parental combinations that produce hybrids with superior performance. While GCA ensures that parents contribute consistent genetic improvements, SCA identifies the unique genetic interactions that result in hybrids with exceptional performance. In crops like **sorghum** and **sunflowers**, breeders use SCA to develop hybrids that exhibit superior traits, such as drought tolerance and disease resistance. In **sorghum breeding**, hybrids developed through high-SCA crosses are particularly valuable for regions facing drought conditions. These hybrids exhibit enhanced drought tolerance, higher biomass production, and improved grain yield, making them ideal for cultivation in arid and semi-arid regions. SCA analysis allows breeders to identify the specific parental combinations that will produce the most resilient and productive hybrids for these challenging environments.

3. GCA and SCA in Practical Breeding Programs

In practical breeding programs, breeders must strike a balance between selecting for both GCA and SCA to develop high-performing hybrids. While GCA provides a reliable measure of a parent's overall genetic contribution, SCA captures the unique interactions between parental lines that result in superior hybrid performance. By combining high-GCA parents with high-SCA crosses, breeders can develop hybrids that not only perform well across a range of environments but also exhibit strong heterosis for key traits.

i) Balancing GCA and SCA in Hybrid Development

In crops like **maize** and **rice**, breeders aim to balance GCA and SCA to develop hybrids that are both genetically stable and exhibit strong heterosis. High-GCA parents are selected for their consistent contribution to traits like yield and disease resistance, while high-SCA crosses are identified to maximize hybrid vigor. This balance ensures that the resulting hybrids are not only high-yielding but also adaptable to diverse growing conditions. In **hybrid rice breeding**, for example, breeders use GCA to select parental lines with strong additive genetic effects for yield and disease resistance. However, they also conduct SCA analysis to identify the specific crosses that exhibit the greatest heterosis. This approach has led to the development of hybrid rice varieties that significantly outperform traditional varieties in terms of yield, resilience, and adaptability to different environments.

ii) GCA and SCA in Synthetic Varieties

In addition to hybrid breeding, GCA and SCA are also important in the development of **synthetic varieties**, where genetic diversity is maintained to ensure long-term productivity. Synthetic varieties are developed by intercrossing multiple parental lines with high GCA and allowing their progeny to cross-pollinate freely. This approach ensures that the resulting varieties maintain a balance between genetic diversity and trait expression. In **alfalfa breeding**, synthetic varieties are developed by intercrossing parental lines with high GCA for traits like forage yield and disease resistance. By allowing these lines to cross-pollinate freely, breeders can maintain genetic diversity while ensuring that the synthetic varieties exhibit high performance across a range of environments. The combination of GCA and SCA ensures that these synthetic varieties are both productive and resilient, making them ideal for forage production in diverse regions.

F. Self-Incompatibility in Crop Plants and Their Commercial Exploitation

Self-incompatibility (SI) is a natural mechanism that prevents self-fertilization in many angiosperms. It is estimated that more than 300 species across 20 different plant families exhibit self-incompatibility, a process that encourages cross-pollination and the maintenance of genetic diversity within plant populations. SI plays an essential role in reproductive biology, particularly for crop plants, where it influences breeding practices and commercial seed production. In plants exhibiting self-incompatibility, pollen from the same plant or genetically similar individuals fails to fertilize the ovules, thus ensuring that only genetically different individuals can contribute to the next generation. The primary purpose of this mechanism is to prevent inbreeding and promote outcrossing, which enhances the plant's ability to adapt to changing environmental conditions and resist diseases. While self-incompatibility is a beneficial trait in nature, it presents both challenges and opportunities for plant breeders, particularly when developing hybrid seeds for commercial purposes. The process of self-incompatibility is controlled by genetic mechanisms that vary depending on the species. There are two primary systems through which self-incompatibility operates: the heteromorphic system and the homomorphic system. Each system can further be divided based on whether the self-incompatibility reaction is controlled by the genotype of the pollen or the genotype of the plant that produced the pollen.

Heteromorphic System of Self-Incompatibility

The heteromorphic system of self-incompatibility is characterized by morphological differences in the flowers of plants within the same species. These differences in floral structure create physical barriers that prevent self-pollination. Heteromorphic self-incompatibility is commonly observed in species such as *Primula* and *Lythrum*, where flowers exhibit different types of floral morphology. In *Primula*, for example, two distinct floral types—pin and thrum—exist, with the pin flowers having long styles and short stamens, while thrum flowers have short styles and long stamens. In the heteromorphic system, successful fertilization occurs only between pin and thrum flowers, while self-pollination or pollination between flowers of the same type (e.g., pin to pin or thrum to thrum) is unsuccessful. This phenomenon is known as distyly. In some species, a more complex system known as tristyly exists, where flowers may have three different morphs: short, medium, and long styles. In such cases, successful fertilization can only occur between morphologically different flowers.

The genetic control of the heteromorphic system is relatively simple and is often governed by a single gene with two alleles. For example, in *Primula*, the gene controlling floral morphology is referred to as the S locus, with the S allele being dominant over the s allele. Thrum flowers have the Ss genotype, while pin flowers have the ss genotype. The pollen produced by thrum flowers contains both S and s alleles, but only the S allele is phenotypically expressed, meaning that thrum pollen is incompatible with thrum flowers. Conversely, the pollen produced by pin flowers carries only the s allele, which can successfully fertilize thrum flowers. This system ensures outcrossing between morphologically distinct flowers and promotes genetic diversity. Although the heteromorphic system of self-incompatibility is an interesting evolutionary adaptation, its application in crop breeding is limited. It is primarily observed in ornamental plants and a few crop species, such as sweet potato (*Ipomoea batatas*) and buckwheat (*Fagopyrum esculentum*). Due to the morphological differences between flowers, heteromorphic self-incompatibility is difficult to manipulate for large-scale commercial seed production. As a result, plant breeders tend to focus on other systems of self-incompatibility, such as the homomorphic system, which is more common in economically important crops.

Homomorphic System of Self-Incompatibility

In contrast to the heteromorphic system, the homomorphic system of self-incompatibility is not associated with any visible morphological differences between flowers. Instead, the self-incompatibility reaction is controlled solely by the genetic makeup of the pollen and the pistil, with no structural barriers involved. Homomorphic self-incompatibility is more widespread and has greater relevance in crop breeding programs due to its occurrence in many commercially important plant species. The homomorphic system can be further divided into two main types based on how the self-incompatibility reaction is controlled: the gametophytic system and the sporophytic system.

Gametophytic Self-Incompatibility (GSI)

Gametophytic self-incompatibility (GSI) is controlled by the genotype of the pollen grain itself. This system was first described in *Nicotiana sanderae* by East and Mangelsdorf in 1925. In GSI, the compatibility of the pollen is determined by its own genetic makeup at the S locus. If the allele carried by the pollen grain matches one of the alleles in the pistil, the self-incompatibility reaction is triggered, preventing fertilization. In plants that exhibit GSI, the pollen carries a single allele from the S locus, and fertilization can only occur if the allele in the pollen is different from both alleles in the pistil. The self-incompatibility reaction typically occurs after the pollen has landed on the stigma and attempted to germinate. If the pollen grain carries the same S allele as the pistil, the pollen tube's growth is either slowed or completely halted, preventing it from reaching the ovule and effecting fertilization.

There are three possible outcomes in gametophytic self-incompatibility:

1. **Fully incompatible:** This occurs when the alleles in the pollen and the pistil are identical (e.g., S1S2 × S1S2). In such cases, no fertilization occurs, and no seeds are produced.
2. **Fully compatible:** This occurs when the alleles in the pollen and pistil are entirely different (e.g., S1S2 × S3S4), resulting in successful fertilization and seed production.

3. **Partially compatible:** This occurs when one allele is shared between the pollen and the pistil (e.g., S1S2 × S2S3), leading to partial fertilization, where only half of the pollen is able to fertilize the ovules.

GSI is found in many economically important crops, including tomatoes (*Solanum lycopersicum*), potatoes (*Solanum tuberosum*), and tobacco (*Nicotiana tabacum*). While GSI can be beneficial for maintaining genetic diversity and preventing inbreeding depression, it presents challenges for plant breeders who need to control the pollen genotype to produce consistent seed sets. Despite these challenges, GSI has been exploited in some crops, such as clover (*Trifolium*), to produce hybrid seeds with enhanced vigor and yield.

Sporophytic Self-Incompatibility (SSI)

Sporophytic self-incompatibility (SSI) is controlled by the genotype of the entire plant (the sporophyte) that produced the pollen, rather than the genotype of the pollen grain itself. This system was first discovered in *Brassica* species, where it has been extensively studied due to its commercial significance in crop breeding. In SSI, the self-incompatibility reaction is determined by the genotype of the plant producing the pollen. Even if the pollen grain's genotype is compatible with the pistil, the sporophyte's genotype may prevent fertilization. The genetic control of SSI is often more complex than GSI, as it involves multiple alleles at the S locus and can result in dominant relationships between alleles. For example, in *Brassica oleracea* (cabbage, broccoli, cauliflower), more than 30 alleles have been identified at the S locus, and these alleles can interact in different ways, leading to varying degrees of self-incompatibility. Some alleles may be dominant over others, while in other cases, alleles may act independently or compete with each other. In SSI, the self-incompatibility reaction occurs at the stigmatic surface, where proteins from the pollen coat interact with the stigma to determine whether fertilization will proceed. If the pollen is incompatible, biochemical signals trigger the formation of callose, a polysaccharide that blocks the growth of the pollen tube, preventing fertilization. SSI is of great importance in commercial hybrid seed production, particularly in crops like *Brassica*. Japanese seed companies have successfully exploited SSI to produce high-quality hybrid seeds with desirable traits, such as disease resistance and improved yield. However, SSI also presents challenges for plant breeders, including the need to maintain inbred lines for hybrid production and the influence of environmental factors, such as temperature and humidity, which can override the self-incompatibility response and lead to self-fertilization.

Mechanism of Self-Incompatibility

The mechanism of self-incompatibility is a complex physiological process that involves interactions between the pollen, stigma, style, and ovule. The process is poorly understood due to its complexity, but it can be broken down into three major categories: pollen-stigma interaction, pollen tube-style interaction, and pollen tube-ovule interaction. These mechanisms operate differently depending on whether the self-incompatibility system is gametophytic or sporophytic.

The process can generally be divided into three main interactions:

1. Pollen-Stigma Interaction
2. Pollen Tube-Style Interaction
3. Pollen Tube-Ovule Interaction

These interactions act as checkpoints, preventing self-fertilization at different stages of the fertilization process. these interactions are key to comprehending how self-incompatibility operates in different plant species.

1. Pollen-Stigma Interaction

The pollen-stigma interaction is the first point at which the self-incompatibility reaction can occur. After the pollen lands on the stigma, it must germinate and begin to grow a pollen tube, which will eventually reach the ovule for fertilization. In self-incompatible plants, this process is often blocked at the pollen-stigma interface, preventing the pollen from even germinating. In gametophytic self-incompatibility (GSI), the pollen-stigma interaction occurs later in the process, as the stigma in GSI plants generally has a wet surface (plumose stigma), allowing for some initial pollen germination. However, the pollen tube's growth is later inhibited.

In sporophytic self-incompatibility (SSI), the interaction is more immediate, as the stigma in SSI plants has a dry surface (papillate stigma) covered by a proteinaceous pellicle. The incompatibility reaction occurs almost instantly when the pollen lands on the stigma, preventing pollen germination in incompatible combinations. Research has shown that within minutes of landing on the stigma, the pollen releases exudates (proteins or glycoproteins) that trigger a response from the stigma, leading to the formation of callose deposits that block the pollen from germinating or penetrating the stigma. Once the pollen tube penetrates the stigma in SSI systems, there are typically no further barriers to its growth, meaning the stigma serves as the primary site of the self-incompatibility reaction. In homomorphic systems, particularly sporophytic self-incompatibility, the rejection of pollen often depends on the proteins present on the surface of the pollen and the stigma. These proteins are coded by the S alleles, and they determine whether the pollen will be accepted or rejected by the stigma. The biochemical signaling involved in these interactions remains an area of active research, but it is clear that the S alleles play a crucial role in controlling the compatibility reaction.

2. Pollen Tube-Style Interaction

If the pollen successfully germinates and penetrates the stigma, the next checkpoint occurs in the style. This is particularly important in gametophytic self-incompatibility, where the pollen tube must grow down through the style to reach the ovule. In incompatible combinations, the growth of the pollen tube is often retarded or entirely blocked within the style, preventing successful fertilization. In the case of gametophytic self-incompatibility, the pollen tube's growth is controlled by its genotype. If the pollen grain shares the same S allele as the style, the growth of the pollen tube is inhibited. For example, in species like *Lycopersicon* (tomato), *Petunia*, and *Lilium*, the pollen tube may start growing but will eventually stop or burst due to the breakdown of its walls. This is typically caused by a cessation of protein and polysaccharide synthesis within the pollen tube, which leads to the degeneration of the tube wall, preventing the pollen tube from reaching the ovule. In species like *Oenothera* and other plants exhibiting gametophytic self-incompatibility, the pollen tube may penetrate the stigma but grow too slowly to reach the ovule before the flower senesces (drops). This is an effective way to prevent self-fertilization, as it ensures that only pollen from genetically different plants will be able to grow quickly enough to reach the ovule and achieve fertilization. In sporophytic self-incompatibility systems, the stigma is the primary site of rejection, and pollen tubes that manage to penetrate the stigma generally face no further barriers in the style. This is because the sporophytic self-incompatibility reaction typically ends at the stigma, meaning the pollen tube can grow freely once it has crossed this initial barrier.

3. Pollen Tube-Ovule Interaction

The final stage of the self-incompatibility process occurs at the ovule. In rare cases, pollen tubes may successfully grow through the style and reach the ovule, but fertilization still fails due to interactions between the pollen tube and the ovule. This type of interaction is less common and is generally observed in species with particularly complex self-incompatibility systems. For example, in *Theobroma cacao* (cocoa), the pollen tube may reach the ovule and affect fertilization, but the embryo degenerates at an early stage of development in incompatible combinations. This failure of embryo development prevents the production of seeds, effectively ensuring that only genetically diverse crosses result in viable offspring. While the pollen tube-ovule interaction is not as common a barrier as the pollen-stigma or pollen tube-style interactions, it demonstrates the multiple layers of protection that some plants have evolved to prevent self-fertilization and encourage outcrossing. These different checkpoints in the self-incompatibility process illustrate the complexity of this mechanism and its importance in maintaining genetic diversity in plant populations.

Exploitation of Self-Incompatibility in Hybrid Production

Self-incompatibility has been successfully exploited in plant breeding programs, particularly for the production of hybrid seeds. Hybrid vigor, or heterosis, results in plants that are more vigorous and productive than their inbred counterparts, making hybrids highly desirable in commercial agriculture. The use of self-incompatibility in hybrid seed production takes advantage of the plant's natural ability to prevent self-fertilization and ensures that only cross-pollination occurs between genetically diverse individuals.

1. Hybrid Seed Production Using Self-Incompatible Lines

In commercial hybrid seed production, breeders often exploit self-incompatibility by interplanting two self-incompatible but cross-compatible lines. When these lines are grown together, cross-pollination occurs naturally, and the resulting seeds are hybrid seeds. These hybrid seeds typically display hybrid vigor, resulting in higher yields, better disease resistance, and improved stress tolerance. This approach has been particularly successful in crops like *Brassica* (cabbage, broccoli, cauliflower), where the sporophytic self-incompatibility system ensures that only cross-pollination occurs between inbred lines. Japanese seed companies have pioneered the use of self-incompatibility for hybrid seed production in *Brassica*, producing high-quality hybrid seeds with improved agronomic traits. Alternatively, breeders can use a self-incompatible line in combination with a self-compatible line. In this scheme, only the self-incompatible line will produce hybrid seeds, while the self-compatible line acts as the pollen donor. This method can be useful when only one parent needs to be hybridized, reducing the complexity of maintaining two self-incompatible lines.

2. Double and Triple Cross Hybrids

In addition to simple hybrid seed production, more complex schemes have been developed for the production of double-cross and triple-cross hybrids. In these systems, two or more self-incompatible lines are crossed to produce hybrids with even greater genetic diversity and hybrid vigor. These schemes have been successfully demonstrated in crops like brassicas, where the gametophytic system has been used to produce hybrid seeds with enhanced traits. Double and triple cross hybrids are typically more expensive to produce, as they require careful management of inbred lines and controlled pollination. However, the resulting plants often display superior agronomic traits, making them highly desirable for commercial agriculture.

4. Challenges in Using Self-Incompatibility for Hybrid Production

While self-incompatibility offers many advantages for hybrid seed production, it also presents several challenges. One of the primary challenges is the need to maintain inbred lines through hand pollination, which is labor-intensive and costly. The production and maintenance of inbred lines are essential for hybrid seed production, but the process is time-consuming and requires significant resources. Additionally, continued selfing (self-fertilization) of inbred lines can lead to depression in self-incompatibility, unintentionally selecting for self-fertility. Over time, this can reduce the effectiveness of the self-incompatibility mechanism, making it more difficult to produce hybrid seeds. In the gametophytic system, continued inbreeding can give rise to new incompatibility reactions, limiting the usefulness of inbred lines as parents for hybrid production.

Environmental factors can also affect the self-incompatibility reaction. High temperatures, humidity, and other environmental conditions can reduce or even overcome self-incompatibility, leading to an increased proportion of selfed seeds. This can reduce the quality of hybrid seed production and increase costs for breeders. Finally, pollinators such as bees can complicate the process of hybrid seed production. Bees often prefer to stay within a single parental line, particularly if the parental lines differ morphologically. This behavior increases the proportion of selfed seeds, further reducing the efficiency of hybrid seed production. Despite these challenges, self-incompatibility remains a valuable tool in plant breeding programs, offering a natural and effective way to produce hybrid seeds with improved traits. By overcoming the limitations of self-incompatibility and refining breeding techniques, plant breeders can continue to exploit this mechanism for the benefit of commercial agriculture.

G. Male Sterility in Crop Plants and Its Commercial Exploitation

Male sterility refers to a condition where plants produce non-functional or aborted pollen grains, rendering them incapable of self-fertilization, while their female reproductive organs remain functional and can participate in cross-pollination. Male sterility is a critical tool in plant breeding, especially in hybrid seed production, as it eliminates the need for manual emasculation (removal of male reproductive organs), which can be labor-intensive and costly. In crop plants, male sterility occurs naturally due to mutations or can be induced through various breeding strategies.

There are three major types of male sterility: genetic, cytoplasmic, and cytoplasmic-genetic, each of which plays a distinct role in facilitating hybrid seed production. the mechanisms of male sterility and its applications in commercial breeding are vital for developing high-yielding and resilient hybrid varieties that are essential for modern agriculture.

Types of Male Sterility

1. Genetic Male Sterility
2. Cytoplasmic Male Sterility
3. Cytoplasmic-Genetic Male Sterility

1. Genetic Male Sterility

Genetic male sterility is determined by nuclear genes and is generally controlled by a single recessive gene, denoted as *ms*. In some species, dominant genes can also cause male sterility, as seen in crops like safflower (*Carthamus tinctorius*). The sterile condition results from mutations that disrupt pollen development, rendering the plant's male reproductive organs non-functional, though the female reproductive organs remain viable. In typical genetic male sterility, when a homozygous male sterile plant (*ms ms*) is crossed with a heterozygous male fertile plant (*Ms ms*), the progeny segregates in a 1:1 ratio of male sterile (*ms ms*) and male fertile (*Ms ms*) individuals. The key to exploiting genetic male sterility in breeding is the ability to maintain and propagate male sterile lines while controlling the occurrence of male fertile plants.

Mechanism and Maintenance of Genetic Male Sterility

In genetic male sterility, the non-functional pollen grains are a direct result of the defective expression of the *ms* gene. Male sterility alleles can arise spontaneously due to natural mutations or can be induced artificially through mutagenesis using chemicals or radiation. These mutations prevent the proper development of pollen grains, but the rest of the reproductive processes in the plant, including the development of viable ovules, remain intact. Maintaining a male sterile line requires crossing it with heterozygous male fertile plants (*Ms ms*), which will produce progeny in a 1:1 ratio of male sterile and male fertile individuals. The heterozygous male fertile plants are known as "maintainer lines" or "B lines." These lines are critical for sustaining the male sterile population, as they ensure that the genetic material necessary for sterility is preserved in the breeding process.

Utilization of Genetic Male Sterility in Plant Breeding

Genetic male sterility has significant applications in hybrid seed production, where the male sterile plants serve as the female parent in crosses. In this breeding scheme, male sterile plants (*ms ms*) are crossed with a homozygous male fertile pollinator (*Ms Ms*), ensuring that the resulting hybrid seeds exhibit the desired traits from both parent lines. These hybrid seeds are used for commercial planting because they exhibit hybrid vigor, which leads to higher yields, improved resistance to diseases, and better adaptability to environmental stresses.

The genotypes of the male sterile (*ms ms*) and the maintainer (*Ms ms*) lines are identical, except for the *ms* locus, meaning they are isogenic. This allows for precise control of breeding outcomes, as the male sterile and fertile lines share the same genetic background, ensuring consistency in the hybrid seed production process. A challenge in using genetic male sterility in breeding is that male fertile plants must be identified and removed before pollen shedding to prevent unintended self-pollination. This process, known as "rouging," involves physically removing the male fertile plants from the breeding population. Identifying male fertile plants at the seedling stage can be facilitated by using pleiotropic effects of the *ms* gene or marker genes closely linked to the *ms* locus. In maize (*Zea mays*), for example, the *pigmented hypocotyl* (P) gene is closely linked to the male sterility locus. Plants with pigmented hypocotyls are sterile, while those with green hypocotyls are fertile, allowing breeders to remove the green plants early in the growing process.

Limitations and Challenges

Despite its utility, genetic male sterility presents several challenges:

1. **Rouging Costs:** The process of identifying and removing male fertile plants before pollen shedding is labor-intensive and increases the cost of hybrid seed production.
2. **Poor Pollen Dispersal:** In many cases, pollen dispersal from the male pollinator line is poor, leading to suboptimal seed set in the female line.
3. **Limited Commercial Use:** Due to these difficulties, genetic male sterility has been commercially exploited in only a few crops, such as castor (*Ricinus communis*) in the United States and safflower in India. Suggestions have been made for its use in crops like cotton, barley, tomato, sunflower, and cucurbits, but practical challenges have prevented its widespread adoption.

2. Cytoplasmic Male Sterility (CMS)

Cytoplasmic male sterility (CMS) is a maternally inherited trait controlled by the cytoplasmic organelles, primarily the mitochondria, rather than by nuclear genes. Because the cytoplasm is inherited exclusively from the female parent (through the egg cell), the progeny of cytoplasmic male sterile plants will also be male sterile. This type of male sterility is of great significance in plant breeding because it eliminates the need for rouging, as all offspring of a CMS plant will be male sterile, making it easier to produce hybrid seeds.

Mechanism of Cytoplasmic Male Sterility

In CMS, defective pollen development is caused by mutations or rearrangements in the mitochondrial genome, which affect the energy production pathways essential for pollen formation. These mutations are often associated with plasmid-like elements in the mitochondria that disrupt normal mitochondrial function, leading to the failure of pollen development. Since the cytoplasm (and hence the mitochondria) is inherited from the egg cell, all progeny of a CMS plant will inherit the same male sterile cytoplasm. The nuclear genotype of a CMS plant is nearly identical to that of its recurrent pollinator, except the sterile cytoplasm. To maintain a CMS line, breeders must continually backcross the CMS plant with a fertile "maintainer" line that has the same nuclear genotype but possesses fertile cytoplasm. The male sterile line is referred to as the "A line," while the maintainer line is referred to as the "B line."

Utilization of Cytoplasmic Male Sterility in Plant Breeding

Cytoplasmic male sterility is particularly useful in producing hybrid seeds, especially in crops where manual emasculation is difficult or impossible. By using CMS lines as the female parent and a male fertile line as the pollinator, breeders can produce hybrid seeds without the need for manual removal of the male reproductive organs. This method has been widely used in crops like maize (*Zea mays*), sorghum (*Sorghum bicolor*), and onions (*Allium cepa*). CMS is not universally applicable to all crops, especially those where seeds are the primary economic product, such as cereals and legumes. In such cases, the hybrid progeny would inherit the male sterile cytoplasm and would be unable to produce functional pollen, making them unsuitable for further seed production. Therefore, CMS is more commonly used in crops where vegetative parts are harvested, such as fodder sorghum, onions, and cabbages.

Challenges and Limitations

1. **Limited Use in Seed-Producing Crops:** CMS cannot be used in crops where seeds are the primary economic product, as the hybrid progeny would also be male sterile, making it impossible to produce subsequent generations of seeds.
2. **Cytoplasmic-Nuclear Interactions:** The expression of CMS can be influenced by nuclear genes, leading to variability in the degree of sterility. This requires careful selection of compatible nuclear and cytoplasmic genotypes to ensure reliable male sterility.
3. **Environmental Sensitivity:** The expression of CMS can sometimes be influenced by environmental factors such as temperature, which may lead to partial fertility restoration under certain conditions.

3. Cytoplasmic-Genetic Male Sterility

Cytoplasmic-genetic male sterility is a combination of cytoplasmic male sterility and nuclear genes that can restore male fertility. In this system, the cytoplasm determines male sterility, but the presence of specific nuclear genes, known as restorer genes (R), can restore male fertility. These restorer genes are dominant and are found in certain strains or related species, allowing breeders to control the sterility or fertility of the plant depending on the desired outcome.

Mechanism of Cytoplasmic-Genetic Male Sterility

In cytoplasmic-genetic male sterility, male sterility is conferred by the sterile cytoplasm, but the fertility of the plant can be restored if the nuclear genome contains one or more dominant restorer genes (*R*). If the plant has the restorer allele (*R*), it will produce functional pollen even in the presence of sterile cytoplasm. Conversely, if the plant lacks the restorer gene (*rr*), it will remain male sterile.

This system is particularly useful in hybrid seed production because it allows breeders to produce male sterile lines for use as female parents and then restore fertility in the hybrid progeny by using pollinators that carry the restorer gene. This ensures that the hybrid plants will be fertile and capable of producing seeds, making the system applicable to a wide range of crops where seeds are the economic product.

Utilization of Cytoplasmic-Genetic Male Sterility in Plant Breeding

Cytoplasmic-genetic male sterility is widely used in hybrid seed production in crops like maize, sorghum, rice, and wheat. In this system, breeders use three lines:

1. **A Line:** The male sterile line, which carries the sterile cytoplasm and lacks the restorer gene.
2. **B Line:** The maintainer line, which has the same nuclear genotype as the A line but carries fertile cytoplasm, allowing the A line to be maintained through backcrossing.

3. **R Line:** The restorer line, which carries the restorer gene and is used as the pollinator to restore fertility in the hybrid progeny.

By crossing the A line with the R line, breeders can produce hybrid seeds that are both fertile and exhibit hybrid vigor. This system has been successfully exploited in crops like sorghum and pearl millet, where hybrid seeds are in high demand for their improved yield, disease resistance, and stress tolerance.

Advantages and Challenges

1. **Restorer Genes:** The presence of restorer genes allows breeders to produce fertile hybrid progeny, making the system applicable to a wide range of seed-producing crops.
2. **Versatility:** Cytoplasmic-genetic male sterility can be used in both seed-producing and vegetatively propagated crops, making it a highly versatile tool for hybrid seed production.
3. **Complexity:** The system requires careful management of the A, B, and R lines to ensure successful hybrid seed production. Additionally, not all crops have identified restorer genes, limiting the application of this system in some species.

Male sterility is an invaluable tool in plant breeding, especially for hybrid seed production. By eliminating the need for manual emasculation, male sterility systems reduce labor costs and improve the efficiency of breeding programs. The three types of male sterility—genetic, cytoplasmic, and cytoplasmic-genetic—each offer unique advantages and challenges, and their use depends on the specific crop and breeding goals. While genetic male sterility has limited commercial application due to the need for rouging, cytoplasmic male sterility, and cytoplasmic-genetic male sterility are widely used in crops like maize, sorghum, and onions. The discovery of restorer genes has expanded the applicability of cytoplasmic-genetic male sterility, allowing breeders to produce fertile hybrid seeds in a wide range of crops. As plant breeding technology continues to advance, the exploitation of male sterility systems will remain a key strategy for developing high-yielding, disease-resistant, and resilient hybrid crops that meet the demands of modern agriculture.

Chapter - 2

Principles of Breeding for Biotic and Abiotic Stresses

A. Principles of Breeding for Biotic and Abiotic Stresses

Breeding for resistance to biotic and abiotic stresses is a key objective in crop improvement programs globally. Biotic stresses refer to damage caused by living organisms such as pathogens (fungi, bacteria, viruses), insects, nematodes, and weeds. Abiotic stresses, on the other hand, include environmental factors like drought, heat, salinity, cold, and nutrient deficiencies that adversely affect plant growth and productivity. Breeding for stress resistance in crops involves selecting and developing varieties that can withstand or minimize the damage caused by these stressors, thereby improving yields, stability, and sustainability in agricultural systems. The principles underlying breeding for biotic and abiotic stress resistance rely on understanding plant-pathogen interactions, genetic control of stress tolerance, mechanisms of resistance or tolerance, and the use of modern technologies such as molecular markers, genetic engineering, and genomics to facilitate breeding programs.

Breeding for Biotic Stress Resistance

Biotic stresses pose a significant threat to agricultural productivity, especially as changing climates and agricultural practices alter pest and disease dynamics. Effective breeding for biotic stress resistance focuses on developing plant varieties that can withstand pathogens and pests, reducing the need for chemical pesticides, and improving long-term crop sustainability. The following principles guide the process of breeding for biotic stress resistance:

1. Host-Pathogen Interactions

Host-pathogen interactions refer to the complex interplay between a plant (host) and a pathogen (such as fungi, bacteria, viruses, or nematodes) that results in either successful infection or resistance. these interactions are fundamental for developing crop varieties that can resist biotic stress effectively. Breeding for biotic stress resistance starts with an in-depth study of how plants respond to pathogen attacks and how pathogens evolve to overcome plant defenses. In essence, the battle between a plant and a pathogen is an ongoing evolutionary arms race, where plants evolve resistance mechanisms, and pathogens develop new strategies to overcome these defenses.

DOI: 10.1201/9781003757412-2

Types of Plant Resistance: Qualitative and Quantitative

Plant resistance to pathogens is categorized into two main types: qualitative resistance and quantitative resistance. Both play distinct roles in protecting crops from pathogen attacks, and understanding the differences between them helps breeders strategize for both short-term and long-term resistance in breeding programs.

1. Qualitative Resistance

Qualitative resistance, also referred to as vertical resistance, is controlled by a small number of genes, often a single major gene, commonly called an *R* (resistance) gene. These *R* genes are highly specific and provide strong resistance to certain pathogen strains, which makes qualitative resistance highly effective but often short-lived.

Mechanism of Qualitative Resistance

Qualitative resistance operates through gene-for-gene interactions between the host plant's *R* gene and the pathogen's corresponding avirulence (*Avr*) gene. When a plant carrying a specific *R* gene recognizes the presence of a pathogen carrying the corresponding *Avr* gene, it triggers a defense response known as the hypersensitive response (HR). This response results in localized cell death around the site of infection, effectively containing the pathogen and preventing it from spreading further.

» **Gene-for-gene interaction:** This is a key concept in qualitative resistance, where each *R* gene in the plant recognizes a specific *Avr* gene in the pathogen. The plant responds by mounting a strong defense, often leading to localized necrosis that restricts pathogen growth.

» **Hypersensitive response (HR):** The HR is one of the most well-known defense mechanisms associated with qualitative resistance. When a plant detects the pathogen, it triggers rapid cell death in the infected area to stop the pathogen from spreading to healthy tissue. This response is fast and effective, leading to strong resistance against the specific pathogen strain carrying the matching *Avr* gene.

Limitations of Qualitative Resistance

While qualitative resistance provides robust protection against specific pathogen strains, it has significant limitations. The main issue is the potential for pathogens to evolve and overcome this resistance. Pathogens are highly adaptable, and under selection pressure from resistant crops, they can mutate or lose their *Avr* genes, rendering the plant's *R* gene ineffective. This results in the breakdown of resistance, where previously resistant crops become susceptible to new strains of the pathogen.

» **Breakdown of resistance:** Pathogens can quickly evolve to overcome the plant's *R* genes. For example, a pathogen may mutate its *Avr* gene, rendering the plant's *R* gene unable to recognize it, allowing the pathogen to infect the plant. This is especially problematic in monoculture farming systems where the same *R* gene is used over large areas, increasing the selection pressure on the pathogen to evolve.

» **Durability concerns:** Because qualitative resistance is based on single or few genes, it is often short-lived, particularly in environments where pathogens have high genetic diversity or mutation rates. As a result, breeders must frequently replace old resistance genes with new ones, making it a labor-intensive and ongoing process to maintain crop protection.

2. Quantitative Resistance

Quantitative resistance, also known as horizontal resistance or partial resistance, is controlled by multiple genes, often referred to as polygenes. This type of resistance provides a broader and more durable defense against a wide range of pathogen strains, making it a valuable long-term strategy for disease management.

Mechanism of Quantitative Resistance

Unlike qualitative resistance, which operates through a single gene-for-gene interaction, quantitative resistance is the result of the cumulative effect of many genes, each contributing a small amount to the overall resistance. These genes collectively reduce the pathogen's ability to infect the plant, but they do not provide complete immunity. Instead, they lower the severity of the disease, slowing down the pathogen's development and spread within the plant.

» **Polygenic control:** In quantitative resistance, multiple genes contribute to the plant's defense, each having a small effect on the overall resistance. This polygenic control makes it harder for pathogens to overcome resistance, as they would need to mutate multiple virulence factors simultaneously to bypass all the resistance genes.

» **Reduction in disease severity:** Quantitative resistance does not prevent the infection entirely but reduces its impact. This means that the plant may still become infected, but the disease develops more slowly, and its severity is reduced. As a result, the crop suffers less damage, and the yield losses are minimized.

Durability of Quantitative Resistance

The durability of quantitative resistance is one of its most significant advantages. Because it is controlled by many genes, it is much more difficult for pathogens to overcome this type of resistance. Even if a pathogen mutates to overcome one of the resistance genes, the other genes continue to provide protection, making quantitative resistance more stable over time.

» **Broad-spectrum resistance:** Quantitative resistance often provides protection against multiple strains or species of pathogens, making it an ideal strategy for breeding crops that will be exposed to diverse pathogen populations. It also tends to be more effective against pathogens that evolve rapidly, as the polygenic nature of quantitative resistance makes it harder for the pathogen to bypass.

» **Longer-lasting protection:** Unlike qualitative resistance, which can break down relatively quickly, quantitative resistance tends to be more durable. This is because pathogens must overcome several genetic defenses simultaneously, which is less likely to happen through simple mutations. As a result, quantitative resistance provides more sustainable long-term protection for crops.

Host-Pathogen Coevolution

In the context of plant breeding, understanding the coevolution of hosts and pathogens is essential. Both plants and pathogens are constantly evolving in response to each other. Plants evolve resistance mechanisms to fend off pathogens, while pathogens evolve new strategies to overcome plant defenses.

- **Pathogen virulence factors:** Pathogens possess virulence factors, such as enzymes, toxins, or effector proteins, that allow them to infect and damage the plant. In response, plants evolve resistance mechanisms that can recognize and neutralize these virulence factors. However, pathogens can mutate their virulence factors to avoid detection by the plant, leading to a continuous arms race between plant defenses and pathogen virulence.
- **Plant defense mechanisms:** In addition to *R* genes, plants have a range of other defense mechanisms, including physical barriers (e.g., cell walls, waxy cuticles) and chemical defenses (e.g., antimicrobial compounds). Breeding programs aim to enhance these defenses by selecting traits that improve the plant's ability to withstand pathogen attacks.

This dynamic is critical for breeders when selecting parent plants for breeding programs. By recognizing how pathogens evolve and how plants respond, breeders can make informed decisions about which resistance genes to incorporate into new varieties. Additionally, using strategies such as gene pyramiding, which involves stacking multiple resistance genes into a single variety, can help enhance the durability of resistance.

Molecular Basis of Host-Pathogen Interactions

Modern advances in molecular biology and genetics have provided deeper insights into host-pathogen interactions. These insights are invaluable for developing disease-resistant crops. Key areas of research include:

- **Effector-triggered immunity (ETI):** This is a type of plant immune response that is activated when a plant recognizes specific pathogen effectors. Effectors are molecules secreted by pathogens that promote infection. When a plant's *R* gene recognizes an effector, it triggers a defense response, often leading to the hypersensitive response.
- **Pathogen-associated molecular pattern (PAMP)-triggered immunity (PTI):** PTI is the first line of defense in plants. It is triggered when plants recognize conserved molecules found in many pathogens, known as PAMPs. This recognition activates a broad-spectrum immune response that slows down the pathogen's ability to infect the plant.
- **Molecular markers and gene editing:** Advances in molecular markers and gene editing technologies, such as CRISPR-Cas9, have revolutionized plant breeding for disease resistance. By identifying specific resistance genes and manipulating them through genetic engineering, breeders can rapidly introduce disease resistance traits into crops without the need for traditional cross-breeding methods.

Implications for Breeding Programs

The insights gained from studying host-pathogen interactions have significant implications for plant breeding programs. The use of molecular markers for tracking resistance genes, gene pyramiding to combine multiple resistance genes, and the development of crops with durable quantitative resistance are all strategies that help breeders develop varieties with improved resistance to pathogens.

- **Marker-assisted selection (MAS):** Molecular markers linked to resistance genes can be used to select plants with desirable resistance traits. This accelerates the breeding process by allowing breeders to track the inheritance of resistance genes without relying on time-consuming field evaluations.

- **Gene pyramiding:** Combining multiple *R* genes in a single plant (gene pyramiding) is an effective strategy for enhancing resistance. Pathogens are less likely to overcome multiple resistance genes simultaneously, leading to more durable protection.
- **Gene editing and transgenic approaches:** Technologies like CRISPR-Cas9 enable precise modification of plant genomes, allowing breeders to introduce or enhance resistance traits more efficiently. These approaches can accelerate the development of disease-resistant crops, especially for pathogens that evolve rapidly.

2. Types of Resistance: Horizontal and Vertical

In plant breeding, resistance to biotic stresses is typically categorized into two main types: vertical resistance and horizontal resistance. These classifications are based on the specificity of the resistance to pathogen races and the underlying genetic control of the resistance mechanisms. Both vertical and horizontal resistance are essential for crop protection, and understanding their differences is critical for developing long-lasting and effective disease management strategies.

1. Vertical Resistance (Race-Specific Resistance)

Vertical resistance, also known as race-specific or qualitative resistance, is the type of resistance that provides strong protection against specific races or strains of a pathogen. It is often governed by single resistance (*R*) genes that interact directly with corresponding avirulence (*Avr*) genes in the pathogen. When the plant's *R* gene recognizes the pathogen's *Avr* gene, a defense response is triggered, which can be very effective in preventing the pathogen from infecting the plant.

Mechanism of Vertical Resistance

Vertical resistance operates through a highly specific interaction between the host plant and the pathogen, known as the gene-for-gene relationship. In this model, for every *R* gene in the plant, there is a corresponding *Avr* gene in the pathogen. When the plant detects the presence of the pathogen's *Avr* gene, it initiates a defense response, typically leading to localized cell death around the infection site, which prevents the pathogen from spreading. This response is often referred to as the **hypersensitive response** (HR).

- **Hypersensitive response (HR):** The HR is a rapid, localized defense mechanism that occurs at the point of pathogen entry. The plant cells surrounding the infection site undergo programmed cell death, forming a barrier that prevents the pathogen from advancing further into the plant tissue. This response is highly effective in containing the pathogen, especially in the early stages of infection.
- **Gene-for-gene interaction:** In vertical resistance, the relationship between the plant's *R* genes and the pathogen's *Avr* genes is highly specific. The plant's immune system can recognize specific molecules produced by the pathogen, triggering the HR and preventing infection. However, this specificity also makes vertical resistance vulnerable to the pathogen's ability to evolve and evade detection.

Advantages of Vertical Resistance

1. **Strong Protection:** Vertical resistance provides a high level of protection against specific races

of a pathogen. Once triggered, the hypersensitive response effectively halts the pathogen's progress and limits the damage to the plant.

2. **Simplicity in Breeding:** Since vertical resistance is often controlled by single major genes, it is relatively easy to transfer these genes into new varieties through traditional breeding or molecular techniques. Once a useful *R* gene is identified, it can be incorporated into crop lines quickly, providing effective short-term protection.

Limitations of Vertical Resistance

1. **Breakdown of Resistance:** The major limitation of vertical resistance is its vulnerability to pathogen evolution. Pathogens can mutate rapidly, and under selective pressure from resistant crop varieties, new virulent strains can emerge that are no longer recognized by the plant's *R* genes. This leads to the breakdown of resistance, rendering the previously resistant crop susceptible to the pathogen. For instance, resistance to wheat rust (*Puccinia spp.*) has broken down multiple times as new pathogen strains evolve.
2. **Narrow Spectrum of Protection:** Vertical resistance is usually effective against only a few specific races of a pathogen. If the pathogen population contains other races that are not recognized by the plant's *R* gene, the crop will remain vulnerable to those races.
3. **Short-Term Solution:** Due to its race-specific nature, vertical resistance is often considered a short-term solution to disease control. It is effective as long as the pathogen does not evolve, but long-term reliance on vertical resistance can lead to the rapid emergence of new virulent pathogen races.

2. Horizontal Resistance (Race-Non-Specific or Durable Resistance)

Horizontal resistance, also referred to as race-non-specific, quantitative, or durable resistance, provides partial protection against a wide range of pathogen races. Unlike vertical resistance, horizontal resistance is typically controlled by multiple genes, each contributing a small effect to the overall resistance. These genes act together to slow the progression of the disease and reduce its severity rather than completely preventing infection.

Mechanism of Horizontal Resistance

Horizontal resistance is based on a more generalized defense mechanism that does not depend on specific gene-for-gene interactions. Instead, it involves multiple genes that contribute to the plant's overall ability to tolerate or resist pathogen infection. This type of resistance usually works by slowing down the pathogen's life cycle, limiting its ability to reproduce, or reducing the amount of damage it can cause to the plant.

- **Polygenic control:** Horizontal resistance is governed by many genes, often referred to as quantitative trait loci (QTLs), which collectively enhance the plant's resistance to a broad spectrum of pathogen races. Each gene involved in horizontal resistance has a small effect, but together, they provide significant protection against multiple strains of the pathogen.
- **Reduction in disease severity:** Horizontal resistance does not typically prevent the pathogen from infecting the plant. Instead, it reduces the severity of the disease by limiting the

pathogen's ability to cause damage. The plant may still become infected, but the infection progresses more slowly, and the overall impact on the plant is reduced.

Advantages of Horizontal Resistance

1. **Durability:** Horizontal resistance is generally more durable than vertical resistance. Since it is controlled by multiple genes, it is much more difficult for the pathogen to overcome this resistance through simple genetic mutations. Even if the pathogen evolves to overcome one gene, the other resistance genes continue to provide protection. This makes horizontal resistance more stable and long-lasting in the field.
2. **Broad-Spectrum Protection:** Horizontal resistance is effective against a wide range of pathogen races, providing a broader level of protection than vertical resistance. This makes it particularly useful in environments where the pathogen population is highly diverse or prone to rapid evolution.
3. **Sustainable Disease Management:** Because of its durability and broad-spectrum protection, horizontal resistance is often preferred for long-term disease management. It reduces the need for chemical controls, such as fungicides or pesticides, and contributes to more sustainable agricultural practices.

Limitations of Horizontal Resistance

1. **Partial Resistance:** Horizontal resistance does not provide complete immunity to the pathogen. The plant may still become infected, and disease symptoms may still be present, though at a reduced severity. In some cases, this partial resistance may not be enough to prevent significant yield losses, especially under conditions of high disease pressure.
2. **Complexity in Breeding:** Because horizontal resistance is controlled by many genes, it can be more challenging to breed for than vertical resistance. Traditional breeding methods may require several generations to accumulate the desired resistance genes, and the effects of individual genes can be difficult to measure. However, advances in molecular markers and genomic selection have made it easier to identify and select for quantitative resistance traits in modern breeding programs.

Comparison of Horizontal and Vertical Resistance

While both types of resistance have their advantages and limitations, they serve complementary roles in crop protection. The choice between vertical and horizontal resistance depends on the specific crop, pathogen, and agricultural environment. In many cases, the best approach is to combine both types of resistance to create a more robust and sustainable disease management strategy.

» **Vertical Resistance:**
  - Provides strong, race-specific protection.
  - Vulnerable to the evolution of virulent pathogen strains.
  - Easy to incorporate into breeding programs due to single-gene control.
  - Short-term solution, with the risk of resistance breakdown.

» **Horizontal Resistance:**

- Provides broad, race-non-specific protection.
- Difficult for pathogens to overcome due to polygenic control.
- Slower to incorporate into breeding programs due to the complexity of multiple genes.
- Long-term, sustainable solution with partial but durable resistance.

Combining Vertical and Horizontal Resistance for Durable Crop Protection

One of the most effective strategies in modern plant breeding is to combine both vertical and horizontal resistance to create a more comprehensive defense against pathogens. This can be achieved through gene pyramiding, where multiple *R* genes (vertical resistance) are combined with quantitative resistance genes (horizontal resistance) in a single plant variety.

» **Gene Pyramiding:** By stacking several *R* genes and quantitative resistance genes into one plant, breeders can provide both strong, race-specific protection and durable, broad-spectrum protection. This strategy makes it harder for the pathogen to evolve and overcome the plant's defenses, as it would need to mutate multiple virulence factors simultaneously to bypass both types of resistance.

» **Integrated Disease Management:** Combining vertical and horizontal resistance can also be part of an integrated disease management strategy, where resistant crop varieties are used alongside other control measures, such as crop rotation, biological control, and judicious use of chemical pesticides. This holistic approach reduces the selection pressure on pathogens, helping to prolong the effectiveness of resistance genes and slow the evolution of virulent pathogen strains.

3. Sources of Resistance

The foundation of any successful breeding program for biotic stress resistance lies in identifying and utilizing diverse sources of resistance. These sources of resistance serve as the genetic pool from which breeders can draw to develop new, resilient crop varieties that can withstand attacks from pests, pathogens, and other biotic stressors. As pathogens evolve and environmental pressures increase, identifying new sources of resistance becomes even more critical for maintaining food security and agricultural productivity.

Breeding programs rely on several key sources of resistance, including wild relatives of crops, landraces and traditional varieties, and induced mutations. Each of these sources offers unique advantages and challenges, and their use in breeding programs is essential for developing disease-resistant crops.

Wild Relatives of Crops

Wild relatives of domesticated crops are some of the most valuable sources of resistance to biotic stress. These species have not undergone the same selective breeding as cultivated varieties, which means they have retained a high level of genetic diversity, including traits that allow them to survive and thrive in the presence of pathogens and pests.

Advantages of Using Wild Relatives

1. **Natural Resistance**: Wild relatives often possess natural resistance to a wide range of diseases and pests. This resistance is the result of millions of years of evolution in challenging environments where pathogens and pests exert strong selective pressure. For example, wild relatives of wheat, such as *Aegilops* species, have been a key source of resistance to diseases like rusts and powdery mildew. These species have been used in breeding programs to transfer resistance genes into cultivated wheat varieties.
2. **Genetic Diversity**: Domesticated crops often have reduced genetic diversity due to selective breeding for specific traits like yield or quality. This makes them more vulnerable to biotic stress. In contrast, wild relatives have not undergone this genetic narrowing and therefore retain a broad range of alleles that can confer resistance to a variety of pathogens and pests. This diversity can be harnessed to improve the resilience of cultivated crops.
3. **Introgression of Resistance Traits**: Through a process known as introgression, breeders can transfer resistance genes from wild relatives into cultivated varieties. Introgression involves crossing a wild relative with a domesticated variety and then backcrossing the offspring with the cultivated parent over several generations to maintain the desired resistance trait while minimizing the influence of undesirable wild traits. For example, resistance to wheat stem rust (*Puccinia graminis*) has been introgressed from wild relatives into commercial wheat varieties, significantly reducing yield losses due to this disease.

Challenges of Using Wild Relatives

1. **Compatibility Issues**: One of the main challenges of using wild relatives in breeding programs is the difficulty in achieving successful crosses between wild and domesticated species. In some cases, wild relatives may belong to different species or genera, making traditional breeding methods less effective. However, advances in techniques such as wide hybridization, embryo rescue, and biotechnology have helped overcome some of these barriers.
2. **Linkage Drag**: While wild relatives may possess valuable resistance traits, they may also carry undesirable traits such as low yield, poor agronomic characteristics, or incompatibility with modern farming practices. When transferring resistance genes from wild relatives, breeders must be careful to minimize "linkage drag," where undesirable traits are co-inherited along with the desired resistance trait. This often requires multiple generations of backcrossing and careful selection to separate the beneficial genes from the undesirable ones.
3. **Adaptation to Cultivated Environments**: Wild relatives are often adapted to environments that are very different from the highly controlled conditions of modern agriculture. This means that while they may possess resistance to pathogens or pests, they may not perform well in terms of yield, quality, or adaptability to the conditions under which commercial crops are grown. Therefore, the challenge for breeders is to transfer the resistance traits while ensuring that the resulting crop variety performs well in cultivated conditions.

Landraces and Traditional Varieties

Landraces and traditional varieties are another important source of resistance to biotic stress. Landraces are locally adapted crop varieties that have evolved over time in specific geographic regions. They have typically been maintained by farmers for generations and are often well-adapted to the local environment, including the pests and diseases that occur in those areas.

Advantages of Using Landraces

1. **Genetic Diversity and Adaptation**: Landraces are often genetically diverse and have evolved in response to the specific challenges of their environment. This includes resistance to local pathogens and pests. Because they have been selected by farmers for traits such as disease resistance, drought tolerance, and pest resistance, they are a valuable reservoir of genetic diversity that can be tapped for breeding programs. For instance, certain landraces of rice (*Oryza sativa*) have been found to possess resistance to bacterial blight (*Xanthomonas oryzae*) and have been used to develop commercial rice varieties with improved resistance to the disease.
2. **Historical Evidence of Resistance**: Landraces have been maintained in farming systems for centuries, often in the face of biotic stresses. This long-term survival suggests that they have developed resistance mechanisms that may not be present in modern commercial varieties. By studying landraces that have been used in regions with a high incidence of certain diseases, breeders can identify resistance traits that are effective against local pathogen populations.
3. **Low Input Systems**: Landraces are often cultivated in low-input farming systems, where farmers use minimal chemical inputs like pesticides and fertilizers. This means that these varieties may have developed natural resistance mechanisms to cope with pests and diseases without relying on external inputs. These traits are especially valuable in sustainable and organic farming systems, where the use of chemical pesticides is restricted.

Challenges of Using Landraces

1. **Lower Yield Potential**. One of the main challenges of using landraces in modern breeding programs is that they often have lower yield potential compared to commercial varieties. While landraces may have valuable resistance traits, they may not meet the yield expectations of modern agriculture. Breeding programs must strike a balance between incorporating disease resistance from landraces while ensuring that the resulting varieties have acceptable levels of yield.
2. **Time-Consuming Selection**: Incorporating resistance traits from landraces into commercial varieties can be a time-consuming process. Because landraces often carry a complex array of traits, breeders must carefully select for the desired resistance trait while eliminating undesirable traits, such as low yield or poor quality. This can take several generations of crossing and backcrossing, which can slow down the breeding process.
3. **Limited Availability**: While landraces are an important source of resistance, many have been lost over time due to the widespread adoption of high-yielding commercial varieties. As modern agriculture has favored the use of uniform, high-yielding crops, traditional landraces have often been abandoned. This loss of genetic diversity makes it more difficult to find suitable landraces for breeding programs. However, efforts to conserve and document landraces through seed banks and genetic repositories are helping to mitigate this issue.

Induced Mutations

Induced mutations are a powerful tool in plant breeding for creating new sources of resistance to biotic stress. Mutations can be induced through the application of chemicals, radiation, or other mutagenic agents, which cause changes in the plant's DNA. These mutations can result in new traits, including resistance to pathogens and pests.

Advantages of Induced Mutations

1. **Creation of Novel Resistance Traits**: Induced mutations can create new alleles that provide resistance to diseases or pests. These mutations may not exist in the natural gene pool or may not have been discovered yet. For example, resistance to stem rust in wheat (*Puccinia graminis*) has been developed through the use of induced mutations that produced new resistance genes.
2. **Broad Application**: Induced mutations can be applied to any crop species, making it a versatile tool for breeding. Whether in cereals, fruits, vegetables, or ornamental plants, breeders can use mutagenesis to develop resistance traits that might not be available in wild relatives or landraces. For instance, in barley (*Hordeum vulgare*), induced mutations have led to the discovery of resistance genes for powdery mildew (*Blumeria graminis*).
3. **Faster Development of Resistant Varieties**: Induced mutations can speed up the process of developing resistant varieties. Traditional breeding methods can take several years to introduce resistance traits through crossing and backcrossing, but mutagenesis can introduce these traits in a shorter time frame. Once a desirable mutation is identified, it can be rapidly incorporated into breeding programs.

Challenges of Induced Mutations

1. **Unpredictability**: One of the main challenges of using induced mutations is that the process is largely random. While mutagenesis can create new sources of resistance, it can also produce unwanted or harmful mutations. This means that breeders must carefully screen large populations of plants to identify those with beneficial mutations while eliminating those with undesirable traits.
2. **Linkage Drag and Off-Target Effects**: Mutations can sometimes affect genes that are linked to important agronomic traits, leading to linkage drag. This occurs when the mutation confers resistance but also negatively affects traits like yield, quality, or plant architecture. In addition, induced mutations can cause off-target effects, where unrelated genes are inadvertently altered, leading to unexpected changes in the plant's performance.
3. **Need for Extensive Screening**: After inducing mutations, breeders must screen large numbers of plants to identify individuals with the desired resistance traits. This process can be labor-intensive and time-consuming, as the majority of mutations will not confer useful traits. However, advances in molecular marker technology have made it easier to identify and select plants with specific mutations, improving the efficiency of the screening process.

4. Breeding Methods for Biotic Stress Resistance

Biotic stresses, such as pests, pathogens, and diseases, pose a significant threat to agricultural productivity and food security. Breeding for biotic stress resistance is a fundamental approach to developing crop varieties that can withstand these threats. Several breeding methods are employed to incorporate resistance traits into crops, each with its advantages, limitations, and specific applications. These methods include conventional breeding, marker-assisted selection (MAS), genetic engineering and gene editing, and pyramiding resistance genes.

1. Conventional Breeding

Conventional breeding is one of the oldest and most widely used methods for developing resistance to biotic stresses. It involves crossing plants with known resistance traits (often landraces, wild relatives, or existing resistant varieties) with high-yielding, susceptible varieties to produce offspring that combine both resistance and desirable agronomic traits.

Process of Conventional Breeding

The typical process for conventional breeding for biotic stress resistance includes several key steps:

- **Crossing resistant and susceptible varieties:** The first step is to cross a plant with strong resistance to a specific pest or pathogen with a high-yielding but susceptible variety. The goal is to combine the resistance traits of the resistant parent with the desirable traits of the susceptible parent (such as high yield, quality, or adaptability).
- **Selection of progeny:** In the next generations, breeders select offspring that exhibit both resistance and good agronomic performance. This selection is based on the phenotype (physical appearance) of the plants, where resistance is visually evaluated through disease screening in field or greenhouse trials. This process requires several generations of crossing and selection to fix the desired traits.
- **Backcrossing:** To ensure that the resistance genes are incorporated into high-yielding commercial varieties without losing the desirable agronomic traits, backcrossing is often employed. Backcrossing involves crossing the selected resistant progeny with the original susceptible parent repeatedly. This process retains the resistance gene while restoring the original characteristics of the high-yielding variety, ensuring that the final product has both improved resistance and agronomic performance.

Advantages of Conventional Breeding

1. **Wide Applicability:** Conventional breeding can be applied to a wide range of crops and resistance traits, making it a versatile method for improving crops globally. It does not require advanced technologies or infrastructure, making it accessible to breeding programs in both developed and developing countries.
2. **Combines Multiple Traits:** This method allows breeders to incorporate not only resistance traits but also other desirable traits such as high yield, drought tolerance, or improved quality. Through careful selection, breeders can create varieties that are well-rounded in terms of agronomic performance and biotic stress resistance.

Limitations of Conventional Breeding

1. **Time-Consuming:** Conventional breeding is a slow process that may take several years or even decades to develop a successful resistant variety. This is because multiple generations of plants must be grown, crossed, and selected to fix the desired traits. Pathogens can evolve rapidly, making it difficult to keep up with new virulent strains through conventional methods alone.
2. **Environmental Sensitivity:** The expression of resistance traits can be influenced by environmental conditions, making it difficult to consistently select resistant plants across different environments. Field evaluations may not always reveal the full spectrum of resistance potential, especially in locations where pathogen pressure is variable.
3. **Limited to Existing Genetic Variation:** Conventional breeding relies on the natural or existing genetic variation within the species or closely related species. If there are no available resistant sources in the gene pool, it becomes challenging to develop resistant varieties through traditional methods.

2. Marker-Assisted Selection (MAS)

Marker-assisted selection (MAS) is a more recent breeding method that uses molecular markers linked to resistance genes to accelerate the breeding process. MAS allows breeders to track the inheritance of specific resistance traits in a breeding population, reducing the time and resources required for phenotypic evaluations in the field.

How MAS Works

Molecular markers are segments of DNA that are tightly linked to resistance genes. By identifying these markers, breeders can determine whether a plant carries the desired resistance genes without having to expose the plant to the pathogen. This process can be done at the seedling stage, allowing breeders to select resistant individuals early in the breeding cycle.

- **DNA extraction and marker analysis:** DNA is extracted from plant tissue (often from leaves), and molecular markers are used to determine whether the plant contains the resistance gene. Plants with the desired markers are selected for further breeding.
- **Combining MAS with conventional breeding:** MAS is often combined with conventional breeding techniques. After performing a cross between a resistant and a susceptible variety, MAS is used to identify the offspring that carry the resistance genes, allowing breeders to focus on those individuals and speed up the selection process.

Advantages of MAS

1. **Accelerates Breeding:** MAS significantly shortens the time needed to develop resistant varieties by allowing breeders to select plants with resistance genes early in the breeding process. This reduces the number of generations required for selection and eliminates the need for repeated field trials under disease pressure.
2. **More Precise Selection:** Unlike conventional breeding, which relies on phenotypic selection (which can be influenced by environmental conditions), MAS provides a more precise and

reliable method of identifying resistant plants. This precision increases the likelihood of selecting plants that carry the desired resistance genes.

3. **Applicable to Complex Traits:** MAS is particularly useful for traits controlled by multiple genes (polygenic traits), such as quantitative resistance to diseases. Molecular markers linked to quantitative trait loci (QTLs) can be used to select plants that have combinations of resistance genes, increasing the durability of resistance.

Limitations of MAS

1. **High Initial Costs:** MAS requires specialized equipment and expertise in molecular biology, which may not be available in all breeding programs, particularly in developing countries. The initial costs of setting up MAS facilities and conducting marker analyses can be prohibitive for small-scale programs.
2. **Limited by Available Markers:** The success of MAS depends on the availability of molecular markers that are closely linked to the resistance genes. In some cases, markers may not be available for important resistance traits, limiting the effectiveness of MAS in those situations.
3. **Requires Phenotypic Validation:** Although MAS is more efficient than conventional breeding, it still requires some level of phenotypic validation to ensure that the selected plants exhibit the desired level of resistance in the field. This is particularly true for quantitative resistance, where multiple genes contribute to the trait.

3. Genetic Engineering and Gene Editing

Genetic engineering and **gene editing** are advanced breeding methods that allow for the direct modification of a plant's genome to introduce or enhance resistance traits. Unlike conventional breeding or MAS, which rely on the natural inheritance of genes, genetic engineering allows breeders to transfer specific genes into a plant's genome, including genes from other species.

Genetic Engineering

In genetic engineering, scientists insert a foreign gene (often from another species) into the plant's genome. This process can introduce novel resistance traits that are not present in the plant's natural gene pool. For example, resistance to bacterial, viral, or fungal pathogens can be introduced through the insertion of *R* genes or genes that produce antimicrobial proteins.

» **Transgenic plants:** One of the most famous examples of genetic engineering for biotic stress resistance is the development of Bt crops (e.g., Bt cotton, Bt maize). These plants have been engineered to express genes from the bacterium *Bacillus thuringiensis* (Bt), which produce proteins toxic to specific insect pests. Bt crops have significantly reduced the need for chemical insecticides, making them an important tool in pest management.

Gene Editing

Gene editing technologies, such as CRISPR-Cas9, allow for precise modifications to be made in the plant's existing genome without introducing foreign DNA. This method can be used to enhance a plant's natural resistance by activating or disabling specific genes.

» **CRISPR-Cas9:** This technology enables breeders to edit the plant's genome with pinpoint accuracy, making it possible to improve existing resistance genes or remove susceptibility genes. For example, gene editing can be used to enhance the plant's immune response to pathogens or improve its ability to detoxify pathogen-produced toxins.

Advantages of Genetic Engineering and Gene Editing

1. **Precision and Speed:** Genetic engineering and gene editing provide a precise and fast method for developing resistance traits. Unlike conventional breeding, which can take years to transfer resistance genes into a new variety, genetic modification allows for the direct and immediate incorporation of resistance traits.
2. **Novel Resistance Sources:** Genetic engineering makes it possible to introduce resistance genes from unrelated species, providing new sources of resistance that would not be available through conventional breeding. For example, resistance to viruses or pests can be introduced from bacteria or other plants.
3. **Avoids Linkage Drag:** By directly inserting or editing the genes responsible for resistance, genetic engineering and gene editing avoid the issue of linkage drag, where undesirable traits are inherited along with the resistance genes in conventional breeding programs.

Limitations of Genetic Engineering and Gene Editing

1. **Regulatory and Public Acceptance Issues:** Genetically modified organisms (GMOs) face significant regulatory hurdles in many countries, as well as opposition from certain segments of the public. The approval process for GMOs can be time-consuming and expensive, limiting their widespread adoption. Gene-edited crops, however, are often subject to fewer regulations, depending on the country and the specific technology used.
2. **High Costs and Technical Expertise:** Both genetic engineering and gene editing require advanced technical expertise and significant investment in laboratory facilities, making them less accessible to small-scale or resource-limited breeding programs.
3. **Ethical and Environmental Concerns:** There are ongoing debates about the ethical and environmental implications of using genetically modified crops. Concerns include the potential for cross-contamination with non-GMO crops, the impact on biodiversity, and the long-term sustainability of genetically engineered resistance.

4. Pyramiding Resistance Genes

Pyramiding resistance genes is a breeding strategy that involves combining multiple resistance genes (often referred to as *R* genes) into a single plant. The goal is to increase the durability of resistance by making it more difficult for pathogens to evolve virulence against multiple resistance genes simultaneously.

Process of Gene Pyramiding

» **Combining multiple *R* genes:** Breeders use traditional breeding or molecular techniques to introduce several *R* genes into the same plant. These genes may provide resistance to different pathogen strains or act against different stages of the pathogen's life cycle, creating a multi-layered defense.

» **Marker-assisted pyramiding:** MAS is often used to facilitate gene pyramiding, as it allows breeders to track the inheritance of multiple *R* genes simultaneously. This reduces the time needed to combine the genes and ensures that the desired combination is present in the final variety.

Advantages of Pyramiding Resistance Genes

1. **Increased Durability:** By combining multiple resistance genes, pyramiding makes it more difficult for pathogens to overcome the plant's defenses. Pathogens would need to evolve virulence against all the *R* genes simultaneously, which is unlikely to happen quickly. This increases the durability of resistance and reduces the likelihood of resistance breakdown.
2. **Broad-Spectrum Protection:** Gene pyramiding can provide resistance against multiple pathogen strains or species, offering broader protection than single-gene resistance. This makes it a valuable strategy in environments where multiple pathogens are present.

Limitations of Pyramiding Resistance Genes

1. **Complex Breeding Process:** Pyramiding multiple genes requires careful planning and execution, as well as the use of molecular markers to track the inheritance of each gene. This adds complexity to the breeding process and requires advanced breeding techniques.
2. **Cost and Time:** Developing varieties with multiple resistance genes through pyramiding can be time-consuming and costly, especially if MAS or other molecular techniques are needed.

Breeding for Abiotic Stress Tolerance

Abiotic stresses such as drought, heat, cold, salinity, and nutrient deficiencies are major challenges in agriculture, particularly in regions where climate change exacerbates these stress conditions. Breeding for abiotic stress tolerance involves selecting and developing varieties that can withstand or perform better under these environmental conditions. The principles guiding breeding for abiotic stress tolerance include understanding plant responses to stress, identifying genes responsible for tolerance, and using breeding methods that enhance stress resilience.

1. Mechanisms of Abiotic Stress Tolerance

Abiotic stresses such as drought, salinity, extreme temperatures, and nutrient deficiencies are major factors limiting crop productivity worldwide. To survive and thrive under these harsh environmental conditions, plants have evolved a wide array of physiological, biochemical, and molecular mechanisms to cope with abiotic stresses. These mechanisms help plants mitigate the damaging effects of stress, ensuring their growth and development even in unfavorable environments. These mechanisms are crucial for breeding stress-tolerant crops that can withstand the challenges posed by climate change and other environmental pressures.

1. Osmotic Adjustment

One of the primary mechanisms plants use to cope with drought and salinity stress is **osmotic**

adjustment, which helps plants maintain water uptake and cell turgor under stressful conditions. Osmotic adjustment occurs when plants accumulate certain organic and inorganic solutes, known as **osmolytes**, to lower their cellular water potential and improve water retention. This allows the plant to continue to function and grow even when water availability is limited.

Key Osmolytes in Plants

» **Proline**: Proline is one of the most common osmolytes that accumulates in plants during drought or salinity stress. It plays multiple roles in stress tolerance, including osmoprotection, stabilization of proteins and membranes, and detoxification of reactive oxygen species (ROS). Proline accumulation helps plants maintain cell turgor by drawing water into the cells, preventing dehydration.

» **Glycine betaine**: Another important osmolyte is glycine betaine, which accumulates in many plant species in response to salt and drought stress. Glycine betaine helps protect cellular structures, such as membranes and proteins, by stabilizing their structure under stress conditions. It also functions as an osmoprotectant by regulating the osmotic balance between the plant cells and their surrounding environment.

» **Sugars (e.g., trehalose, sucrose)**: Sugars such as trehalose and sucrose also play a role in osmotic adjustment by serving as osmoprotectants. These sugars help maintain the structural integrity of cell membranes and proteins under osmotic stress and act as energy reserves that can be utilized by the plant when normal photosynthetic activity is reduced due to stress.

Role of Osmotic Adjustment in Abiotic Stress Tolerance

The accumulation of osmolytes during osmotic adjustment serves multiple purposes:

» **Maintaining water uptake**: By lowering the water potential inside cells, osmolytes enable plants to continue absorbing water from the soil even when external water availability is low (as in drought) or when soil salinity is high (as in salt stress).

» **Protecting cellular structures**: Osmolytes protect cellular membranes, proteins, and enzymes from damage caused by dehydration and high salt concentrations. This helps preserve cellular function during periods of stress.

» **Enabling recovery**: Osmotic adjustment not only helps plants survive stress but also allows them to recover more quickly once normal conditions are restored. The ability to bounce back from stress is critical for maintaining crop productivity in environments where periodic drought or salinity stress is common.

2. Antioxidant Defense Systems

Abiotic stresses often lead to the overproduction of **reactive oxygen species (ROS)**, such as hydrogen peroxide (H_2O_2), superoxide radicals (O_2^-), and hydroxyl radicals (OH^-), which are highly reactive molecules that can cause oxidative damage to lipids, proteins, and nucleic acids in plant cells. To protect themselves from oxidative damage, plants have developed complex **antioxidant defense systems** that can detoxify ROS and prevent cellular injury.

Key Components of the Antioxidant Defense System

- **Superoxide dismutase (SOD):** This enzyme is the first line of defense against ROS. It catalyzes the conversion of superoxide radicals (O_2^-) into hydrogen peroxide (H_2O_2) and oxygen (O_2), which are less harmful to the cell. SOD is present in various cellular compartments, including chloroplasts, mitochondria, and the cytosol.
- **Catalase (CAT)**: Catalase is responsible for breaking down hydrogen peroxide (H_2O_2) into water (H_2O) and oxygen (O_2). This enzyme is particularly important because hydrogen peroxide can be highly toxic to cells at elevated levels. Catalase is primarily found in peroxisomes, where it detoxifies hydrogen peroxide produced during various metabolic processes.
- **Ascorbate peroxidase (APX):** This enzyme is part of the ascorbate-glutathione cycle, a key antioxidant pathway in plants. APX reduces hydrogen peroxide to water using ascorbate (vitamin C) as an electron donor. This process occurs in several organelles, including chloroplasts and mitochondria, helping to protect the plant's photosynthetic machinery from oxidative damage.
- **Glutathione (GSH)**: Glutathione is a tripeptide that plays a central role in the detoxification of ROS. It acts as a substrate for enzymes like glutathione reductase (GR) and glutathione peroxidase, which are involved in the regeneration of antioxidants and the neutralization of ROS.

Role of Antioxidant Defense Systems in Abiotic Stress Tolerance

- **Protection from oxidative stress**: Under abiotic stress conditions such as drought, salinity, or extreme temperatures, the production of ROS increases dramatically. The antioxidant defense system helps protect cells by neutralizing these harmful molecules, preventing oxidative damage to proteins, lipids, and DNA.
- **Maintenance of metabolic processes**: By detoxifying ROS, antioxidant enzymes ensure that critical metabolic processes, such as photosynthesis and respiration, can continue even under stress. This is particularly important in maintaining plant growth and productivity during periods of abiotic stress.
- **Improvement of stress recovery**: Plants with more robust antioxidant systems are better able to recover from periods of stress, as they sustain less damage during the stress period and can resume normal metabolic activities more quickly once the stress is alleviated.

3. Stress-Responsive Gene Expression

Under abiotic stress conditions, plants activate a suite of **stress-responsive genes** that regulate physiological, biochemical, and molecular responses. These genes encode proteins that help the plant adapt to stress, such as enzymes involved in detoxification, signal transduction, and osmoprotection. The ability to activate these genes is crucial for plant survival and stress tolerance.

Key Categories of Stress-Responsive Genes

- **Transcription factors**: Transcription factors are proteins that regulate the expression of other genes in response to stress. Many stress-responsive transcription factors belong to families such as **DREB** (dehydration-responsive element-binding proteins), **NAC**, and **bZIP**. These transcription factors control the expression of downstream genes involved in stress tolerance, including those that regulate osmotic balance, ROS detoxification, and stress signaling pathways.

» **Heat shock proteins (HSPs)**: HSPs are a family of proteins that help maintain protein stability and function during stress conditions such as high temperatures, drought, and salinity. HSPs act as molecular chaperones, preventing the aggregation of misfolded proteins and facilitating their refolding or degradation. This helps protect the plant's cellular machinery from damage caused by stress-induced protein denaturation.

» **Aquaporins**: Aquaporins are membrane proteins that facilitate the transport of water across cell membranes. During drought or salinity stress, aquaporins play a key role in maintaining water balance within cells by regulating water uptake and movement. By modulating aquaporin activity, plants can improve their ability to retain water and avoid dehydration under stress conditions.

» **Late embryogenesis abundant (LEA) proteins**: LEA proteins are involved in protecting cells from damage caused by dehydration during drought or salt stress. These proteins stabilize membranes and other cellular structures, helping to prevent damage from water loss and ensuring that cells can maintain their integrity under stressful conditions.

Role of Stress-Responsive Genes in Abiotic Stress Tolerance

» **Regulation of stress responses**: The activation of stress-responsive genes is a key part of the plant's overall stress response. These genes regulate a wide range of processes that allow the plant to adapt to and survive abiotic stress. By controlling osmotic balance, ROS detoxification, and protein stability, stress-responsive genes help plants minimize the damage caused by drought, salinity, or extreme temperatures.

» **Signal transduction**: Many stress-responsive genes are involved in signal transduction pathways that detect environmental stress signals and transmit them to the appropriate cellular machinery. These pathways enable plants to rapidly adjust their physiology and biochemistry in response to changing environmental conditions.

» **Development of stress-tolerant varieties**: Understanding the roles of stress-responsive genes has opened up new opportunities for breeding stress-tolerant crop varieties. By identifying key genes involved in stress tolerance and incorporating them into breeding programs, breeders can develop crops that are better equipped to cope with abiotic stresses. For example, the overexpression of transcription factors like DREB has been shown to improve drought tolerance in crops like rice and wheat.

2. Sources of Abiotic Stress Tolerance

Abiotic stress tolerance is a crucial trait in plants, especially given the increasing challenges posed by climate change, such as drought, salinity, temperature extremes, and nutrient deficiencies. To develop crop varieties that can thrive under such harsh conditions, plant breeders must identify and incorporate genetic sources of abiotic stress tolerance into breeding programs. These sources can come from wild relatives of crops, landraces, and mutations. Each of these sources provides a reservoir of valuable traits that can be harnessed to improve the resilience of cultivated crops to environmental stresses.

1. Wild Relatives of Crops

Wild relatives of domesticated crops are some of the most important sources of abiotic stress tolerance. These species, which have evolved in natural environments without human intervention, have often developed mechanisms to cope with extreme environmental conditions such as drought, high salinity, and nutrient-poor soils. Unlike domesticated varieties, which have been selectively bred for traits such as yield or quality, wild relatives retain a greater diversity of stress tolerance traits.

Advantages of Using Wild Relatives

- **Natural Adaptation to Harsh Environments:** Wild relatives have evolved over millennia in challenging environments, making them inherently tolerant to various abiotic stresses. For example, wild relatives of barley (*Hordeum vulgare*) and rice (*Oryza spp.*) have been found to possess natural tolerance to drought and salinity. These wild relatives have survived in regions with limited water resources or saline soils, developing physiological and genetic mechanisms that allow them to thrive where domesticated crops would fail.
- **Diverse Genetic Pool:** One of the primary benefits of wild relatives is their rich genetic diversity. Domesticated crops often have a narrow genetic base due to selective breeding, which limits their ability to adapt to changing environmental conditions. Wild relatives, in contrast, possess a wide range of alleles for traits like drought tolerance, cold hardiness, and salt tolerance. This diversity makes them an invaluable resource for breeding programs aimed at improving stress tolerance.
- **Introgression of Stress Tolerance Genes:** Through a process known as **introgression**, breeders can transfer stress tolerance traits from wild relatives into cultivated varieties. Introgression involves crossing the wild relative with the domesticated crop and then repeatedly backcrossing the offspring with the cultivated parent to retain the stress tolerance trait while minimizing undesirable traits from the wild relative. This technique has been successfully used to transfer drought tolerance from wild relatives of barley and rice into modern high-yielding varieties.

Examples of Wild Relatives Used in Breeding for Abiotic Stress Tolerance

- **Barley (Hordeum vulgare)**: Wild barley species have been identified as sources of drought and salt tolerance. These wild relatives possess traits such as deeper root systems, osmotic adjustment mechanisms, and more efficient water use, which help them survive in arid and saline environments. Breeding programs have utilized wild barley to introduce these traits into commercial barley varieties.
- **Rice (Oryza spp.)**: Wild relatives of rice, such as *Oryza rufipogon* and *Oryza nivara*, have been a key source of tolerance to abiotic stresses, including drought, flooding, and salinity. These wild species possess adaptive traits that allow them to survive in extreme conditions. For example, genes for submergence tolerance (*Sub1*) and drought resistance have been introgressed from wild relatives into cultivated rice, resulting in improved stress tolerance in high-yielding varieties.

» **Wheat (Triticum spp.)**: Wild relatives of wheat, such as *Aegilops* species, have been used to introduce traits like drought tolerance and salinity resistance into modern wheat varieties. These wild species are adapted to semi-arid environments and have evolved mechanisms to cope with water scarcity, making them valuable resources for breeding wheat varieties that can withstand dry and saline conditions.

Challenges of Using Wild Relatives

» **Compatibility Issues:** Crossing wild relatives with domesticated crops can be challenging due to genetic incompatibility between species. In some cases, wild relatives may belong to different species or genera, making it difficult to achieve successful crosses through conventional breeding methods. However, advances in biotechnology, such as wide hybridization, embryo rescue, and molecular markers, have helped overcome these barriers.

» **Linkage Drag:** Wild relatives often carry undesirable traits, such as poor yield, low quality, or undesirable plant architecture. When introgressing stress tolerance genes from wild relatives, breeders must be careful to minimize "linkage drag," where these undesirable traits are co-inherited along with the desired stress tolerance traits. This requires several generations of backcrossing and selection to separate the beneficial genes from the unwanted ones.

» **Time-Consuming Process:** The process of introgressing traits from wild relatives into domesticated crops can be time-consuming, as it involves multiple generations of crossing, backcrossing, and selection. While the benefits of stress tolerance are significant, the time and resources required to achieve successful introgression can be a limiting factor for breeding programs.

2. Landraces and Traditional Varieties

Landraces are locally adapted crop varieties that have been cultivated by farmers over many generations. These varieties have evolved in specific geographic regions and are often well-suited to the local environmental conditions, including abiotic stresses like drought, salinity, and poor soil fertility. Landraces are an important source of abiotic stress tolerance because they have retained a high level of genetic diversity and possess traits that have been naturally selected for survival in challenging environments.

Advantages of Using Landraces

» **Local Adaptation to Environmental Stress:** Landraces are often better adapted to the local climate and soil conditions than modern, high-yielding varieties. They have evolved over time to cope with the specific abiotic stresses present in their environment. For example, certain landraces of maize (*Zea mays*) grown in dry, mountainous regions of Mexico have developed traits such as drought tolerance and efficient water use, making them valuable for breeding programs targeting arid environments.

» **Genetic Diversity:** Landraces often exhibit greater genetic diversity than modern varieties, which have undergone genetic bottlenecks due to selective breeding. This diversity provides a wide range of alleles that can be used to improve stress tolerance in crops. For instance, drought-resistant landraces of millet (*Pennisetum glaucum*) and sorghum (*Sorghum bicolor*) have been used in breeding programs to enhance the resilience of these crops to water scarcity.

- **Resilience in Low-Input Systems:** Many landraces are cultivated in low-input farming systems, where chemical fertilizers, pesticides, and irrigation are rarely used. As a result, these varieties have evolved natural mechanisms to cope with nutrient deficiencies, drought, and other abiotic stresses without relying on external inputs. This makes landraces particularly valuable for breeding stress-tolerant varieties that can thrive in low-input or organic farming systems.

Examples of Landraces Used in Breeding for Abiotic Stress Tolerance

- **Rice (Oryza sativa):** Landraces of rice from regions prone to drought or flooding, such as the traditional varieties grown in the rain-fed lowlands of South Asia, have been used to introduce traits for water stress tolerance. For example, the landrace *Khao Dawk Mali*, grown in Thailand, has been used in breeding programs to develop drought-tolerant rice varieties.
- **Maize (Zea mays):** Landraces of maize from Mexico, the center of origin for the crop, are known for their tolerance to drought and poor soil conditions. These landraces possess traits such as deep root systems and efficient nutrient uptake, which make them resilient to abiotic stresses. Breeding programs have used these landraces to improve the drought tolerance of modern maize varieties.
- **Millet (Pennisetum glaucum):** Millet landraces grown in the arid regions of Africa and India have been an important source of drought tolerance. These landraces have adapted to survive with minimal water, making them valuable for breeding drought-resistant millet varieties that can be cultivated in water-scarce regions.

Challenges of Using Landraces

- **Lower Yield Potential:** One of the main challenges of using landraces in modern breeding programs is their lower yield potential compared to high-yielding commercial varieties. While landraces may possess valuable stress tolerance traits, they often have lower productivity, which can be a limiting factor for their use in large-scale agriculture. Breeders must carefully balance the incorporation of stress tolerance traits with the need to maintain or improve yield.
- **Limited Availability and Conservation Issues:** As modern agriculture has increasingly favored uniform, high-yielding crop varieties, many landraces have been abandoned, leading to the loss of genetic diversity. Efforts to conserve landraces through seed banks and in situ conservation are critical for ensuring that these valuable sources of abiotic stress tolerance remain available for future breeding programs.
- **Complex Genetic Background:** Landraces often have a complex genetic background, with multiple genes contributing to their stress tolerance. This can make it challenging for breeders to isolate and select the specific genes responsible for stress tolerance without inadvertently losing other desirable traits.

3. Induced Mutations

Induced mutations are another valuable source of abiotic stress tolerance. Mutations can be induced in plants through the application of chemicals, radiation, or other mutagenic agents, leading to changes in the plant's DNA. These mutations can result in new traits, including tolerance to abiotic stresses such as drought, salinity, or extreme temperatures. Induced mutations are particularly useful when natural sources of stress tolerance are limited or unavailable in the existing gene pool.

Advantages of Induced Mutations

» **Creation of Novel Traits:** Induced mutations can create new alleles that provide tolerance to abiotic stresses. These traits may not exist in the natural gene pool or may be difficult to identify through conventional breeding. For example, mutations that improve water-use efficiency or enhance salt tolerance can be induced in crops like wheat or rice, providing new sources of stress tolerance.

» **Applicability to Any Crop:** Induced mutations can be applied to a wide range of crops, making it a versatile tool for breeding. Whether in cereals, legumes, fruits, or vegetables, mutagenesis can create valuable traits for abiotic stress tolerance. For example, in barley, induced mutations have led to the discovery of drought-tolerant lines that can be used in breeding programs.

» **Faster Development of Tolerant Varieties:** Induced mutations can accelerate the development of stress-tolerant varieties by directly altering the plant's genome to introduce new traits. Once a desirable mutation is identified, it can be rapidly incorporated into breeding programs, bypassing the need for cross-breeding with wild relatives or landraces.

Challenges of Induced Mutations

» **Unpredictability:** Induced mutations are largely random, and most mutations do not result in beneficial traits. As a result, breeders must screen large populations of mutant plants to identify individuals with the desired traits, which can be time-consuming and labor-intensive.

» **Potential for Negative Traits:** While induced mutations can create valuable traits for abiotic stress tolerance, they can also produce undesirable traits, such as reduced yield or poor plant architecture. Breeders must carefully select plants with the desired stress tolerance traits while eliminating those with negative characteristics.

» **Extensive Screening Required:** After inducing mutations, breeders must conduct extensive screening to identify plants with the desired abiotic stress tolerance traits. This process often requires several generations of selection to stabilize the mutation and ensure that it is consistently expressed in the plant.

3. Breeding Methods for Abiotic Stress Tolerance

Breeding for abiotic stress tolerance often involves more complex traits that are controlled by multiple genes. This makes the breeding process more challenging. Several methods are employed, including:

» **Conventional breeding:** Traditional crossing and selection methods are used to combine stress tolerance with high yield potential. Progeny is evaluated under controlled stress conditions, such as water-limited environments for drought tolerance.

» **QTL mapping and marker-assisted selection**: Quantitative trait loci (QTL) associated with stress tolerance are identified and linked to molecular markers. Marker-assisted selection (MAS) can then be used to select for plants carrying the desired QTL in breeding populations.

» **Genetic engineering**: Advances in biotechnology have made it possible to directly introduce genes responsible for stress tolerance into crops. For example, genes involved in osmotic adjustment, ROS detoxification, or stress signaling pathways can be inserted into crop genomes to enhance their performance under stress conditions.

» **Breeding for root architecture**: Root system architecture plays a critical role in abiotic stress tolerance, especially under drought and nutrient-limited conditions. Breeding efforts focus on selecting plants with deeper or more efficient root systems that can access water and nutrients in suboptimal environments.

3. Breeding Methods for Abiotic Stress Tolerance

Breeding for abiotic stress tolerance presents a unique challenge because abiotic stress traits are typically complex, polygenic, and significantly influenced by environmental factors. Unlike traits controlled by single genes (e.g., disease resistance), abiotic stress tolerance often involves the coordinated action of multiple genes and biochemical pathways, making the breeding process more intricate. However, advances in breeding techniques and molecular biology have provided powerful tools to enhance the efficiency and precision of breeding programs aimed at improving abiotic stress tolerance. Below are some of the key breeding methods used to develop crops with improved tolerance to abiotic stresses like drought, salinity, heat, cold, and nutrient deficiencies.

1. Conventional Breeding

Conventional breeding is the oldest and most widely practiced method for improving abiotic stress tolerance in crops. It involves the traditional crossing of plants with desirable traits and selecting offspring that exhibit both stress tolerance and high yield potential.

Process of Conventional Breeding for Abiotic Stress Tolerance

The basic process of conventional breeding for abiotic stress tolerance includes the following steps:

» **Identifying parental lines**: Breeders start by identifying parent plants with desirable traits. One parent may have strong abiotic stress tolerance, while the other parent may possess high yield potential or other agronomic traits like disease resistance or quality.

» **Crossing**: The parents are crossed to produce offspring, known as the F1 generation. These offspring inherit a mix of traits from both parents, including stress tolerance traits.

» **Evaluation under stress conditions**: The offspring are evaluated under controlled or natural stress conditions. For example, drought-tolerant varieties may be evaluated in water-limited environments, while salt-tolerant varieties are grown in soils with high salinity. Plants that perform well under these conditions are selected for further breeding.

» **Backcrossing**: To ensure that the stress tolerance traits are integrated into high-yielding varieties without losing important agronomic traits, breeders may employ backcrossing. This involves repeatedly crossing the selected offspring back to the high-yielding parent while retaining the stress tolerance traits. Backcrossing ensures that the offspring retain the desirable traits of the high-yielding parent while incorporating the stress tolerance traits from the other parent.

Advantages of Conventional Breeding

1. **Wide Applicability**: Conventional breeding can be applied to almost any crop and stress type, making it a versatile method for improving abiotic stress tolerance. It is especially useful for complex traits like drought or heat tolerance that are controlled by multiple genes.
2. **Combines Multiple Traits**: This method allows breeders to combine stress tolerance with other desirable traits, such as high yield, disease resistance, or improved quality, by careful selection over several generations.

Limitations of Conventional Breeding

1. **Time-Consuming**: Conventional breeding is a slow process that can take several years or even decades to develop a successful stress-tolerant variety. This is because multiple generations are required to combine stress tolerance with other agronomic traits and to ensure stability in the new variety.
2. **Environmental Sensitivity**: The expression of stress tolerance traits is often influenced by environmental factors. This means that a plant may perform well under controlled conditions but fail to exhibit the same level of tolerance in different environments, making the selection process more challenging.
3. **Limited to Available Genetic Variation**: Conventional breeding relies on the natural or existing genetic variation within a species or its close relatives. If there are no available sources of stress tolerance within the crop's gene pool, conventional breeding may not be sufficient to develop stress-tolerant varieties.

2. QTL Mapping and Marker-Assisted Selection (MAS)

Quantitative Trait Loci (QTL) mapping and **marker-assisted selection (MAS)** are modern breeding methods that use molecular markers to identify and select plants with desirable traits, such as abiotic stress tolerance. QTL mapping allows breeders to locate specific regions of the genome that are associated with stress tolerance traits, while MAS enables the selection of plants carrying those QTLs without the need for labor-intensive phenotypic screening.

QTL Mapping for Abiotic Stress Tolerance

Abiotic stress tolerance traits are often controlled by multiple genes, each contributing a small effect. QTL mapping helps breeders identify these genes and understand their contribution to stress tolerance. The process involves:

» **Crossing two genetically diverse parents**: One parent may have strong abiotic stress tolerance, while the other may be susceptible but have desirable agronomic traits.

» **Generating a mapping population**: The offspring of this cross are used to create a population that segregates for the trait of interest (e.g., drought tolerance). The population is then evaluated under stress conditions, and their genetic makeup is analyzed to identify markers linked to the stress tolerance trait.

» **Identifying QTLs**: Statistical analysis is used to associate specific molecular markers with the trait of interest. These regions of the genome, known as QTLs, are believed to contain the genes responsible for stress tolerance.

Marker-Assisted Selection (MAS)

Once QTLs for stress tolerance have been identified, molecular markers linked to these QTLs can be used to select plants that carry the desired traits in subsequent generations. MAS allows breeders to:

» **Select plants early**: MAS can be used to select plants at the seedling stage, long before they express the stress tolerance trait under field conditions. This accelerates the breeding process by allowing breeders to focus only on plants carrying the desired QTLs.

» **Combine multiple traits**: MAS can also be used to pyramid multiple QTLs for stress tolerance into a single plant, increasing its resilience to abiotic stresses. For example, QTLs for both drought and heat tolerance can be combined in a single variety using MAS.

Advantages of QTL Mapping and MAS

1. **Precision in Selection**: MAS provides a more precise method of selecting plants with stress tolerance traits compared to conventional breeding, where selection is based solely on phenotypic evaluation.
2. **Accelerates Breeding**: MAS reduces the time required for selection by allowing breeders to identify and select plants with the desired traits at the seedling stage, rather than waiting for the plant to fully mature and express the traits in the field.
3. **Applicable to Complex Traits**: QTL mapping and MAS are particularly useful for traits like drought or salinity tolerance, which are controlled by multiple genes. MAS allows breeders to select for these traits even when their phenotypic expression is highly variable or influenced by environmental conditions.

Limitations of QTL Mapping and MAS

1. **High Initial Costs**: QTL mapping and MAS require specialized equipment and expertise in molecular biology, making them costly to implement. However, the long-term benefits often outweigh the initial costs.
2. **Dependence on Identified Markers**: MAS is only effective if molecular markers linked to the desired traits have been identified. In some cases, the genetic basis of stress tolerance may be poorly understood, limiting the effectiveness of MAS.

3. Genetic Engineering

Genetic engineering allows breeders to directly introduce genes responsible for abiotic stress tolerance into crop varieties. Unlike conventional breeding, which relies on natural or existing genetic variation, genetic engineering makes it possible to introduce novel genes from other species or improve the function of native genes to enhance stress tolerance.

Process of Genetic Engineering for Abiotic Stress Tolerance

- **Gene identification**: The first step is to identify genes involved in abiotic stress tolerance, such as genes responsible for osmotic adjustment, ROS detoxification, or stress signaling pathways. These genes may come from other plant species, bacteria, or even animals.
- **Gene insertion**: Once the gene of interest is identified, it is inserted into the target crop's genome using methods like **Agrobacterium-mediated transformation** or **particle bombardment**. This process creates a transgenic plant that expresses the new gene and, consequently, exhibits improved stress tolerance.
- **Selection and testing**: Transgenic plants are selected based on their ability to express the introduced gene and are then tested under stress conditions to evaluate their performance.

Examples of Genetic Engineering for Abiotic Stress Tolerance

- **Drought tolerance in maize**: Transgenic maize varieties have been developed by inserting genes involved in osmotic adjustment and water-use efficiency, leading to improved performance under drought conditions.
- **Salt tolerance in rice**: Genes from salt-tolerant plants or bacteria have been inserted into rice varieties, enabling them to grow in saline soils by enhancing their ability to exclude or sequester salt.

Advantages of Genetic Engineering

1. **Direct Introduction of Traits**: Genetic engineering allows breeders to introduce specific genes responsible for stress tolerance directly into crop varieties, bypassing the need for traditional crossing and backcrossing.
2. **Novel Sources of Resistance**: Genetic engineering makes it possible to introduce genes from unrelated species, providing new sources of stress tolerance that may not be available within the crop's natural gene pool.

Limitations of Genetic Engineering

1. **Regulatory and Public Acceptance Issues**: Genetically modified organisms (GMOs) face significant regulatory hurdles and public opposition in many countries. The approval process for GMOs can be time-consuming and expensive, limiting their widespread adoption.
2. **High Costs and Expertise Required**: Developing and commercializing genetically engineered crops is costly and requires advanced scientific expertise and infrastructure.

4. Breeding for Root Architecture

Root system architecture plays a critical role in abiotic stress tolerance, particularly under conditions of drought, salinity, and nutrient deficiency. Plants with deeper or more efficient root systems are better able to access water and nutrients in suboptimal environments. Breeding for improved root architecture is therefore a key strategy for enhancing abiotic stress tolerance.

Traits Related to Root Architecture

- **Deep root systems**: Plants with deep roots are better able to access water stored deeper in the soil profile, making them more resilient to drought conditions.
- **Root density and branching**: A more branched and dense root system increases the surface area for water and nutrient absorption, improving the plant's ability to uptake resources under stress conditions.
- **Root-to-shoot ratio**: A higher root-to-shoot ratio allows plants to allocate more resources to root growth, enhancing their ability to survive in water-limited environments.

Breeding Methods for Root Architecture

- **Phenotypic selection**: Breeders select plants with desirable root traits by evaluating their performance under stress conditions in field or controlled environments. Modern imaging techniques, such as root scanning and X-ray tomography, allow breeders to visualize and measure root systems more accurately.
- **Marker-assisted selection**: Molecular markers linked to traits associated with root architecture can be used to select plants with deeper or more efficient root systems.

Advantages of Breeding for Root Architecture

1. **Improves Resource Efficiency**: Breeding for improved root architecture allows plants to access water and nutrients more efficiently, improving their resilience to abiotic stresses.
2. **Applicable to Multiple Stresses**: Root traits related to drought tolerance are often also beneficial for nutrient uptake, making this approach useful for improving tolerance to multiple abiotic stresses.

Limitations of Breeding for Root Architecture

1. **Difficult to Measure**: Root traits are often difficult to measure, particularly in field conditions, making phenotypic selection challenging. However, advances in imaging technologies are helping to overcome this limitation.

Integrated Breeding for Both Biotic and Abiotic Stress Tolerance

Crops in agricultural systems are often subjected to both biotic and abiotic stresses simultaneously, such as drought, salinity, heat, pests, and diseases. These stresses can interact and exacerbate each other's effects, making it essential to adopt an integrated approach in breeding programs. An integrated breeding strategy that focuses on both biotic and abiotic stress tolerance is crucial for developing resilient crop varieties capable of withstanding the complex challenges of modern agriculture. Advances in molecular biology, genomics, and phenotyping technologies have greatly enhanced the ability of breeders to tackle multiple stresses simultaneously.

1. Combining Resistance/Tolerance to Multiple Stresses

One of the most significant challenges in plant breeding is combining resistance or tolerance to both biotic (pests, pathogens) and abiotic (drought, salinity, heat) stresses. These stress responses are typically controlled by different sets of genes, and the process of breeding crops that can withstand multiple stresses requires careful selection and combination of these genetic traits.

Pyramiding Traits for Multiple Stress Tolerance

Pyramiding refers to the process of combining multiple genes or traits into a single plant, thereby enhancing its ability to cope with multiple stresses. This is particularly important in the case of biotic and abiotic stress tolerance, where different genes control different responses. For example, resistance to fungal pathogens may be governed by one set of genes, while drought tolerance might involve a completely different set of genes. The challenge lies in ensuring that both sets of genes are inherited and expressed effectively in the same variety.

Advances in Molecular Marker Technology

Advances in **molecular marker technology** have made it possible to track multiple genes and traits simultaneously in breeding populations. **Marker-assisted selection (MAS)** allows breeders to identify plants carrying specific genes for biotic and abiotic stress tolerance, ensuring that these traits are combined in the next generation of crops. MAS provides precision in selection, allowing for the integration of complex traits without relying solely on phenotypic evaluations.

» **Gene stacking:** Through MAS, breeders can combine multiple *R* genes for resistance to different pathogens (e.g., resistance to fungal and bacterial diseases) with genes for abiotic stress tolerance (e.g., drought or heat resistance). This technique allows breeders to "stack" genes, ensuring that the resulting varieties are resilient to a wide range of stresses.

Challenges in Combining Multiple Traits

Combining resistance to multiple stresses poses several challenges, including:

» **Gene interactions:** In some cases, genes for biotic and abiotic stress tolerance may interact in ways that affect their overall expression. For example, selecting for drought tolerance may inadvertently reduce the plant's ability to resist certain pests or pathogens. Breeders must carefully evaluate the genetic interactions between the traits they are selecting for to avoid unintended consequences.

» **Complex inheritance patterns:** Traits for abiotic stress tolerance, such as drought or salinity resistance, are often controlled by multiple genes (polygenic traits) and can be influenced by environmental factors. This makes it more challenging to fix these traits in breeding populations compared to simpler, single-gene traits like resistance to certain pathogens.

» **Trade-offs between stress tolerance and yield:** Breeding for multiple stress tolerances must also consider potential trade-offs between stress resistance and yield. For example, plants with high levels of resistance to drought may allocate more energy to root growth, which could reduce the energy available for grain production. Balancing stress tolerance with yield potential is a key consideration in integrated breeding programs.

2. Genomic Selection and Phenotyping Platforms

The integration of **genomic selection (GS)** and advanced **phenotyping platforms** has revolutionized the breeding of crops for both biotic and abiotic stress tolerance. These tools allow breeders to make more informed decisions by predicting the performance of plants based on genetic data and real-time monitoring of their responses to stress.

Genomic Selection (GS)

Genomic selection is a modern breeding method that uses genome-wide marker data to predict the performance of complex traits, such as stress tolerance, in breeding populations. Unlike marker-assisted selection, which focuses on a few specific genes, GS considers the cumulative effects of many genes across the genome.

» **Genome-wide prediction:** In GS, breeders analyze the entire genome of a plant to identify markers associated with stress tolerance traits. This information is then used to predict which plants are most likely to perform well under stress conditions, allowing breeders to select the best candidates for further breeding without waiting for them to mature.

» **Efficient selection of complex traits:** Genomic selection is particularly useful for traits like drought tolerance, heat resistance, or resistance to multiple pathogens, which are controlled by many genes and are difficult to assess through traditional phenotypic methods alone. By predicting a plant's potential for stress tolerance based on its genetic profile, breeders can accelerate the development of resilient varieties.

High-Throughput Phenotyping Platforms

High-throughput phenotyping (HTP) platforms use advanced imaging and remote sensing technologies to collect precise data on plant growth, physiology, and stress responses. These platforms allow breeders to monitor plant performance under stress conditions in real time, providing valuable insights into how different genotypes respond to environmental challenges.

» **Remote sensing technologies:** Drones, satellites, and ground-based sensors equipped with multispectral and hyperspectral cameras can capture detailed information on plant health, growth, and stress responses. These technologies allow breeders to assess large populations of plants efficiently and non-invasively, even under field conditions.

» **Imaging technologies:** Advanced imaging techniques, such as thermal imaging and chlorophyll fluorescence imaging, can be used to detect physiological changes in plants under stress, such as changes in water use efficiency or photosynthetic activity. This data helps breeders identify genotypes that are better able to cope with abiotic stress.

» **Data integration:** By combining genomic selection with phenotypic data from high-throughput platforms, breeders can make more accurate predictions about a plant's potential for stress tolerance. This integration of genomic and phenotypic data accelerates the breeding process and increases the likelihood of developing varieties that are resilient to both biotic and abiotic stresses.

3. Role of Biotechnology in Stress Breeding

Biotechnological tools, such as transgenics and gene editing, are increasingly being used to enhance both biotic and abiotic stress tolerance in crops. These technologies provide a way to directly modify the plant's genome, introducing new traits or improving existing ones to enhance stress resilience.

Transgenics for Stress Tolerance

Transgenic technologies involve the introduction of genes from other species into a plant's genome to confer new traits, such as resistance to pests, pathogens, or environmental stresses. In the context of abiotic and biotic stress tolerance, transgenic approaches have been used to introduce genes that improve the plant's ability to withstand stress.

» **Abiotic stress tolerance:** Genes involved in osmotic adjustment, such as those encoding for proline or glycine betaine synthesis, have been introduced into crops to improve drought and salinity tolerance. These genes help plants maintain water balance and protect cellular structures during stress conditions.

» **Biotic stress resistance:** Transgenic crops expressing genes from *Bacillus thuringiensis* (Bt) have been developed to provide resistance to insect pests, reducing the need for chemical pesticides. Other transgenic approaches have introduced genes for resistance to fungal, bacterial, and viral pathogens.

Gene Editing for Precision Breeding

Gene editing technologies, such as **CRISPR-Cas9**, have revolutionized plant breeding by allowing for precise modifications to be made in the plant's genome. Unlike transgenics, which introduce foreign genes, gene editing allows breeders to modify the plant's existing genes to enhance stress tolerance or resistance.

» **CRISPR-Cas9**: This powerful tool enables breeders to target specific genes involved in stress responses and make precise changes to improve the plant's performance. For example, CRISPR can be used to knock out susceptibility genes, enhancing the plant's resistance to certain pathogens. It can also be used to enhance genes involved in abiotic stress tolerance, such as those regulating water use efficiency or ROS detoxification.

» **No foreign DNA:** One of the key advantages of gene editing over transgenics is that it does not involve the introduction of foreign DNA. This can help overcome regulatory and public acceptance issues associated with genetically modified organisms (GMOs). Gene-edited crops may also be subject to fewer regulatory hurdles, depending on the country and the specific technology used.

Integration of Biotechnology in Breeding Programs

The integration of biotechnological tools, such as gene editing and transgenics, into traditional breeding programs allows breeders to combine the strengths of both approaches. For example:

» **CRISPR and MAS:** Gene editing can be used to create new alleles for stress tolerance, which can then be tracked using marker-assisted selection in breeding populations. This combination of precise genome editing and molecular markers accelerates the development of multi-stress-tolerant crops.

» **Pyramiding transgenic traits:** Transgenic approaches can be used to introduce multiple genes for both biotic and abiotic stress tolerance into a single variety, enhancing its overall resilience. For example, a crop could be engineered to express Bt genes for pest resistance while also being modified for improved drought tolerance.

B. Breeding Self-Pollinated Crops

Breeding self-pollinated crops is a crucial area of agricultural science, as it deals with the enhancement of crop species that primarily undergo self-pollination. This type of breeding focuses on improving traits such as yield, disease resistance, stress tolerance, and quality traits like grain size or nutritional content. Since self-pollinated crops tend to be highly homozygous and genetically uniform, breeding methods for these crops differ from those applied to cross-pollinated species. Examples of self-pollinated crops include rice, wheat, barley, beans, peas, and tomatoes.

In the context of plant breeding, self-pollinated crops present both opportunities and challenges. The ease of maintaining genetic purity is an advantage, but it also limits genetic variability, which makes the introduction of new traits more complex. The breeding of self-pollinated crops primarily relies on techniques such as **pure-line selection**, **mass selection**, **pedigree breeding**, **backcross breeding**, and **mutation breeding**.

Characteristics of Self-Pollinated Crops

Self-pollinated crops possess several defining characteristics that make them distinct from cross-pollinated species. These characteristics influence their genetic makeup, breeding methods, and overall resilience to environmental challenges. One of the most important traits of self-pollinated crops is their tendency to become highly homozygous, meaning they exhibit uniform genetic material across generations. This homozygosity has both advantages and challenges, which must be carefully managed in breeding programs to enhance crop performance.

1. High Homozygosity

High homozygosity is a key feature of self-pollinated crops. In self-pollination, the same plant or flower provides both the pollen and the ovule, resulting in fertilization that occurs without genetic material from a different plant. Over time, as self-pollination continues generation after generation, the genetic variation within the crop diminishes, leading to a state of homozygosity. In this state, nearly all of the plant's genes are represented by identical alleles, which causes the plants to be genetically uniform. The high degree of homozygosity in self-pollinated crops has significant implications for both their agricultural performance and the strategies used to breed them.

Advantages of Homozygosity

1. Uniformity in Traits

One of the major advantages of high homozygosity in self-pollinated crops is the **uniformity** that results. When crops are homozygous, they exhibit consistent traits across multiple generations, meaning that farmers can expect predictable performance in terms of yield, quality, and resistance to environmental or biotic stresses. For instance, in a field of self-pollinated wheat, each plant will likely have the same response to drought, pests, or diseases, and their grain size, shape, and nutritional quality will also be highly uniform. The uniformity provided by homozygosity is particularly valuable for commercial agriculture, where consistency is key to achieving efficient management, harvest, and marketability. Seed producers, in particular, benefit from this genetic consistency, as they can maintain the same variety over time without significant variation in performance or quality.

Simplification of the Breeding Process: Since the genetic information is consistent from one generation to the next, breeders have an easier task when it comes to selecting and fixing desirable traits. For example, once a plant line has been developed with high yield or disease resistance, breeders can be confident that these traits will continue to be expressed in future generations. This stability in traits greatly simplifies the breeding process, as it eliminates the variability seen in cross-pollinated crops.

2. Fixed Desirable Traits

Another advantage of high homozygosity is that once a desirable trait is introduced into a self-pollinated crop, it becomes "fixed" and remains stable over time. **Fixed traits** are those that are consistently passed down to future generations without alteration, ensuring that beneficial characteristics such as disease resistance, drought tolerance, or increased yield are preserved. In breeding programs, this stability is crucial for developing resilient crops. For example, when breeders introduce a gene for drought tolerance into a self-pollinated crop like rice, the trait becomes fixed after a few generations of self-pollination. This means that the offspring will inherit the trait uniformly, allowing farmers to reliably plant drought-tolerant varieties year after year. The **stability of traits** is particularly important for crops grown in environments with consistent stress factors, such as regions prone to drought or disease outbreaks. By ensuring that desirable traits are fixed in the crop's genome, breeders can provide farmers with a dependable tool for maintaining productivity in challenging conditions.

3. Easier Inbreeding

Unlike cross-pollinated crops, which often suffer from **inbreeding depression** when subjected to self-pollination, self-pollinated crops are naturally adapted to inbreeding. Inbreeding depression refers to the reduction in vigor, yield, and overall health that occurs when genetically similar individuals mate, leading to the expression of harmful recessive alleles.

Self-pollinated crops, such as rice, wheat, and beans, have evolved to tolerate inbreeding, allowing them to be self-pollinated repeatedly without experiencing a loss of vigor or a reduction in yield. In fact, self-pollination is a natural mechanism for these crops, and repeated inbreeding tends to stabilize their genetic makeup, leading to high levels of homozygosity and uniform performance. **Self-Pollination as a Breeding Tool**: The natural tolerance for inbreeding in these crops makes them ideal for **pure-line breeding**, where breeders aim to develop highly uniform varieties that exhibit specific desirable traits. In this context, inbreeding is not only tolerated but leveraged to ensure that the selected traits are consistently expressed in future generations.

Challenges of Homozygosity

1. Limited Genetic Variation

While high homozygosity provides advantages like trait uniformity and stability, it also introduces a significant challenge: **limited genetic variation**. Genetic variation is the raw material for evolution and adaptation, and when crops have a limited pool of genetic diversity, they may struggle to adapt to new or changing environmental conditions. In self-pollinated crops, the process of repeated self-pollination reduces the genetic variation within the population. This lack of variation can make it difficult for breeders to introduce new traits, such as improved disease resistance or tolerance to emerging environmental stresses like climate change. The absence of genetic diversity also limits the ability of the crop to respond to novel threats, as there are fewer alleles available to confer resistance to new pests, diseases, or abiotic stresses.

Challenges in Breeding for Adaptability: Because self-pollinated crops lack the genetic diversity seen in cross-pollinated species, breeders must rely on external sources of variation, such as wild relatives or induced mutations, to introduce new traits. This makes breeding for adaptability a more complex and time-consuming process.

Example: For instance, in self-pollinated crops like rice, the genetic uniformity that makes the crop stable in one environment might limit its ability to thrive in a new environment where different stressors are present, such as increased salinity or temperature fluctuations. To overcome this, breeders may need to cross the crop with wild rice species that possess the necessary genetic variation for stress tolerance.

2. Potential for Vulnerability

Another significant challenge of high homozygosity is the **potential for vulnerability** to diseases, pests, or other stress factors. When a crop population is genetically uniform, it is more susceptible to being wiped out by a single pathogen or environmental stressor. This is because if one plant is susceptible, the entire population is likely to be susceptible due to the shared genetic makeup.

Increased Risk from Pathogens: In a diverse population, some plants may carry genes that provide resistance to a particular pathogen, allowing them to survive even if others succumb to the disease. In a homozygous, self-pollinated population, however, if the genetic makeup does not include resistance genes, the entire crop may be at risk. This can lead to devastating crop losses, as seen in the history of agricultural monocultures where uniformity made crops vulnerable to specific diseases.

Pathogen Evolution and Resistance Breakdown: Pathogens and pests evolve over time, and if a pathogen develops a way to overcome the resistance mechanisms of a genetically uniform crop, the entire crop could be at risk of destruction. This is particularly concerning in the context of climate change, where the range and virulence of pests and pathogens are expected to shift.

Example: A classic example of this vulnerability occurred with the Irish potato famine in the 1840s, where the reliance on a single potato variety, which was genetically uniform, allowed a pathogen (late blight, *Phytophthora infestans*) to wipe out the entire crop, leading to widespread famine. Similarly, modern monocultures of self-pollinated crops could face similar risks if genetic diversity is not maintained or introduced through breeding programs.

Strategies to Mitigate Vulnerability: To address this challenge, breeders often work to introduce genetic diversity into self-pollinated crops by crossing them with wild relatives or by employing techniques like **mutation breeding** to create new sources of genetic variation. This helps to ensure that the crops are more resilient to evolving pathogens and environmental changes. Additionally, breeders may pyramid multiple resistance genes into a single variety to provide broader and more durable protection against biotic and abiotic stresses.

2. True Breeding

True breeding is a fundamental concept in the breeding of self-pollinated crops. These crops are known as **true-breeding crops**, meaning that their offspring will consistently exhibit the same traits as the parent generation. This consistency is a direct result of the high homozygosity achieved through repeated self-pollination. In true-breeding crops, the genetic makeup becomes fixed, with little to no genetic variation present between generations. This trait makes self-pollinated crops highly valuable in agricultural systems, especially for seed production and commercial farming, where uniformity and reliability are key. The stability that true-breeding provides is both a strength and a limitation. While true breeding ensures that desirable traits are passed down through generations, it also presents challenges when breeders need to introduce new traits or adapt to changing environmental conditions. Below is a detailed discussion of the benefits and limitations of true breeding in self-pollinated crops.

Benefits of True Breeding

1. Consistency Across Generations

One of the greatest benefits of true breeding in self-pollinated crops is the consistency it provides across generations. In true-breeding crops, desirable traits such as high yield, disease resistance, or specific grain quality traits are consistently expressed in every generation. This is because the plants are highly homozygous, meaning that they carry identical alleles for most of their genes. As a result, the offspring of true-breeding crops will inherit the same genetic makeup as the parent generation, ensuring that the traits selected by the breeder or farmer remain stable over time.

Importance of Consistency in Agriculture:

- » **Commercial Agriculture**: In large-scale commercial agriculture, consistency in crop performance is crucial for efficient management, harvesting, and marketing. Farmers need to be confident that the seeds they plant will produce crops with predictable traits, such as uniform plant height, grain size, or flowering time. This predictability allows for better planning and reduces the risk of unexpected crop failures or variations in yield.
- » **Seed Production**: True-breeding crops are particularly valuable in seed production. Seed companies rely on the fact that true-breeding plants will consistently produce offspring with the same traits as the parent plants. This allows them to produce and sell genetically uniform seed varieties, ensuring that farmers receive seeds with reliable performance year after year.
- » **Uniformity in Traits**: The uniformity provided by true breeding is also important for traits like disease resistance or stress tolerance. In a true-breeding population, if the parent plants are resistant to a particular pathogen, the offspring will also exhibit the same level of resistance, reducing the likelihood of disease outbreaks in the field.

Example: An example of this can be seen in self-pollinated crops like wheat or rice, where true breeding ensures that the entire crop matures at the same time, making it easier to harvest mechanically and ensuring uniform grain quality. This consistency is vital for maintaining high levels of productivity in large-scale agricultural operations.

2. Stable Varieties

The true-breeding nature of self-pollinated crops allows breeders to develop **stable varieties** that can be reliably used by farmers over many seasons. Stability in plant breeding refers to the ability of a variety to consistently express the same traits under varying environmental conditions. True-breeding crops achieve this stability through their high levels of homozygosity, which ensures that genetic variation is minimized, and the desired traits remain fixed.

Benefits of Stable Varieties:

» **Long-Term Use**: Stable varieties are particularly valuable for farmers who operate in regions with consistent environmental conditions. For example, a farmer in a region that experiences regular droughts may rely on a true-breeding, drought-tolerant variety to provide consistent yields year after year. Because the variety is stable, the farmer does not need to replace the seed frequently or worry about variations in the plant's performance.

» **Reduced Need for Re-Selection**: In cross-pollinated crops, breeders often need to re-select for desirable traits in each generation, as genetic variation can lead to the loss of important traits. In contrast, true-breeding crops do not require frequent re-selection, as the traits are already fixed in the population. This reduces the time and cost associated with maintaining the variety.

» **Predictability for Farmers**: The stability provided by true breeding means that farmers can plan their planting, management, and harvesting strategies with greater confidence. They can predict how the crop will respond to environmental conditions, how it will perform in terms of yield, and how it will resist pests or diseases. This predictability is crucial for minimizing risk and optimizing productivity in farming operations.

Example: In the case of crops like beans or peas, which are often grown in smallholder farming systems, the stability provided by true-breeding varieties allows farmers to save seeds from one season to the next, knowing that the plants grown from these seeds will perform similarly to the parent plants. This ability to save and reuse seeds is an important factor in promoting food security and reducing the cost of inputs for small-scale farmers.

Limitations of True Breeding

1. Difficulty in Introducing New Traits

While true breeding provides consistency and stability, it also introduces challenges when it comes to the **introduction of new traits**. Once a crop is true-breeding for a set of traits, such as disease resistance or high yield, it becomes more difficult to introduce additional traits into the population. This limitation arises because self-pollinated crops have a fixed genetic makeup, with little genetic variation present to facilitate the incorporation of new traits.

Challenges of Introducing New Traits:

» **Limited Genetic Variation**: In true-breeding crops, the genetic diversity within the population is reduced due to the repeated self-pollination process. This limited genetic variation makes it more difficult for breeders to combine multiple desirable traits, such as drought tolerance, disease resistance, and high yield. Unlike cross-pollinated crops, where genetic recombination introduces new variations, self-pollinated crops lack the same opportunities for new trait combinations to emerge naturally.

» **Slower Breeding Progress**: When breeders aim to introduce new traits into a true-breeding crop, such as resistance to a new pathogen or improved stress tolerance, they must cross the crop with a genetically diverse donor plant. This process can be time-consuming, as it often requires several generations of crossing and backcrossing to introduce the new trait while maintaining the crop's original characteristics. The need to reintroduce and stabilize traits slows down the overall breeding progress.

» **Combining Multiple Traits**: In many modern breeding programs, the goal is to develop crop varieties that combine multiple beneficial traits, such as resistance to multiple diseases, tolerance to environmental stresses, and high nutritional content. In true-breeding crops, achieving this goal can be more challenging due to the genetic uniformity of the population. Breeders must carefully manage the crossing and selection process to ensure that all desired traits are successfully combined without losing any of the original characteristics of the crop.

Example: In a true-breeding wheat variety that has been selected for high yield and disease resistance, introducing a new trait for heat tolerance may require crossing the wheat with a wild relative that possesses the desired trait. This process could take several breeding cycles, and breeders would need to carefully select offspring that retain the high yield and disease resistance of the original variety while also incorporating the new heat tolerance trait.

2. Potential for Reduced Adaptability

Another limitation of true breeding is the potential for **reduced adaptability** to changing environmental conditions or evolving pest and disease pressures. Because true-breeding crops are genetically uniform, they may lack the diversity needed to adapt to new challenges. In cross-pollinated crops, genetic variation within the population provides a buffer against environmental changes, as some individuals may carry alleles that confer resistance to new stresses. In true-breeding crops, this variation is absent, making the crop more vulnerable to shifts in environmental conditions or the emergence of new pests or diseases.

Challenges of Reduced Adaptability:

» **Environmental Changes**: As global climates continue to change, agricultural systems are facing new challenges, such as increased temperatures, changing rainfall patterns, and the spread of salinity. True-breeding crops, which are genetically uniform and adapted to specific conditions, may struggle to adapt to these changes. Without the genetic diversity needed to evolve in response to new stressors, these crops may experience reduced yields or increased susceptibility to environmental extremes.

» **Pest and Disease Evolution**: Pathogens and pests are constantly evolving, and they can develop new virulent strains that overcome the resistance mechanisms present in crops. In true-breeding crops, once a pest or pathogen finds a way to bypass the plant's defenses, the entire population is at risk because all the plants share the same genetic makeup. In contrast, crops with greater genetic diversity are more likely to contain individuals with natural resistance to the evolving threat, providing a safeguard against widespread crop failure.

Example: A true-breeding variety of tomato that has been selected for resistance to a specific fungal disease may lose its effectiveness if the pathogen evolves a new strain that can infect the plant. In this scenario, the entire tomato crop could be at risk, as the genetic uniformity of the true-breeding variety limits its ability to adapt to the evolving pathogen.

Strategies to Overcome Limitations: To overcome the limitations of true breeding, breeders often use techniques such as **backcrossing** or **mutation breeding** to introduce new genetic variation into self-pollinated crops. Backcrossing involves crossing the true-breeding crop with a genetically diverse donor plant that carries the desired trait, followed by repeated backcrossing to retain the original characteristics of the crop while introducing the new trait. Mutation breeding involves inducing genetic mutations to create new sources of variation, which can then be selected for desirable traits.

3. Genetic Purity

Genetic purity is a critical characteristic of self-pollinated crops, referring to the uniformity and consistency within a plant population. In self-pollinated crops, due to the inherent process of self-pollination, a high degree of homozygosity is achieved, meaning that the plants in the population are either genetically identical or very similar. This high genetic purity results from repeated self-pollination over generations, which stabilizes the genetic makeup of the crop, fixing desirable traits. In the context of agriculture, genetic purity is particularly important for seed production and commercial farming. Farmers and seed producers rely on genetically pure seeds to ensure that the plants grown from those seeds exhibit the exact traits selected for, such as high yield, disease resistance, or specific quality characteristics. While genetic purity offers many advantages, it also presents some challenges, particularly regarding the crop's vulnerability to biotic and abiotic stresses.

Advantages of Genetic Purity

1. Uniform Crop Performance

One of the major advantages of genetic purity in self-pollinated crops is the **uniformity in crop performance** that results from the genetic consistency within the population. In a genetically pure population, all the plants exhibit the same traits, leading to uniformity in growth patterns, yield, and responses to environmental factors like pests, diseases, and abiotic stresses. This uniformity is highly desirable for commercial farming, where predictability and consistency are key to efficient crop management and production.

Importance of Uniformity in Crop Performance:

» **Predictability in Management**: When crops are genetically pure, farmers can predict their growth and development more easily. For example, if all plants in a wheat field are genetically identical, they will likely flower, mature, and ripen at the same time, making it easier to plan for activities like fertilization, irrigation, and harvesting. Uniform maturity ensures that the entire crop is ready for harvest simultaneously, which is especially important for mechanized farming, where efficient harvesting requires crops to reach the same growth stage at once.

» **Consistent Yield**: In genetically pure crops, yield performance is more consistent across the field, reducing variability in the amount of produce harvested. This consistency is crucial for maximizing productivity and ensuring that farmers can meet market demands reliably. For example, in rice production, a genetically pure variety will produce grains of similar size, weight, and quality, allowing for easier processing and marketing.

- **Improved Resistance**: Crops with genetic purity often have uniform resistance to specific pests and diseases if they are bred for these traits. This uniformity in resistance ensures that the entire crop is protected from a particular pathogen or pest, providing a safeguard against widespread infection or damage.

Example of Uniform Crop Performance:

In commercial maize production, genetic purity allows farmers to plant hybrid maize varieties that exhibit uniform growth, ensuring that all plants reach maturity at the same time. This is particularly important in large-scale maize farming, where uniformity in plant height, ear size, and kernel quality leads to easier mechanical harvesting and processing. The result is increased efficiency in crop management and higher overall productivity.

2. Easier Selection

Another significant advantage of genetic purity in self-pollinated crops is the **ease of selection** for breeders. In self-pollinated crops, breeders aim to fix desirable traits such as disease resistance, drought tolerance, or increased yield. Once these traits are introduced into a genetically pure population, they remain stable across generations, making it easier to select and propagate plants with those traits.

Advantages of Easier Selection:

- **Stable Inheritance of Traits**: The high degree of homozygosity in genetically pure crops ensures that desirable traits are consistently passed down to subsequent generations. This stability simplifies the breeding process, as breeders can be confident that the selected traits will remain fixed in the population. This is especially important in seed production, where genetic purity ensures that the seeds produced will reliably exhibit the desired traits.
- **Efficient Breeding Programs**: In genetically pure populations, breeders can focus on fine-tuning specific traits without worrying about variability within the population. This makes breeding programs more efficient, as less time is spent on reselecting for traits that may be lost due to genetic variability. In contrast, in cross-pollinated crops, breeders must deal with greater genetic diversity, which can complicate the selection process.
- **Maintenance of Genetic Integrity**: For seed producers, genetic purity ensures the **genetic integrity** of the variety. When producing seeds for commercial sale, seed companies must ensure that the seeds meet certain standards of purity and that the plants grown from those seeds will perform as expected. Genetic purity simplifies this process, as the high homozygosity of the crop guarantees that the offspring will be genetically uniform and exhibit the same traits as the parent generation.

Example of Easier Selection:

In self-pollinated crops like tomatoes, breeders can easily select for specific traits such as fruit size, color, or resistance to certain fungal diseases. Once these traits are fixed in the population, they remain stable across generations, making it easier to produce seeds that consistently express the desired characteristics. This simplifies the breeding process and allows seed companies to maintain the genetic purity of their tomato varieties.

Challenges of Genetic Purity

1. Vulnerability to Biotic and Abiotic Stresses

While genetic purity is advantageous for maintaining desirable traits, it can also lead to increased **vulnerability to biotic and abiotic stresses**. When a plant population is genetically uniform, it lacks the genetic diversity needed to adapt to changing environmental conditions or evolving pathogens. This lack of diversity makes the crop more susceptible to widespread damage if a new pest, disease, or environmental stressor emerges.

Biotic Stress Vulnerabilities:

» **Pathogen and Pest Evolution**: Pathogens and pests are constantly evolving, and they can develop new strains that overcome the resistance mechanisms of genetically uniform crops. In a genetically pure population, if one plant is susceptible to a new pathogen or pest, the entire crop is likely to be at risk, as all plants share the same genetic makeup. This lack of diversity can lead to significant crop losses, as there are no individuals within the population that possess natural resistance to the new threat.

» **Example**: One of the most famous examples of this vulnerability is the Irish Potato Famine in the 1840s, where a genetically uniform potato crop was devastated by the late blight fungus (*Phytophthora infestans*). The lack of genetic diversity in the potato population meant that the entire crop was susceptible to the disease, leading to widespread famine.

Abiotic Stress Vulnerabilities:

» **Environmental Changes**: Crops with high genetic purity are also more vulnerable to abiotic stresses such as drought, salinity, and temperature extremes. If the crop is not bred for tolerance to these stresses, the uniform genetic makeup limits its ability to adapt to sudden changes in environmental conditions. In contrast, crops with greater genetic diversity are more likely to contain individuals with natural tolerance to abiotic stresses, providing a buffer against environmental fluctuations.

» **Climate Change Impact**: As climate change leads to more frequent and severe environmental stress events, crops with high genetic purity may struggle to adapt. For example, a genetically pure wheat variety that is highly susceptible to heat stress may experience significant yield reductions in regions where temperatures are rising due to climate change. The lack of genetic diversity means that the population cannot evolve in response to the changing climate, increasing the risk of crop failure.

2. Reduced Genetic Flexibility

Genetic purity, while beneficial for maintaining desirable traits, can also reduce the **genetic flexibility** of a crop population. In self-pollinated crops, the process of repeated self-pollination fixes the genetic makeup of the population, reducing the potential for new genetic combinations to arise. This can be a disadvantage when breeders need to introduce new traits or adapt the crop to changing environmental conditions.

Challenges of Reduced Genetic Flexibility

- **Introduction of New Traits**: In self-pollinated crops with high genetic purity, introducing new traits, such as resistance to a new pathogen or tolerance to a new environmental stress, can be more difficult. Breeders must cross the crop with a genetically diverse donor plant that possesses the desired trait. This process can be time-consuming, as it often requires several generations of crossing and backcrossing to introduce the new trait while maintaining the original characteristics of the crop.
- **Limited Ability to Adapt**: The lack of genetic flexibility also limits the crop's ability to adapt to new challenges. In a genetically diverse population, there may be individuals that carry alleles for resistance to a new disease or tolerance to a new environmental condition. In a genetically pure population, these alleles are absent, making the crop more vulnerable to changing conditions.

Example of Reduced Genetic Flexibility:

In self-pollinated crops like barley, high genetic purity can limit the crop's ability to adapt to new environmental stresses such as increased salinity in soils. If the population is genetically uniform and lacks alleles for salinity tolerance, the entire crop may be at risk of reduced yields or even crop failure in areas where soil salinity is increasing. To overcome this challenge, breeders may need to introduce new genetic variation through crosses with wild relatives or through mutation breeding.

Breeding Methods for Self-Pollinated Crops

Breeding self-pollinated crops focuses on fixing desirable traits within genetically uniform populations. Self-pollinated crops undergo repeated self-fertilization, which naturally leads to high levels of homozygosity over time, stabilizing the genetic material. The primary goal in breeding these crops is to enhance specific traits like yield, disease resistance, abiotic stress tolerance, and quality characteristics, such as nutritional content. To achieve these goals, several breeding methods have been developed and refined over the years, each suited to the genetic characteristics of self-pollinated crops.

The most commonly used methods in breeding self-pollinated crops include **pure-line selection, mass selection, pedigree breeding, backcross breeding**, and **mutation breeding**. Below is a more detailed expansion of one of these key methods used in the breeding of self-pollinated crops: **pure-line selection**.

1. Pure-Line Selection

Pure-line selection is one of the most common and effective methods used in breeding self-pollinated crops. This technique involves selecting a single superior plant from a genetically variable population and propagating it through successive generations of self-pollination until a homozygous line is developed. The end result is a genetically uniform line (pure line) that consistently expresses desirable traits. This method is particularly effective for stabilizing traits like disease resistance, high yield, or specific quality characteristics in self-pollinated crops such as rice, wheat, barley, and beans. Pure-line selection offers a streamlined approach to achieving genetic uniformity while ensuring that the selected plants carry traits that improve agricultural productivity and resilience. Below are the steps involved in pure-line selection, along with its advantages and limitations.

Steps in Pure-Line Selection:

1. Selection of Superior Plants

The first step in pure-line selection is to identify and select superior individual plants from a genetically diverse population. This population could be a landrace, a traditional variety, or even a population generated through mutation breeding or hybridization. The goal is to choose plants based on observable traits, such as high yield, resistance to diseases, tolerance to drought, or desirable grain quality.

Process of Selection:

» **Phenotypic Evaluation**: Breeders evaluate the plants for specific traits through phenotypic analysis. This evaluation might include examining traits like plant height, flowering time, seed size, grain quality, disease resistance, and abiotic stress tolerance. Advanced breeding programs also use molecular markers to aid in identifying plants with the desired genetic traits.

» **Initial Selection**: From this evaluation, the best-performing individual plants are selected for further self-pollination. For example, if the breeding objective is to improve drought tolerance, breeders would select plants that show the best performance under drought stress conditions. Similarly, if disease resistance is the target, plants that remain unaffected by common pathogens would be prioritized for selection.

» **Environmental Conditions**: The selection process is often carried out across multiple environments to ensure that the selected traits are stable and can perform well under different conditions. For example, a breeder might test wheat plants for drought tolerance in both controlled environments and in regions prone to drought. This ensures that the selected plants have a broad adaptability to various growing conditions.

2. Self-Pollination and Evaluation

Once the superior plants are selected, they undergo **self-pollination**, where the pollen from the same plant fertilizes its own ovules. Self-pollination ensures that the genetic material from the selected plant is passed down to the next generation with minimal genetic variation. This process is repeated over several generations until the progeny become genetically uniform, or homozygous.

Process of Evaluation:

» **Successive Generations**: Over multiple generations, breeders closely monitor the progeny for the desired traits. The goal is to identify plants that continue to express the selected traits, such as disease resistance or high yield, consistently across generations. In each generation, breeders select the best-performing individuals and discard plants that do not exhibit the desired characteristics.

» **Trait Fixation**: As self-pollination continues, the plants become increasingly homozygous, meaning that the traits selected for become genetically "fixed" in the population. For instance, if the breeding objective was to develop a wheat variety resistant to rust disease, the progeny in each generation would be evaluated for their resistance to rust. By the fourth or fifth generation, the rust-resistance trait would be fixed in the population, ensuring that all plants express this trait consistently.

» **Field Trials**: As part of the evaluation process, breeders conduct field trials to observe how the selected lines perform under real-world agricultural conditions. These trials provide important data on how the plants respond to environmental factors like soil type, climate, and irrigation practices. The most promising lines are advanced for further selection.

3. Selection of the Best Pure Lines

After several generations of self-pollination and evaluation, breeders select the most superior lines that demonstrate uniformity in the desired traits. These **pure lines** are genetically homogeneous and consistently express the traits that were selected for, making them ideal for large-scale seed production and commercial farming.

Final Selection:

» **Uniformity**: One of the key goals of pure-line selection is to achieve **genetic uniformity**. The pure lines selected at this stage should show uniformity not only in yield and resistance traits but also in other agronomic characteristics like plant height, maturity, and grain quality. Uniformity is crucial for mechanical harvesting, consistent processing, and predictable market performance.

» **Propagation for Seed Production**: Once the best pure lines are identified, they are propagated for seed production. These pure lines are multiplied and distributed to farmers or seed companies. Since the plants in these lines are genetically identical, they provide uniform performance in the field, making them highly reliable for commercial agriculture.

» **Commercial Release**: Before being released for commercial use, pure lines often undergo additional testing by national or international agricultural research organizations. These tests ensure that the lines meet the required standards for traits like disease resistance, stress tolerance, and yield potential. Once a line is approved, it can be marketed as a new variety for widespread agricultural use.

Advantages of Pure-Line Selection:

1. Genetic Uniformity:

One of the most significant advantages of pure-line selection is that it results in **genetically uniform populations**. This uniformity ensures that all plants in the population express the same traits consistently, which is particularly valuable for commercial farming, where predictability and consistency are essential. Crops with genetic uniformity are easier to manage, harvest, and market because they mature at the same time and exhibit similar growth patterns.

» **Predictable Performance**: Genetic uniformity ensures that crops exhibit **predictable performance** across growing seasons. Farmers can rely on these crops to provide consistent yields and maintain stable resistance to diseases or pests. For example, a pure-line variety of rice that has been selected for resistance to blast disease will exhibit uniform disease resistance in every growing season, making it easier for farmers to manage disease outbreaks.

» **Efficient Management**: In genetically uniform crops, inputs like fertilizers, pesticides, and irrigation can be applied more efficiently because the plants respond similarly to these interventions. This improves overall farm management, reduces costs, and increases productivity.

2. **Stable Traits**:

Pure-line selection allows breeders to **fix desirable traits** within the population, ensuring that these traits are passed down to future generations without variability. Once a trait is fixed in a pure line, it remains stable, reducing the need for ongoing re-selection or modification.

» **Reliability in Seed Production**: For seed producers, stability is a critical advantage. Pure-line varieties provide consistent seed quality and performance, ensuring that farmers who purchase seeds can expect reliable results in the field. This stability also allows seed companies to maintain the genetic purity of their varieties over many production cycles.

» **Long-Term Performance**: Pure-line varieties offer **long-term performance stability**. Once traits like high yield, disease resistance, or drought tolerance are fixed in a population, farmers can use the variety for many seasons without worrying about losing these traits.

Limitations of Pure-Line Selection:

1. **Limited Genetic Diversity:**

One of the drawbacks of pure-line selection is that it leads to a reduction in **genetic diversity**. Since pure-line selection results in genetically uniform populations, there is little genetic variation within the crop. This lack of diversity can make the crop more vulnerable to new pathogens, pests, or environmental changes, as the entire population may lack the genetic resources needed to adapt.

» **Risk of Disease Outbreaks**: If a new disease emerges that the pure-line population is not resistant to, the entire crop could be wiped out due to its uniform susceptibility. For example, if a pure-line wheat variety is susceptible to a newly evolved strain of rust disease, the entire population could suffer significant losses because all the plants share the same genetic vulnerability.

2. **Time-Consuming:**

The process of pure-line selection can be **time-consuming**, often requiring multiple growing seasons to stabilize traits and ensure genetic uniformity. This prolonged breeding cycle can slow down the introduction of new varieties, especially when breeders need to address urgent challenges like the emergence of new diseases or changing environmental conditions.

» **Generational Time**: Depending on the crop, it can take several years to complete the pure-line selection process. For crops like rice or wheat, which have long growing seasons, the time required to develop pure lines can be particularly extended, making it difficult to respond quickly to new challenges.

2. Mass Selection

Mass selection is a traditional breeding method widely used for improving self-pollinated crops. Unlike pure-line selection, where a single plant is selected and propagated, mass selection involves choosing a large number of individuals that exhibit desirable traits and allowing them to self-pollinate. The seeds from these selected plants are then **bulked together** to form the next generation. Over successive generations, breeders aim to improve the overall population by continuously selecting individuals that perform well in terms of traits like yield, disease resistance, and stress tolerance.

Mass selection has been employed for centuries, especially for improving traditional landraces and variable populations. The simplicity and low-cost nature of this method make it accessible for breeders working with limited resources or in environments where advanced breeding technologies are unavailable. Below is a detailed exploration of the mass selection process, its advantages, and limitations.

Process of Mass Selection

The process of mass selection involves several key steps, including the initial selection of plants, self-pollination and seed collection, and the planting of the next generation. This cycle is repeated over several generations to enhance the traits of the overall population.

1. Selection of Plants

The first step in mass selection is the **selection of plants** based on phenotypic traits. Breeders visually inspect a large population of plants and choose individuals that exhibit desirable characteristics such as high yield, disease resistance, or tolerance to abiotic stresses like drought or salinity. The selection is based on observable traits rather than on genetic markers or molecular techniques, making this method highly dependent on the breeder's experience and environmental conditions.

Phenotypic Evaluation:

» **Visual Assessment**: Breeders carefully examine the crop population, assessing plants for traits like plant height, grain size, flowering time, and resistance to pests or diseases. For example, in a rice field, breeders may select plants that show minimal signs of blast disease or plants that produce larger grains.

» **Environmental Factors**: The selection process is often influenced by the specific environmental conditions in which the plants are grown. For instance, breeders selecting for drought tolerance would assess plants under water-limited conditions to ensure that the selected individuals are those that perform best in such an environment. Similarly, plants grown in regions prone to pests or diseases would be selected based on their resistance to those particular biotic stresses.

» **Diverse Populations**: One of the key aspects of mass selection is that it is typically used in genetically diverse populations, such as traditional landraces or open-pollinated varieties. These populations often contain a wide range of genetic variability, allowing breeders to select individuals that display the most favorable traits for the target environment.

2. Self-Pollination and Seed Collection

Once the superior plants are selected, they are allowed to **self-pollinate**. In self-pollinated crops, this ensures that the genetic traits of the selected plants are passed down to the next generation. After pollination, breeders collect the seeds from these plants, which are then **bulked together** to form the seed stock for the next generation.

Self-Pollination:

» **Preservation of Desirable Traits**: By allowing the selected plants to self-pollinate, breeders ensure that the desirable traits, such as high yield or disease resistance, are preserved in the seeds. This is particularly important in self-pollinated crops, where the goal is to fix these traits in the population over time.

» **Seed Collection**: After self-pollination, the seeds are harvested from the selected plants. Instead of keeping the seeds from each plant separate, breeders **bulk the seeds together**, meaning that the seeds from all the selected plants are combined into a single seed lot. This bulked seed represents the genetic makeup of the entire selected population, and it will be used to plant the next generation.

3. Planting the Next Generation

In the next step, the **bulked seeds** are planted to grow the next generation of plants. This new generation is a direct result of the self-pollination and selection process from the previous generation. The selection process is then repeated, with breeders once again identifying and selecting individuals that exhibit the best performance for the target traits.

Repetition of Selection:

» **Continuous Improvement**: Over successive generations, breeders continue to select the best-performing plants and bulk their seeds to form the next generation. Each cycle of selection and bulking helps to improve the overall population, gradually increasing the frequency of desirable traits. For instance, if the goal is to improve disease resistance, breeders will continue to select plants that show the highest resistance to diseases, gradually building a population that is more resilient to pathogens.

» **Field Trials and Environmental Adaptation**: As part of the selection process, breeders may conduct field trials in different environments to ensure that the selected population performs well under various conditions. For example, if breeders are selecting for drought tolerance, they may test the plants in both controlled environments and real-world drought-prone areas to ensure that the selected plants exhibit stable performance.

Advantages of Mass Selection

1. Simplicity

One of the main advantages of mass selection is its **simplicity**. This method does not require advanced technology, molecular markers, or extensive field trials, making it accessible to breeders with limited resources or those working in regions without access to modern breeding technologies.

» **Cost-Effective**: Mass selection is a cost-effective method because it relies on visual assessment of traits rather than expensive equipment or laboratory techniques. This makes it particularly suitable for small-scale breeding programs or for farmers who want to improve their own crop varieties through selection.

» **Ease of Implementation**: The straightforward nature of mass selection allows breeders to implement it in a wide range of settings, from traditional farming systems to research stations. The lack of complex equipment or specialized knowledge makes this method easy to use, even in resource-constrained environments.

- **Traditional Use**: Mass selection has been used for centuries by farmers and breeders alike, particularly in traditional agricultural systems. For example, farmers in developing countries often use mass selection to improve local landraces, selecting plants that perform best under the specific environmental conditions of their region.

2. Genetic Improvement in Diverse Populations

Another significant advantage of mass selection is that it allows for **genetic improvement in diverse populations**. Unlike pure-line selection, which focuses on a single individual, mass selection works with a large number of plants, ensuring that genetic diversity is maintained within the population. This is particularly valuable when improving traditional landraces or open-pollinated varieties.

- **Genetic Diversity**: Because mass selection involves selecting multiple individuals, it helps preserve a wider range of genetic diversity within the population. This diversity can be beneficial for adapting to changing environmental conditions or evolving pest and disease pressures. For example, a genetically diverse population is more likely to contain individuals with natural resistance to new pathogens, providing a safeguard against disease outbreaks.
- **Improvement of Landraces**: Mass selection is commonly used to improve traditional landraces, which are often genetically diverse populations that have evolved under specific environmental conditions. By selecting individuals with superior traits, breeders can improve the overall performance of the landrace while maintaining its genetic diversity. For instance, farmers in drought-prone areas might use mass selection to enhance the drought tolerance of their local landrace while preserving its adaptability to the region's other environmental conditions.
- **Sustainable Breeding**: The genetic diversity preserved through mass selection can contribute to the long-term sustainability of crop populations. By maintaining a broad genetic base, mass-selected populations are more resilient to environmental stresses, ensuring that they remain productive even in the face of climate change or new pests and diseases.

Limitations of Mass Selection

1. Less Precision

While mass selection offers simplicity and cost-effectiveness, it is generally **less precise** than other breeding methods, such as pure-line selection. Because mass selection relies on visual assessment of traits, there is a higher likelihood of selecting plants based on observable characteristics rather than their underlying genetic potential. This can result in slower genetic progress compared to more precise methods.

- **Phenotypic Selection**: In mass selection, breeders select plants based on their observable traits (phenotypes), without directly assessing the plants' genotypes. This means that some selected plants may not carry the desired genetic traits for long-term improvement, leading to slower progress in achieving the breeding goals.
- **Environmental Influence**: The visual assessment of traits in mass selection can be influenced by environmental factors, which may not always accurately reflect the plant's genetic potential.

For example, a plant that performs well in a particular environment might be selected based on its superior phenotype, but it may not necessarily carry the genetic traits needed to perform well in other environments.

- **Limited Use of Genetic Markers**: Unlike methods such as marker-assisted selection (MAS), mass selection does not involve the use of molecular markers to track the inheritance of specific traits. This can limit the breeder's ability to make precise selections, particularly for complex traits controlled by multiple genes.

2. Limited Genetic Gains

Another limitation of mass selection is that it tends to result in **limited genetic gains** compared to more focused breeding methods. Because mass selection involves selecting and bulking seeds from multiple plants, the genetic progress achieved in each generation may be smaller than methods that concentrate on selecting single superior individuals.

- **Bulked Seeds:** In mass selection, the seeds from multiple selected plants are bulked together, meaning that the genetic gains from each individual plant are diluted in the next generation. This can slow down the overall breeding process, as the population evolves more gradually compared to methods like pure-line selection or pedigree breeding.
- **Longer Time Frame**: Achieving significant genetic improvement through mass selection may take more generations compared to more precise breeding methods. For example, while pure-line selection might stabilize desirable traits within a few generations, mass selection may require more generations to achieve the same level of genetic improvement.

3. Pedigree Breeding

Pedigree breeding is a sophisticated method employed by plant breeders to combine desirable traits from two selected parents and fix these traits in the offspring through successive generations of self-pollination. This method is particularly effective for self-pollinated crops, where the goal is to develop pure lines that carry the best traits of both parent plants, such as higher yield, disease resistance, or stress tolerance. Pedigree breeding is widely used in crop improvement programs for crops like wheat, rice, and beans, where the desired genetic gains can be achieved through careful selection and evaluation over multiple generations.

Pedigree breeding is considered more precise than mass selection, as it allows breeders to systematically track the inheritance of specific traits. By maintaining detailed records of each cross and the subsequent generations, breeders can ensure that desirable traits are successfully transferred and stabilized in the progeny. However, the process is time-consuming and requires extensive field trials to ensure that the traits remain stable across different environments.

Steps in Pedigree Breeding

1. Crossing Selected Parents

The first step in pedigree breeding is **crossing two selected parent plants** that exhibit complementary traits. These parents are carefully chosen based on their genetic characteristics, with the aim of combining the best traits of each parent in the progeny. For example, one parent might have a high yield potential, while the other might be resistant to a specific disease. By crossing these two plants, breeders hope to create offspring that inherit both traits, leading to improved overall performance.

Selection of Parents:

- **Complementary Traits:** The parents selected for crossing should possess traits that complement each other. For instance, if one parent is known for its resistance to a particular pest but has low yield potential, it may be crossed with a high-yielding plant that is susceptible to pests. The goal is to combine these traits in the offspring to produce plants that have both high yield and pest resistance.
- **Diverse Genetic Backgrounds**: Breeders often choose parents with diverse genetic backgrounds to increase the likelihood of combining multiple desirable traits in the progeny. This diversity can also help prevent issues such as inbreeding depression, where the genetic uniformity of self-pollinated crops can lead to reduced vigor and performance.
- **Controlled Crosses:** The crossing process is typically performed in controlled environments to ensure that the pollen from one parent fertilizes the ovules of the other parent without contamination from external sources. This may involve manually transferring pollen between flowers or using controlled pollination techniques in greenhouses or field settings.

2. Selfing and Selection

After the initial cross, the **F1 progeny** (the first generation of offspring) are self-pollinated to produce the **F2 generation**. The F2 population is genetically diverse because it results from the combination of alleles from both parents. This generation provides a wide range of genetic variability, which breeders can exploit to select individuals that exhibit the best combination of traits.

Selfing:

- **Genetic Segregation:** In the F2 generation, the genes from both parents segregate, meaning that some individuals will carry a combination of desirable traits from both parents, while others may only carry traits from one parent or exhibit unfavorable traits. This genetic variability provides breeders with the opportunity to select individuals that express the ideal combination of traits.
- **Evaluation of Traits:** The F2 population is carefully evaluated for the desired traits. This may involve assessing traits such as yield, plant height, disease resistance, or drought tolerance. For example, breeders may look for plants that exhibit high yield potential while also showing strong resistance to common diseases. Plants that meet the selection criteria are chosen for further self-pollination and evaluation.
- **Selection Process:** The selection process is repeated in each subsequent generation (F3, F4, etc.), with breeders choosing the best-performing plants in each generation. The goal is to **fix the desirable traits** in a homozygous state, meaning that the selected traits become genetically stable and are consistently passed on to future generations. By the later generations (e.g., F6 or F7), the plants should be highly homozygous, with little genetic variation remaining for the selected traits.

3. Generation of Pure Lines

After several generations of self-pollination and selection, the goal of pedigree breeding is to develop **pure lines** that combine the desirable traits of both parents. These pure lines are genetically uniform and consistently express the selected traits, making them ideal for large-scale cultivation or seed production.

Development of Pure Lines:

- **Homozygosity**: Over successive generations of self-pollination, the plants become increasingly homozygous, meaning that they carry identical alleles for the traits being selected. This homozygosity ensures that the traits are fixed and will not segregate in future generations. For example, if a breeder is selecting for drought tolerance, the final pure lines will consistently exhibit this trait, regardless of environmental conditions.
- **Field Trials**: Before releasing the pure lines for commercial use, breeders often conduct field trials to ensure that the lines perform well in different environments. These trials help confirm that the selected traits are stable and that the pure lines are adaptable to a variety of growing conditions.
- **Seed Multiplication**: Once the pure lines are developed and tested, they are propagated for seed production. These pure lines are valuable because they provide uniformity in the field, ensuring that farmers can expect consistent performance in terms of yield, disease resistance, and other key traits.

Advantages of Pedigree Breeding

1. Combining Multiple Traits

One of the key advantages of pedigree breeding is its ability to **combine multiple desirable traits** from different parents. By selecting parents with complementary traits, breeders can create offspring that exhibit the best characteristics of both plants. This makes pedigree breeding particularly useful for improving complex traits, such as yield, disease resistance, and stress tolerance.

- **Combining Yield and Resistance**: For example, a breeder might cross a high-yielding rice variety with a variety that is resistant to bacterial leaf blight. Through pedigree breeding, the breeder can select offspring that combine both high yield and disease resistance, resulting in a more robust variety that performs well in the field.
- **Complex Traits**: Pedigree breeding is also well-suited for improving complex traits that are controlled by multiple genes. For example, traits like drought tolerance or nutrient efficiency often involve several genes working together. By carefully selecting and crossing parents, breeders can create offspring that inherit the full range of genetic factors necessary for expressing these complex traits.

2. Documentation of Genetic Progress

Another major advantage of pedigree breeding is the ability to **document genetic progress** throughout the breeding process. Breeders maintain detailed records of the pedigree of each plant, including information on which parents were crossed, the traits selected in each generation, and the performance of the progeny. This documentation allows breeders to track the inheritance of traits and monitor the progress of the breeding program over time.

» **Tracking Inheritance**: By recording the pedigree of each plant, breeders can track how traits are inherited across generations. For example, if a disease resistance gene is introduced into the breeding program, breeders can document how this gene is passed on to subsequent generations and ensure that it remains present in the final pure lines.

» **Informed Decision-Making**: The detailed documentation also helps breeders make informed decisions about which plants to select in each generation. For example, if a particular trait is not being expressed as expected, breeders can refer to the pedigree records to identify potential issues and adjust the selection process accordingly.

Limitations of Pedigree Breeding

1. Time-Consuming

Like pure-line selection, pedigree breeding can be a **time-consuming** process. The development of pure lines through pedigree breeding requires multiple generations of self-pollination and selection, each of which can take several growing seasons to complete. This slow process makes pedigree breeding less suitable for situations where rapid genetic gains are needed.

» **Generational Time**: For crops like rice or wheat, which have relatively long growing seasons, it can take several years to develop pure lines through pedigree breeding. Each generation must be carefully evaluated and selected, meaning that the breeding program may need to run for five to seven years before the desired pure lines are developed.

» **Slow Progress**: The slow nature of pedigree breeding means that genetic progress can be incremental. While pedigree breeding allows for the combination of multiple traits, achieving these gains may take several generations, especially for complex traits that involve multiple genes.

2. Requires Extensive Field Trials

Pedigree breeding requires **extensive field trials** to ensure that the selected traits are stable and consistent across different environments. These trials are essential for confirming that the traits remain expressed in a wide range of conditions, including different soil types, climates, and levels of biotic and abiotic stress.

» **Environmental Variation**: One of the challenges of pedigree breeding is ensuring that the traits selected in one environment remain stable in other environments. For example, a variety selected for drought tolerance in a dry region may not perform as well in regions with higher rainfall. Extensive field trials help breeders identify lines that are adaptable to multiple environments, ensuring that the pure lines will perform well under diverse agricultural conditions.

» **Resource-Intensive**: Conducting field trials across different environments requires significant resources, including land, labor, and time. This makes pedigree breeding a resource-intensive process, particularly for large-scale breeding programs that need to evaluate plants in multiple locations.

4. Backcross Breeding

Backcross breeding is a targeted breeding method used to introduce specific traits, such as disease resistance, pest tolerance, or abiotic stress tolerance, into an already well-established elite variety. The key advantage of this method is that it allows breeders to improve a specific trait in a high-performing variety without compromising its other desirable characteristics, such as high yield or grain quality. Backcross breeding is especially useful for traits that are governed by one or a few major genes, such as resistance to a particular pathogen or pest. The essence of backcross breeding is that breeders repeatedly cross the hybrid offspring (F1) with the elite variety (recurrent parent) while selecting for the desired trait from the donor parent. Over several generations, the genetic makeup of the hybrid approaches that of the recurrent parent, while the desired trait from the donor parent is retained. The result is an elite variety that retains its original superior traits but now carries the additional trait for which the breeding process was initiated.

Process of Backcross Breeding

The process of backcross breeding involves three primary steps: the initial cross between the donor parent and the recurrent parent, repeated backcrossing of the progeny to the recurrent parent, and the final selection of plants that exhibit the desired trait along with the genetic background of the recurrent parent.

1. Initial Cross

The first step in backcross breeding is the **initial cross** between the donor parent and the recurrent parent. The donor parent carries the specific trait that breeders want to introduce into the elite variety, while the recurrent parent is the high-performing variety that lacks the desired trait.

Selection of Parents:

- » **Donor Parent**: The donor parent is chosen for its possession of a single trait that is not present in the recurrent parent. For example, the donor parent may carry a gene for disease resistance, drought tolerance, or improved nutrient use efficiency. The donor parent may come from a wild relative, a landrace, or another crop variety that has been identified for its specific advantageous trait.
- » **Recurrent Parent**: The recurrent parent is usually an elite variety that is already well-established for its agronomic performance. This parent is chosen because it exhibits superior characteristics such as high yield, quality, adaptability, and stability. The goal of backcross breeding is to improve this elite variety by incorporating the desirable trait from the donor parent without disturbing its genetic background.
- » **Initial Cross**: The initial cross between the donor and recurrent parents produces the **F1 hybrid**, which carries a mix of genetic material from both parents. At this stage, the F1 hybrid may show partial expression of the desired trait, but it also carries many other genes from the donor parent that are not needed in the final product. The next step is to backcross the F1 progeny to the recurrent parent to reduce the influence of the donor parent while retaining the desired trait.

2. Backcrossing

After the initial cross, the **F1 progeny** are backcrossed to the recurrent parent to progressively recover the genetic makeup of the elite variety while selecting for the desired trait. The backcrossing process is repeated over several generations (usually five or more) to ensure that the progeny increasingly resemble the recurrent parent.

Process of Backcrossing:

» **First Backcross (BC1)**: In the first backcross generation (BC1), the F1 progeny are crossed with the recurrent parent. The BC1 progeny will carry about 75% of the recurrent parent's genes and 25% of the donor parent's genes. Breeders select individuals that show expression of the desired trait from the donor parent, such as disease resistance or drought tolerance. These selected plants are then backcrossed again to the recurrent parent.

» **Subsequent Backcrosses (BC2, BC3, etc.)**: The backcrossing process continues with the second backcross (BC2), where the BC1 progeny are crossed again with the recurrent parent. With each subsequent backcross, the proportion of the recurrent parent's genes increases, while the genetic influence of the donor parent decreases. By the fifth backcross generation (BC5), the progeny will carry more than 96% of the recurrent parent's genetic background, with only the desired trait from the donor parent retained.

» **Selection for Desired Trait**: In each backcross generation, breeders must carefully select plants that express the desired trait from the donor parent while eliminating individuals that do not show the trait. For example, if the goal is to introduce resistance to a particular disease, breeders will inoculate the plants with the disease pathogen and select individuals that show strong resistance. These resistant plants are then used for the next backcross generation.

3. Re-establishing the Elite Variety

After several backcross generations, the resulting plants will closely resemble the recurrent parent in terms of their genetic makeup and agronomic traits. However, they will now possess the additional trait introduced from the donor parent, such as disease resistance or stress tolerance.

Final Selection and Testing:

» **Restoring Elite Performance**: By the fifth or sixth backcross generation, the progeny will have regained almost all the genetic characteristics of the recurrent parent, ensuring that the variety still performs at the high level expected of an elite variety. Breeders will conduct extensive field trials to confirm that the newly developed plants exhibit the same high yield, quality, and other desirable traits as the original recurrent parent, with the added benefit of the newly introduced trait.

» **Elimination of Unwanted Traits**: During the backcrossing process, some undesirable traits from the donor parent may be inadvertently transferred along with the desired trait. This phenomenon is known as **linkage drag**, where genes that are closely linked to the desired gene are also inherited. Breeders must carefully evaluate the progeny for any negative traits and conduct additional backcrossing or selection if necessary to eliminate these unwanted characteristics.

- » **Commercial Release**: Once the new variety is developed, it undergoes further testing and evaluation to ensure that it is suitable for large-scale cultivation. The final variety will be released as an improved version of the original elite variety, with the added trait that enhances its performance in specific environments or against certain stresses.

Advantages of Backcross Breeding

1. Preservation of Elite Varieties

One of the major advantages of backcross breeding is that it allows breeders to **preserve the genetic background of elite varieties** while introducing new traits. This is particularly valuable in situations where farmers have become reliant on a well-performing variety, but there is a need to introduce a new trait, such as resistance to a newly emerging pest or disease.

- » **Maintaining Yield and Quality**: By using backcross breeding, breeders can improve specific traits without affecting the core characteristics of the variety, such as its high yield, grain quality, or adaptability to local conditions. This ensures that farmers can continue to grow a familiar variety with confidence, while also benefiting from improved resistance to biotic or abiotic stresses.
- » **Minimal Genetic Disruption**: Because the backcrossing process focuses on introducing a single trait while maintaining the rest of the genetic makeup of the recurrent parent, there is minimal disruption to the variety's overall performance. For example, if an elite maize variety is backcrossed with a donor carrying drought tolerance, the final variety will retain all of the elite maize's high-yielding and disease-resistant traits, while now also being more resilient to drought.

2. Rapid Improvement of Specific Traits

Backcross breeding is particularly useful for the **rapid improvement of specific traits** in established varieties. By focusing on the transfer of one or a few genes, breeders can quickly enhance the variety's performance in specific areas, such as pest resistance or stress tolerance, without having to undergo lengthy recombination processes.

- » **Targeted Trait Improvement**: The method is ideal for traits that are controlled by single genes, such as resistance to a specific disease or herbicide tolerance. By introducing these genes into an elite variety, breeders can rapidly improve the crop's performance against these specific stresses, while leaving other agronomic traits unchanged.
- » **Faster Results**: Compared to methods like pedigree breeding, which can take many generations to combine multiple traits, backcross breeding is faster when only a single trait is targeted. This makes it particularly useful in urgent situations, such as when a new pest or disease threatens crop production, and rapid genetic improvement is necessary.

Limitations of Backcross Breeding

1. Limited to Single Traits

One of the primary limitations of backcross breeding is that it is typically used to introduce **only a**

single trait into a variety. This method is not as effective for improving multiple traits simultaneously, as the focus is on transferring one specific gene from the donor parent to the recurrent parent.

» **Single-Gene Focus**: Backcross breeding is most effective for traits that are controlled by a single gene, such as resistance to a specific pathogen or herbicide. However, for more complex traits, such as yield potential, drought tolerance, or nutrient efficiency, which are governed by multiple genes, backcross breeding is less effective. These traits often require more comprehensive breeding approaches, such as pedigree breeding or genomic selection.

» **Additional Breeding Required**: If breeders want to introduce multiple traits into a variety, they may need to perform several rounds of backcross breeding, each focusing on a different trait. This can be time-consuming and may require additional breeding strategies to combine the desired traits into a single variety.

2. Linkage Drag

Another limitation of backcross breeding is the risk of **linkage drag**, where undesirable traits from the donor parent are inherited along with the desired trait. This occurs when genes that are closely linked to the target gene are also transferred during the backcrossing process.

» **Undesirable Characteristics**: For example, if a breeder is introducing disease resistance from a wild relative, the progeny may also inherit unwanted traits from the wild parent, such as poor grain quality or reduced yield potential. These undesirable traits can negatively affect the performance of the final variety, requiring additional backcrosses or selection to eliminate them.

» **Need for Further Breeding**: To overcome linkage drag, breeders may need to conduct additional generations of backcrossing or use other breeding techniques, such as marker-assisted selection, to separate the desired trait from the unwanted linked traits.

5. Mutation Breeding

Mutation breeding is a pivotal tool in modern plant breeding programs that introduces genetic variability through induced mutations. Unlike traditional breeding methods, which rely on naturally occurring genetic diversity, mutation breeding involves deliberately altering the plant's genetic code to produce new or enhanced traits that can be useful in improving crop varieties. These mutations, when screened and selected, can lead to the development of plants with improved yield, enhanced resistance to diseases and pests, better nutritional qualities, or increased tolerance to abiotic stresses such as drought or salinity. Mutation breeding is particularly valuable when there is limited genetic diversity for a specific trait within a crop's gene pool. It provides breeders with a method to introduce variability in traits like disease resistance, herbicide tolerance, or drought resistance, even when natural sources of these traits are unavailable. This approach has been instrumental in developing thousands of improved crop varieties globally, contributing significantly to food security and agricultural productivity.

Process of Mutation Breeding

Mutation breeding involves a structured and systematic approach that includes several key steps: inducing mutations, screening for beneficial traits, and stabilizing the desired mutations through self-pollination and evaluation. Each step requires careful management to ensure that beneficial traits are identified and stabilized without introducing undesirable characteristics.

1. Inducing Mutations:

The cornerstone of mutation breeding lies in the **induction of mutations**. Mutations can be induced using chemical mutagens, radiation, or modern gene-editing technologies, each of which has its own strengths and application areas. The goal of this step is to create random changes in the plant's genetic material, which can result in novel traits.

» **Chemical Mutagens**: Chemicals like **ethyl methanesulfonate (EMS)** and **sodium azide** are commonly used to induce point mutations, which involve changes in single nucleotide bases of the DNA sequence. EMS, for example, is effective in inducing subtle but important changes that can lead to improved agronomic traits, such as early maturity, enhanced seed quality, or increased resistance to environmental stresses. Chemical mutagens are particularly useful for crops where precise changes in specific genes can lead to significant improvements.

» **Radiation Mutagenesis**: **Gamma rays**, **X-rays**, and **fast neutrons** are powerful radiation sources used to induce large-scale mutations in plants. Radiation can cause breaks in the DNA, leading to deletions, insertions, or rearrangements of genetic material. This method has been successfully used to develop many improved crop varieties, including mutant rice varieties with enhanced resistance to blast disease and drought-tolerant barley varieties. **Gamma radiation** is one of the most widely used methods in mutation breeding, offering a broad spectrum of mutations that can affect various traits.

» **Advanced Gene-Editing Techniques**: While traditional mutagenesis methods rely on random mutations, modern tools like **CRISPR-Cas9** have revolutionized mutation breeding by allowing for targeted genetic alterations. CRISPR can induce precise mutations in specific genes responsible for traits such as disease resistance or drought tolerance, significantly speeding up the breeding process. Although gene editing offers a more controlled way to induce mutations, it is often used alongside traditional mutagenesis to expand the range of genetic variability.

» **Target Material**: Mutations can be induced in various plant tissues, including seeds, pollen, embryos, or vegetative tissues like leaves and stems. **Seeds** are the most commonly used target material because they can be treated in bulk, allowing for the efficient creation of large populations of mutants. In certain cases, vegetative tissues are treated, and the mutated cells are regenerated into whole plants using tissue culture techniques, especially for crops that are difficult to propagate through seeds.

2. Selection of Mutants: Screening for Desirable Traits

Once mutations have been induced, the next crucial step is to select the mutants that display desirable traits. Given that most mutations are either neutral or detrimental, breeders must carefully screen large populations of plants to identify the few individuals that exhibit beneficial changes.

» **Visual Screening**: One of the primary methods of screening is through **visual observation** of plant traits. Breeders look for improvements in traits such as growth habit, flowering time, seed size, or resistance to pests and diseases. For example, in **mutation breeding of rice**, breeders may identify plants that show improved resistance to rice blast or bacterial blight, while in **barley**, they may screen for plants with enhanced drought tolerance or better grain quality.

- **Phenotypic Evaluation**: Beyond visual observation, phenotypic evaluation under specific stress conditions is critical for selecting mutants with desirable traits. Plants may be grown under conditions of water deficit to identify those that perform better in drought, or they may be exposed to pathogens to select for disease resistance. These tests help ensure that the selected traits are not merely environmental responses but are due to stable genetic changes induced by mutation.
- **Molecular Screening**: Advanced breeding programs increasingly use **molecular markers** and **genomic tools** to identify desirable mutations. Molecular markers can be used to detect specific changes in the plant's DNA that are associated with the traits of interest. For instance, breeders may use molecular markers linked to **disease-resistance genes** to confirm that a mutation has introduced the desired genetic variation. This approach significantly enhances the efficiency of the selection process by allowing breeders to identify promising mutants at an early stage, even before the trait is visually observable.

3. Self-Pollination and Evaluation: Stabilizing Mutations

After selecting the most promising mutants, the next step is to **self-pollinate** them to stabilize the desired traits. This step is crucial for fixing the mutations in the population and ensuring that they are heritable.

- **Self-Pollination**: By self-pollinating the selected mutants, breeders ensure that the mutation is passed on to the next generation. The progeny are grown and observed for the expression of the desired traits. Over several generations of self-pollination, the mutation becomes fixed, meaning that the trait will be consistently expressed in all subsequent generations. For example, if a mutant rice plant shows improved grain size due to a mutation, the goal is to ensure that this trait remains stable in the progeny.
- **Field Trials**: Before releasing a mutant variety for commercial cultivation, breeders conduct **field trials** to assess the performance of the new variety under different environmental conditions. These trials are critical for ensuring that the mutant plants perform well not only in controlled conditions but also in real-world agricultural settings. The trials help breeders evaluate the crop's yield potential, adaptability, and resistance to biotic and abiotic stresses. For example, a mutant wheat variety developed for **drought tolerance** would be tested in drought-prone regions to ensure its effectiveness in improving yield under water-limited conditions.

Advantages of Mutation Breeding

1. Creation of New Genetic Variability

Mutation breeding is particularly advantageous for its ability to **create new genetic variations** that can be harnessed for crop improvement. Unlike conventional breeding, which is limited by the natural genetic diversity available within a crop species, mutation breeding generates novel traits that may not have existed before.

- **Introduction of Unique Traits**: Mutation breeding has led to the development of new traits such as **herbicide resistance**, improved grain quality, and enhanced nutritional content. For instance, mutation breeding has been used to create **high-lysine corn** varieties with improved protein quality, addressing malnutrition in regions where corn is a staple food.
- **Overcoming Genetic Bottlenecks**: In some crops, there may be little natural variation for important traits, such as disease resistance or tolerance to environmental stresses. Mutation breeding provides a way to overcome this limitation by introducing new genetic diversity that can be used to enhance these traits. For example, mutation breeding has been used to develop **salt-tolerant rice varieties**, which are crucial for cultivation in saline-affected areas.

2. Useful for Traits with No Natural Variation

In cases where **natural genetic variation** for a trait is absent or insufficient, mutation breeding offers a solution. This is especially useful for introducing traits like **herbicide tolerance**, **resistance to new pests**, or **improved seed oil content**, where conventional breeding methods may be limited.

- **Expanding Trait Availability**: For example, in crops like **peanut** and **soybean**, where natural sources of resistance to certain pathogens are rare, mutation breeding has been successfully used to introduce resistance to fungal diseases like rust and early leaf spot. These traits have made significant contributions to reducing crop losses in regions affected by these diseases.

Limitations of Mutation Breeding

1. Random Nature of Mutations

While mutation breeding is effective in creating genetic variation, it also suffers from the **random nature** of induced mutations. Most mutations are neutral or harmful, meaning that breeders must screen a large number of plants to find the few individuals that carry beneficial traits.

- **Labor-Intensive Process**: Because many mutations are either deleterious or have no visible effect, breeders often need to handle thousands or even tens of thousands of plants to identify the few mutants with desirable traits. This makes mutation breeding a **labor-intensive and time-consuming** process.

2. Unpredictable Results

The **unpredictability** of mutation breeding is another challenge. While the goal is to improve specific traits, the random nature of mutations can sometimes lead to undesirable changes that affect other important traits, such as yield or quality.

- **Unintended Consequences**: For example, a mutation that improves **disease resistance** may also reduce grain size or affect plant height, requiring additional rounds of breeding to correct these issues. This phenomenon, known as **pleiotropy**, occurs when a single mutation affects multiple traits, leading to a trade-off between beneficial and undesirable characteristics.

Breeding Cross-Pollinated Crops

Breeding cross-pollinated crops presents unique challenges and opportunities compared to self-pollinated crops. In cross-pollinated species, genetic exchange occurs between different plants, resulting in higher genetic diversity within a population. This genetic variability allows for the potential of greater adaptability to environmental changes and stresses, but it also requires breeders to employ specialized techniques to stabilize desirable traits. Cross-pollinated crops include important agricultural species such as maize, sunflower, alfalfa, and many forage grasses. Unlike self-pollinated crops, where a high level of homozygosity is achieved over generations, cross-pollinated crops naturally maintain a high degree of heterozygosity. This genetic diversity is both a strength and a challenge, as it can provide resilience to changing environments but also makes it more difficult to fix desirable traits in a stable population.

Characteristics of Cross-Pollinated Crops

Cross-pollinated crops exhibit several key characteristics that distinguish them from self-pollinated species. These traits influence the breeding strategies employed and determine the approach breeders take in developing improved varieties.

1. High Genetic Diversity

One of the defining features of cross-pollinated crops is their **high genetic diversity**. Since these crops depend on cross-pollination, where pollen from one plant fertilizes another, the resulting offspring contain a mixture of genetic material from both parents. This leads to increased heterozygosity and a wide range of genetic combinations within a population.

» **Adaptability to Environmental Changes**: The genetic diversity in cross-pollinated crops provides an advantage in terms of adaptability. Because these crops are not genetically uniform, they are better able to survive under variable environmental conditions, such as drought, pest infestations, or soil nutrient deficiencies. The genetic diversity acts as a buffer against environmental stresses, allowing the population to contain individuals that may be better suited to new or changing conditions.

» **Heterozygosity**: Heterozygosity is a key factor in cross-pollinated crops, where the presence of different alleles at a gene locus provides genetic variability. This heterozygosity contributes to hybrid vigor or **heterosis**, where offspring from genetically diverse parents exhibit superior traits compared to either parent, such as increased yield or improved stress tolerance.

2. Hybrid Vigor (Heterosis)

Hybrid vigor, or **heterosis**, is a powerful biological phenomenon observed in many cross-pollinated crops, where the offspring of two genetically diverse individuals exhibit superior traits compared to either parent. This enhancement of traits in hybrids is a key tool in modern agriculture, enabling breeders to create varieties with improved growth rates, yield potential, disease resistance, and overall resilience to environmental stresses. Hybrid vigor is especially valuable in crops like maize, sorghum, and sunflower, where the development of hybrid varieties has significantly boosted agricultural productivity across the world.

Mechanism of Heterosis

The exact biological mechanisms behind heterosis are still being studied, but several hypotheses have been proposed. One widely accepted explanation is the **dominance hypothesis**, which suggests that when two genetically diverse parents are crossed, the offspring inherit dominant alleles from each parent that mask the effects of deleterious recessive alleles, leading to superior performance. Another explanation is the **overdominance hypothesis**, which posits that the hybrid's performance is greater than either parent due to the interaction between different alleles at the same gene loci. Regardless of the underlying mechanism, the result of heterosis is clear: hybrids outperform both parental lines in a wide range of traits, leading to better crop performance in the field.

Superior Traits Through Hybrid Vigor

1. Increased Yield

One of the most significant advantages of hybrid vigor is the **increase in crop yield**. In crops like maize, hybrid varieties consistently outperform open-pollinated varieties in terms of grain yield. Hybrid maize varieties have been a major contributor to the **Green Revolution**, especially in countries like the United States, China, and India, where maize is a staple food crop. By combining the best traits of two diverse parent lines, hybrids can achieve a level of productivity that surpasses either parent.

- **Yield Improvements in Maize**: Hybrid maize varieties typically produce up to 25-50% higher yields than open-pollinated varieties. This increase is due to a combination of factors, including enhanced plant vigor, better utilization of water and nutrients, and greater resistance to lodging (the bending or breaking of plants). Hybrid maize varieties are now widely grown in regions with intensive agricultural production, including the United States Corn Belt, Europe, and parts of Africa and Asia.
- **Other Crops**: Hybrid vigor is not limited to maize; it has also been successfully exploited in other crops such as **rice**, **sorghum**, **sunflowers**, and **tomatoes**. In hybrid rice, for example, breeders have developed varieties that are more productive and disease-resistant than their parental lines. These varieties have been particularly beneficial in improving food security in Asia, where rice is a primary staple.

2. Enhanced Disease and Pest Resistance

Another key benefit of heterosis is the ability to **combine disease and pest resistance traits** from different parent lines. By crossing a disease-resistant parent with a high-yielding parent, breeders can produce hybrids that are not only more productive but also better able to withstand common crop diseases and pest attacks.

- **Resistance in Maize Hybrids**: For instance, maize hybrids often show enhanced resistance to diseases such as **maize streak virus**, **leaf blight**, and **rusts**, as well as increased tolerance to pests like **corn borers**. This is particularly important in tropical and subtropical regions where pest and disease pressures are high. Hybrid maize varieties have helped to mitigate crop losses in sub-Saharan Africa, where smallholder farmers face frequent challenges from diseases and pests.
- **Durability of Resistance**: Hybrid vigor also contributes to the **durability** of resistance traits, as hybrids typically contain a broader genetic base compared to pure lines. This diversity allows hybrids to adapt better to evolving pathogen populations, reducing the likelihood of resistance breakdown.

3. Stress Tolerance

Hybrids also demonstrate improved tolerance to **abiotic stresses**, such as drought, high temperatures, and poor soil conditions. This is especially critical in regions where climate variability poses a significant challenge to agriculture. The ability to withstand these stresses while maintaining high productivity makes hybrids an essential component of climate-resilient farming systems.

- **Drought Tolerance in Maize**: In the case of drought-tolerant maize hybrids, breeders have combined traits from different parent lines that enable the plants to use water more efficiently and maintain growth even under water-limited conditions. This trait is vital in regions like sub-Saharan Africa and parts of Asia, where erratic rainfall patterns frequently lead to drought stress during critical growing periods.
- **Tolerance to Poor Soils**: Additionally, hybrids have been developed to tolerate **nutrient-poor soils** or soils with high salinity. For example, hybrid rice varieties that can grow in saline soils have been introduced in coastal regions, allowing farmers to reclaim otherwise unproductive land for agriculture.

Combining Traits for Optimal Performance

The ability to **combine traits** from different parents is one of the most important advantages of hybrid breeding. Breeders often select parents based on their complementary strengths—one parent may have strong disease resistance, while the other may be highly productive in terms of yield. When these traits are combined through hybridization, the resulting offspring exhibit both attributes, providing significant advantages over either parent.

- **Maize Breeding Example**: In hybrid maize breeding, breeders often cross a parent with strong resistance to **northern leaf blight** with another parent that has high grain production potential. The hybrid offspring not only exhibit resistance to northern leaf blight but also maintain or exceed the high yield potential of the second parent. This **synergy** between the two parents results in hybrids that perform better in both yield and disease resistance.
- **Sunflower Hybrids**: In sunflower breeding, hybrids have been developed that combine resistance to **downy mildew** with **tolerance to drought** conditions. The resulting sunflower hybrids not only exhibit better resistance to soil-borne fungal pathogens but also produce high-quality seeds under water-limited conditions.

Challenges in Maintaining Hybrid Vigor

Despite the advantages of hybrid vigor, maintaining and reproducing it requires careful breeding strategies. Once a hybrid variety is created, its vigor is typically **lost in the next generation** (F2) if the hybrid seeds are saved and planted again. This is because the genetic recombination that occurs in the second generation leads to a breakdown of the heterozygosity that provides the hybrid vigor.

1. Controlled Pollination and Hybrid Seed Production

To maintain hybrid vigor, **controlled pollination** and **hybrid seed production** are necessary. This involves producing hybrid seeds each season by crossing the two original parent lines, rather than allowing the hybrid plants to self-pollinate. Hybrid seed production is typically carried out in controlled environments, such as seed farms, where pollination is carefully managed to ensure the desired cross between the parent lines.

» **Male Sterility**: One technique used to ensure controlled pollination is the use of **male-sterile lines**, which cannot produce viable pollen. By planting male-sterile plants alongside pollen-producing plants (the male parent), breeders can ensure that all seeds are hybrids, as the male-sterile plants can only be fertilized by pollen from the male parent.

» **High Seed Production Costs**: However, the production of hybrid seeds is more expensive than open-pollinated varieties due to the need for controlled pollination and the use of specialized techniques like male sterility. This often results in higher seed costs for farmers, although the increased yield and performance of hybrids typically compensate for the higher initial investment.

2. Maintenance of Parental Lines

Maintaining the **genetic purity** of the parental lines used to produce hybrids is also a critical challenge. Over time, these parent lines can accumulate mutations or lose their vigor, which can reduce the quality of the hybrid offspring. To mitigate this, breeders must regularly **refresh and rejuvenate** the parental lines through backcrossing and selection, ensuring that the traits contributing to hybrid vigor are preserved.

» **Inbreeding Depression**: Additionally, because the parental lines are often inbred to achieve homozygosity, they may suffer from **inbreeding depression**, which can reduce their overall fitness. Despite this, the inbred lines are essential for producing hybrids that exhibit heterosis when crossed.

3. The Role of Pollinators in Cross-Pollinated Crops

Cross-pollinated crops rely on pollinators—whether they are wind, insects, or animals—to facilitate the transfer of pollen from one plant to another. This reliance on external pollinators introduces a dynamic variable into the breeding process, as factors such as weather, availability of pollinators, and geographic location significantly influence the success of pollination. Pollination in cross-pollinated crops is essential for ensuring genetic diversity, enhancing crop yield, and maintaining ecosystem health. However, this dependency also poses challenges in controlled breeding environments, where external factors can affect the consistency and success of pollination events.

Importance of Pollinators in Crop Breeding

The role of pollinators is central to the **reproductive success** of many crop species. Pollination is not just a means of ensuring seed formation but also critical in maintaining the genetic diversity necessary for healthy, productive plants. In cross-pollinated species, unlike self-pollinated ones, genetic recombination between different plants promotes heterozygosity, leading to robust populations capable of adapting to environmental stresses and pests.

» **Enhancing Genetic Diversity**: In cross-pollinated crops, the mixing of pollen from different plants results in offspring with a broader genetic base. This genetic diversity is beneficial for maintaining adaptability to changes in environmental conditions, pathogens, or pests.

Breeders leverage this natural diversity to select individuals that combine desirable traits like disease resistance, yield stability, or stress tolerance.

- **Environmental and Ecological Factors**: Because cross-pollination is heavily reliant on external factors like wind and insects, environmental changes can have a profound impact on the breeding process. Poor weather conditions, habitat loss, or a reduction in pollinator populations can severely reduce pollination success, which may, in turn, affect the quality and yield of the crop. As such, understanding and managing the role of pollinators in agricultural ecosystems is critical for successful crop breeding.

Wind-Pollinated Crops

Wind-pollination (anemophily) is a common mechanism in many cross-pollinated crops, where pollen grains are carried by wind currents to nearby plants. Wind-pollinated crops such as maize, wheat, rye, and barley produce large amounts of lightweight pollen that can be transported over long distances, promoting wide genetic mixing within populations.

1. Process and Advantages of Wind Pollination

Wind-pollination is a **passive process** driven by environmental forces, making it one of the most widespread forms of pollination in agricultural systems. In wind-pollinated crops, the male reproductive organs release pollen into the air, and the wind transports these pollen grains to the female reproductive organs of nearby plants, where fertilization occurs.

- **Genetic Mixing and Diversity**: Because wind can transport pollen over great distances, it promotes significant genetic mixing between different plants within a population. This wide dispersal increases the likelihood of **outcrossing**, which results in greater genetic variation and heterozygosity in the offspring. In crops like maize, this genetic diversity is a crucial factor in maintaining robust populations that can adapt to changing environmental conditions or pathogen pressures.
- **Adaptability to Changing Conditions**: The random nature of wind-pollination also allows for greater adaptability within a crop population. For example, in regions prone to fluctuating temperatures, wind-pollinated crops may be more resilient because their genetic diversity allows for a broader range of plant responses to stress. In crops like wheat, this adaptability contributes to greater stability in yield and performance across different growing environments.

2. Challenges and Limitations of Wind Pollination

While wind-pollination is efficient for promoting genetic diversity, it is also **unpredictable** and dependent on environmental conditions. Factors like wind speed, humidity, and temperature can significantly influence the success of pollination, introducing variability into breeding programs and crop production.

- **Variable Pollination Success**: One of the main challenges with wind-pollination is its dependence on favorable environmental conditions. In the absence of sufficient wind or during periods of high humidity or rainfall, pollen may not be effectively dispersed, leading to **poor fertilization rates** and reduced seed set. This is particularly problematic in areas where weather patterns are inconsistent, making it difficult to predict pollination outcomes.

» **Pollen Drift and Contamination**: Another challenge associated with wind-pollination is **pollen drift**, where pollen from one field can travel to adjacent fields, potentially leading to unintended cross-pollination. This can be problematic in cases where farmers are growing different varieties of the same crop, or where genetically modified (GM) crops are grown alongside non-GM varieties. Pollen drift can result in **contamination** of seed stock, reducing the purity of breeding lines or complicating the production of certified organic crops.

Insect-Pollinated Crops

Insect pollination (entomophily) is another major pollination mechanism in cross-pollinated crops. Some of the most economically important crops, including **alfalfa**, **sunflowers**, and **fruits** like apples and pears, rely on insect pollinators—primarily bees—to transfer pollen from one plant to another. The health and activity of insect pollinators are directly linked to the success of these crops, making pollinator management a crucial aspect of agricultural production.

1. Importance of Insect Pollinators

Insect pollinators, particularly **bees**, play a critical role in the success of insect-pollinated crops. Unlike wind-pollinated species, where pollen is passively carried by air currents, insect-pollinated crops depend on **active pollinator behavior** to transfer pollen. As bees and other insects move from flower to flower, collecting nectar and pollen, they inadvertently transfer pollen from the male parts of one flower to the female parts of another, enabling fertilization.

» **Efficient Pollination**: Insect pollination is often more targeted and efficient than wind-pollination because insects can actively seek out and visit flowers, ensuring that pollen is delivered to the right locations. For crops like alfalfa, sunflower, and clover, which rely on insect pollinators for successful fertilization, the presence of a healthy pollinator population can dramatically improve seed set and crop yield.

» **Commercial Pollination**: In some agricultural systems, commercial beekeeping operations are used to provide **managed pollination services**. Beekeepers transport hives of honeybees to fields of insect-pollinated crops during flowering periods, ensuring adequate pollination. This practice is especially common in high-value crops like almonds and apples, where reliable pollination is critical for maximizing fruit set and quality.

2. Challenges of Insect Pollination

While insect pollinators are invaluable for the success of many cross-pollinated crops, their populations are increasingly under threat from environmental pressures such as **habitat loss**, **pesticide exposure**, and **climate change**. These factors can reduce the availability of pollinators and jeopardize the productivity of insect-pollinated crops.

» **Pollinator Decline**: One of the most significant challenges in insect-pollinated crops is the global decline of pollinator populations, particularly bees. This decline is driven by a combination of factors, including the widespread use of **pesticides**, which can kill pollinators or reduce their foraging efficiency, and the destruction of natural habitats due to urbanization and intensive agriculture. Without a healthy pollinator population, crops that rely on insect pollination may experience **reduced yields** and **lower-quality produce**.

» **Dependency on Environmental Conditions**: Like wind-pollinated crops, the success of insect-pollinated crops is highly dependent on environmental factors. Poor weather conditions, such as **heavy rain** or **high winds**, can reduce insect activity, limiting the number of pollination events that occur. In regions with extreme weather patterns, farmers may struggle to achieve consistent pollination from year to year.

» **Management of Pollinators**: Managing pollinator populations is increasingly necessary to ensure the success of insect-pollinated crops. In addition to protecting wild pollinators by reducing pesticide use and preserving natural habitats, many farmers rely on **managed pollination services** to supplement natural pollination. However, the logistics and costs of transporting bees and maintaining hives can be a challenge, particularly for small-scale or resource-limited farmers.

The Role of Other Pollinators

While wind and insects are the primary pollinators in most agricultural systems, other animals, such as **birds** (e.g., hummingbirds) and **bats**, also play important roles in the pollination of certain crops. These pollinators are particularly important in tropical and subtropical regions, where they pollinate crops like bananas, guava, and certain species of cactus. Their unique role in ecosystems highlights the diversity of pollination strategies and the need to preserve natural habitats that support these pollinator species.

Breeding Methods for Cross-Pollinated Crops

Breeding cross-pollinated crops requires different strategies than those used for self-pollinated species. The genetic diversity inherent in cross-pollinated species provides both challenges and opportunities for breeders, as they seek to improve traits like yield, disease resistance, and environmental adaptability while maintaining genetic variability.

1. Mass Selection

Mass selection is a breeding method that has been traditionally used for improving cross-pollinated crops. In this method, breeders select a large number of individuals based on phenotypic traits and allow them to cross-pollinate freely. The seeds from the selected individuals are then bulked together to form the next generation.

» **Selection Process**: Breeders observe a large population of plants and select individuals that exhibit desirable traits, such as high yield, pest resistance, or drought tolerance. These selected plants are then allowed to cross-pollinate, and their seeds are collected and bulked together to form the next generation.

» **Maintaining Genetic Diversity**: One of the advantages of mass selection in cross-pollinated crops is that it maintains genetic diversity within the population. By selecting multiple individuals and allowing natural cross-pollination, breeders ensure that the population retains a wide range of genetic combinations, which can be useful for adapting to changing environments.

» **Gradual Improvement**: Mass selection is a gradual process that requires multiple generations to achieve significant genetic gains. While the process is simple and inexpensive, it is less precise than other breeding methods, as it relies on phenotypic traits rather than direct genetic selection. Nonetheless, it has been successfully used in crops like maize, forage grasses, and legumes.

2. Recurrent Selection

Recurrent selection is a more advanced form of selection used in breeding cross-pollinated crops. This method involves selecting individuals with desirable traits and interbreeding them over multiple generations to improve the frequency of the desired alleles in the population.

» **Continuous Improvement**: Recurrent selection is designed to be a **cyclical process**. Breeders begin by selecting individuals with superior traits from a genetically diverse population. These selected individuals are interbred, and the next generation is evaluated for the same traits. The process is repeated over several cycles, with each cycle gradually improving the overall population.

» **Incorporating New Traits**: Recurrent selection allows breeders to introduce new genetic material into the population while maintaining genetic diversity. For example, a breeder might introduce a new source of disease resistance into a maize population by crossing resistant individuals with the existing population and then selecting for resistance in subsequent generations.

» **Multiple Trait Selection**: This method is particularly useful for selecting multiple traits simultaneously. For instance, breeders may select for both yield and disease resistance in maize. Over several cycles of recurrent selection, they can increase the frequency of individuals that carry both traits, leading to a more resilient and high-performing population.

3. Hybrid Breeding

Hybrid breeding is one of the most effective methods used to exploit hybrid vigor (heterosis) in cross-pollinated crops. In this approach, breeders develop hybrids by crossing two genetically distinct parent lines. The resulting hybrids often exhibit superior traits compared to their parents, such as higher yield, better stress tolerance, and enhanced disease resistance.

» **Inbred Line Development**: Hybrid breeding begins with the development of **inbred lines**, which are created by self-pollinating plants for several generations to produce genetically uniform lines. These inbred lines are homozygous and may exhibit reduced vigor due to inbreeding depression. However, when two distinct inbred lines are crossed, the resulting hybrid offspring show improved performance due to the complementary combination of genes from both parents.

» **Production of Hybrid Seeds**: Hybrid seed production involves carefully controlled pollination to ensure that the desired cross takes place. For example, in maize hybrid seed production, the male flowers of one parent line are removed to prevent self-pollination, and the female plants are pollinated with pollen from the other parent line. This ensures that the seeds produced are hybrids that will exhibit heterosis.

- **Commercial Importance**: Hybrid varieties are widely used in many cross-pollinated crops, including maize, sunflowers, and sorghum. Hybrid maize, in particular, has revolutionized agricultural productivity, leading to significant increases in yield. Farmers prefer hybrids because they offer consistent and predictable performance, making them valuable for large-scale commercial farming.

4. Synthetic Varieties

Synthetic varieties are another breeding strategy used for cross-pollinated crops, particularly in crops like forage grasses, legumes, and alfalfa. Synthetic varieties are developed by interbreeding a selected group of plants that exhibit desirable traits. These plants are chosen based on their ability to perform well both individually and in combination with other plants.

- **Selection of Parents**: In developing synthetic varieties, breeders select a group of plants that show superior performance in traits like yield, disease resistance, or forage quality. These plants are then allowed to cross-pollinate freely, and the seeds from this interbreeding form the synthetic variety.
- **Maintaining Heterozygosity**: One of the key advantages of synthetic varieties is that they maintain a high level of heterozygosity, which provides genetic diversity and adaptability. This makes synthetic varieties well-suited for crops that are grown in variable environments, where adaptability is important for maintaining productivity.
- **Long-Term Stability**: Synthetic varieties can be propagated over multiple generations without losing their genetic diversity, unlike hybrids, which need to be recreated each generation. This makes synthetic varieties a more sustainable option for some crops, especially in regions where farmers may not have access to hybrid seeds every year.

Chapter - 3

Selection Methods in Plant Breeding

Selection in Plant Breeding

Selection is one of the most critical processes in plant breeding, as it involves the identification and isolation of desirable plant types from a population to improve specific traits. This method serves as the foundation of all crop improvement programs. Selection is essential for distinguishing and enhancing traits that are valuable for agriculture, such as yield, disease resistance, stress tolerance, and quality attributes. It is one of two fundamental steps in breeding programs, the other being the creation of genetic variation.

Two primary agents drive the selection process: **natural selection**, which operates in nature, and **artificial selection**, directed by human intervention. While these processes can sometimes align, they usually work in opposite directions due to their different objectives. In natural selection, traits that improve a plant's ability to survive and reproduce in a particular environment are favored. In contrast, artificial selection focuses on traits valuable to humans, such as increased yield, improved flavor, or resistance to pests. Before the domestication of crops, plants were primarily shaped by natural selection, which promoted their adaptation to the environment. However, with the rise of agriculture, humans began to influence plant evolution through artificial selection, either consciously or unconsciously. Over time, crop species were exposed to both natural and artificial selection, which helped shape their modern forms. However, in contemporary plant breeding, artificial selection has become the primary force driving crop improvement, while natural selection plays a minor role.

Basic Principles of Selection

Despite the complex genetic interactions in plants, the principles of selection are relatively straightforward. Understanding these principles is crucial for applying effective selection techniques in crop breeding. The three core principles of selection, as proposed by Walker (1969), are as follows:

1. Selection Operates on Existing Genetic Variation

One of the most fundamental principles is that **selection works by acting on existing genetic variation** within a population. The main function of selection is to identify and discriminate between individuals based on desirable traits. However, for this process to be effective, sufficient variability

DOI: 10.1201/9781003757412-3

must be present in the population. Selection cannot create new genetic variation; it merely utilizes the variation that already exists.

» **The Role of Genetic Diversity**: The variation in traits like plant height, flowering time, or disease resistance within a crop population forms the foundation for selection. In a population of wheat plants, for example, there may be variation in traits such as drought tolerance. By selecting the most drought-tolerant plants, breeders can improve the population's overall ability to withstand water scarcity. However, if the population lacks variation for this trait, no progress can be made through selection alone.

» **Introduction of New Variation**: While selection cannot generate new variation, breeders can introduce additional genetic diversity through methods such as hybridization, mutation breeding, or genetic engineering. These techniques create new combinations of alleles or introduce novel genes into the population, providing more options for selection. For instance, hybridizing two different maize varieties with complementary traits can introduce new variability in traits such as yield or resistance to pests, providing breeders with a broader pool of genetic material to work with.

2. Selection Acts on Heritable Traits

The second principle is that selection is effective only when it targets **heritable traits**, meaning those traits that are genetically controlled and can be passed on to the next generation. Traits influenced by environmental factors, such as variations in soil quality or irrigation, are not heritable and cannot be reliably selected. For selection to be successful, the differences between individuals must be genetic in nature, allowing for the trait to be transmitted from parent to offspring.

» **Genetic vs. Environmental Variation**: In any population, phenotypic variation (observable traits) can result from both genetic factors and environmental influences. For selection to be effective, breeders must ensure that the traits they are selecting for are primarily driven by genetics, not by temporary environmental conditions. For example, if one plant appears to have a higher yield because it was grown in a more fertile area of the field, this difference may not be due to its genetic makeup. To accurately assess genetic variation, breeders often conduct trials in multiple environments or control environmental factors as much as possible.

» **Phenotype and Genotype Relationship**: The closer the relationship between a plant's **phenotype** (its observable characteristics) and its **genotype** (its genetic makeup), the more effective selection will be. Breeders often use **heritability** as a measure of how much of the variation in a trait is due to genetics rather than environmental factors. A high heritability value indicates that the trait is largely controlled by genetics, making it a good candidate for selection. For example, if drought tolerance has high heritability in a rice population, breeders can confidently select for this trait, knowing that it will be passed on to future generations.

3. Selection Favors Superior Individuals

The third principle is that selection operates by favoring certain individuals over others based on their reproductive success. In a population, some individuals will naturally be more successful at surviving and reproducing due to their superior genetic traits. These individuals contribute more offspring to the next generation, thereby increasing the frequency of favorable alleles in the population.

- **Superior Individuals**: In the context of plant breeding, breeders choose individuals with desirable traits—such as disease resistance, higher yield, or faster maturity—to serve as the parents for the next generation. By selecting these superior individuals, breeders ensure that the population evolves in the desired direction over time. For example, in a breeding program for disease resistance in tomatoes, breeders might select the plants that show the strongest resistance to a fungal pathogen and use them to produce the next generation of seeds.
- **Differential Reproduction**: Selection works by allowing superior individuals to reproduce and pass on their genes, while inferior individuals are removed from the breeding population. This process is known as **differential reproduction** and is the driving force behind genetic improvement in crop populations. Over multiple generations, selection increases the frequency of desirable traits while reducing the prevalence of undesirable characteristics.

Characteristics of Selection

Selection in plant breeding has two defining characteristics that are crucial to its success:

1. Selection Acts Only on Heritable Traits

Selection can only be effective if it targets **heritable traits**, meaning traits that are controlled by the plant's genetic makeup. Traits that are influenced primarily by environmental factors cannot be reliably passed on to the next generation, making them poor candidates for selection. For instance, if a plant has a high yield due to favorable environmental conditions (such as extra water or nutrients), this is not an indication of its genetic potential. Breeders focus on selecting traits that are controlled by genetics, ensuring that these traits will appear in future generations.

2. Selection Does Not Create New Genetic Variation

Selection does not generate new genetic variability; it only works on the **existing variation** in a population. This means that the effectiveness of selection is limited by the amount of genetic diversity present. If a population lacks sufficient genetic variation for a particular trait, breeders must introduce new variability through methods like crossing or mutagenesis before they can make progress through selection.

Requirements for Effective Selection

For selection to be effective in a breeding program, two fundamental conditions must be met:

1. Genetic Variation Must Be Present

The first requirement for effective selection is the presence of **genetic variation** within the population. Without sufficient variation, there is no basis for selection, and breeders will be unable to distinguish between superior and inferior individuals. Genetic variation is the raw material that allows selection to improve a population over time.

- **Sources of Variation**: Genetic variation can arise naturally within a population, or it can be introduced by breeders through hybridization, mutation, or genetic engineering. For example, in a breeding program for drought resistance, if there is natural variation in how different plants respond to water stress, breeders can select the most drought-tolerant individuals. However, if the population is highly uniform and lacks variation, selection will not be effective, and breeders may need to introduce new sources of genetic diversity.

2. The Variation Must Be Heritable

The second requirement is that the variation must be **heritable**, meaning it is controlled by the plant's genetic makeup and can be passed on to future generations. Non-heritable variation, caused by environmental factors, cannot be used to make long-term genetic improvements in a population. For example, if one plant in a population appears to have superior growth due to better access to nutrients or water, this advantage may not be passed on to its offspring. To ensure that selection leads to meaningful improvements, breeders focus on traits that are determined by genetics.

Selection Intensity

Selection intensity refers to the proportion of individuals selected to advance to the next generation from a population. Higher selection intensity means that only the top-performing individuals are selected, leading to faster genetic progress. However, this can also reduce genetic diversity in the population. For example, if a breeder selects only 5% of the plants from a large population, those plants are likely to have superior traits, but the genetic diversity of the population may decrease, potentially leading to inbreeding problems over time.

» **Balancing Selection Intensity**: Breeders must strike a balance between selection intensity and maintaining genetic diversity. While high selection intensity can lead to rapid improvements in traits like yield or disease resistance, it can also reduce the overall genetic variability in the population. This loss of diversity can make the population more vulnerable to new pests, diseases, or environmental stresses. To avoid this, breeders may use methods like **recurrent selection** to maintain genetic diversity while still making progress on improving specific traits.

A. Pureline Theory by Johannsen (1903):

Pureline theory, as proposed by Wilhelm Johannsen in 1903, forms a cornerstone of classical genetics and plant breeding. Johannsen's research on **Phaseolus vulgaris** (beans) revolutionized the understanding of heredity, particularly in self-pollinated crops, where genetic uniformity plays a central role. His work differentiated the concepts of **genetic variation** (heritable) and **environmental variation** (non-heritable), introducing the term **pureline** and illustrating the limits of selection in self-fertilized populations.

What is Pure Line?

A **pureline** is defined as the progeny of a single, self-fertilized homozygous plant. This means that all individuals within a pureline share the same genotype and are, genetically, identical to each other. Purelines are significant in plant breeding because they offer stability and uniformity in traits across generations, provided the breeding environment remains consistent. Johannsen introduced this concept based on experiments that showed the difference between genetic and environmental variation, allowing breeders to understand what traits could be selected and enhanced across generations.

Johannsen's Experiment with Beans:

Johannsen's experiments began with the **Princess variety** of beans, which he obtained from a commercial seed lot. He observed that the seeds varied in size, with some being larger and others smaller. This prompted him to explore whether the variation was genetic or simply due to environmental influences. He divided the seeds based on size (larger vs. smaller) and grew them separately. His experiment had the following key observations:

Observation of Genetic Variation in Johannsen's Experiment

Johannsen's experiment with beans revealed significant insights into the genetic basis of seed size variation. By selecting and growing seeds of different sizes, Johannsen was able to demonstrate how seed size is influenced by genetic factors, thereby laying the foundation for the pureline theory in plant breeding.

Genetic Basis of Seed Size

In his experiment, Johannsen observed that when seeds from larger beans were planted, they consistently produced progenies with larger seeds. Similarly, smaller seeds gave rise to progenies that exhibited smaller seed sizes. This observation provided direct evidence that the variation in seed size was heritable and controlled by genetic factors.

» **Inheritance of Traits**: The fact that the size of the seeds was maintained from one generation to the next indicated that the variation was **genetic** rather than purely environmental. The large seeds inherited traits from their parent plants that enabled them to continue producing large seeds, while the smaller seeds passed on their traits for smaller seed size.

» **Implication for Plant Breeding**: This finding was critical for understanding how selection could be used to improve crop traits. If traits like seed size are genetically determined, breeders can select for these traits to enhance crop performance. Johannsen's experiment showed that seed size, like many other traits, can be reliably passed down through generations, enabling plant breeders to target specific characteristics in their breeding programs.

Purelines in Seed Lots

Johannsen concluded from his experiment that the commercial seed lot of beans was not a homogeneous population but rather a mixture of multiple **purelines**, each representing a genetically distinct group. A pureline, by definition, is the progeny of a single self-fertilized homozygous plant. In this case, each pureline had its own characteristic seed size, and the variation observed in the commercial lot was the result of a mixture of these purelines.

» **Mixture of Purelines**: The commercial seed lot contained beans with different seed sizes because it consisted of several genetically distinct purelines. Each pureline was stable, producing progenies that consistently resembled their parent in terms of seed size. For example, larger seeds were part of a pureline that consistently produced larger seeds, while smaller seeds belonged to a pureline that produced smaller seeds.

» **Significance for Crop Improvement**: Johannsen's discovery that a commercial seed lot could contain multiple purelines has important implications for crop improvement. It highlights the importance of identifying and selecting the best purelines for breeding purposes. In a diverse seed lot, breeders can select plants from the purelines that exhibit the most desirable traits, such as larger seed size or higher yield, to use in breeding programs.

» **Genetic Homogeneity in Purelines**: Once a pureline is identified and isolated, it is genetically homogeneous, meaning that all plants within the pureline will exhibit the same traits. This homogeneity makes purelines valuable for breeders because they ensure consistent performance across generations, reducing the variability that can occur in genetically diverse populations.

Purelines and Environmental Variation

In an extension of his foundational study, Johannsen further explored the concept of variation within purelines by examining the influence of environmental factors on seed weight. This part of his research was critical for distinguishing between **genetic variation** and **environmental variation**. Johannsen's experiment with pureline no. 13 provided valuable insights into how environmental conditions, rather than genetic factors, could influence the appearance of variability in a pureline.

Non-Heritable Variation

Johannsen selected seeds from pureline no. 13 and divided them into four distinct weight classes: 200 mg, 300 mg, 400 mg, and 500 mg. These seeds were then planted and their progeny were analyzed to determine the average seed weight in each class. The results demonstrated that despite the differences in seed weight among the parent seeds, the offspring produced seeds of remarkably similar weights. The mean seed weights for the progeny were as follows:

» Seeds from the 200 mg class produced progeny with an average weight of 475 mg.
» Seeds from the 300 mg class produced progeny with an average weight of 450 mg.
» Seeds from the 400 mg class produced progeny with an average weight of 451 mg.
» Seeds from the 500 mg class produced progeny with an average weight of 458 mg.

These findings revealed that **the variation within the pureline was non-heritable**. In other words, the differences in seed weight observed in the parent generation did not translate into significant differences in the offspring. The relatively consistent seed weights in the progeny, regardless of the weight of the parent seed, indicated that the variation was **environmental** in nature, not genetic.

» **Environmental Influence**: The experiment showed that environmental factors such as soil quality, water availability, and nutrient levels were responsible for the initial differences in seed weight. These environmental conditions caused variation in the physical appearance of the seeds but did not alter the underlying genetic makeup of the plants. As a result, the progeny from seeds of varying weights were genetically uniform, leading to minimal differences in seed size across the next generation.

Genetic Uniformity of Purelines

Johannsen's findings on environmental variation further reinforced the concept of **genetic uniformity** in purelines. Since purelines are derived from a single homozygous plant, they consist of individuals that are genetically identical. This genetic uniformity means that any observable variation within a pureline is not due to differences in genetic material but rather to **environmental factors**.

» **Genetically Identical Progeny**: The progeny derived from seeds of different weights were genetically similar because they all originated from the same pureline. The minimal variation in seed weight among the offspring, despite the differences in parent seed size, highlighted the genetic uniformity of the pureline. Since purelines are composed of homozygous individuals, the offspring inherit identical genetic material, resulting in consistent traits across generations.

» **Impact of Environment on Phenotype**: The experiment demonstrated that the **phenotypic variation** (observable traits such as seed weight) in purelines is strongly influenced by the environment. For example, seeds that develop in nutrient-rich conditions may appear larger than those grown in nutrient-poor conditions, but this difference is not passed on to the next generation because it is not genetically determined. This phenomenon highlights the distinction between **phenotypic plasticity**, where the phenotype can change in response to environmental conditions, and **genetic inheritance**, where traits are passed down through genetic material.

Implications for Plant Breeding

Johannsen's findings on environmental variation within purelines have significant implications for **plant breeding**. Understanding that variation within purelines is not heritable allows breeders to focus their efforts on selecting traits that are truly genetic and, therefore, can be passed on to future generations.

» **Selection Within Purelines**: Johannsen's study confirmed that selection within a pureline is **ineffective** in producing genetic improvements. Since all individuals within a pureline are genetically identical, selecting plants based on phenotypic differences caused by environmental factors will not result in meaningful changes. For example, selecting larger seeds within a pureline will not lead to an increase in seed size in future generations because the variation in size is due to environmental influences, not genetic differences.

» **Breeding Programs**: For breeders working with purelines, it is crucial to recognize that once a pureline is established, the genetic traits of that line are fixed. Breeders must introduce new genetic variability through methods such as **crossbreeding** or **mutation breeding** if they wish to improve traits like yield, disease resistance, or stress tolerance. Otherwise, the focus should shift to managing environmental conditions to optimize the expression of desirable traits within the existing genetic framework.

Confirming the Ineffectiveness of Selection Within Purelines

To rigorously test his **pureline theory**, Johannsen conducted a series of additional experiments aimed at determining whether selection could alter traits within a pureline. His findings conclusively showed that **selection within a pureline is ineffective**. This revelation was pivotal in plant breeding, particularly for self-pollinated crops, as it demonstrated the limitations of selection in genetically homogeneous populations. Johannsen's conclusions were based on three critical pieces of evidence, which are elaborated below:

1. Seed Weight Classes Within a Pureline

In his initial investigation, Johannsen sought to explore whether selecting seeds based on size within a pureline could result in progeny with different seed sizes. He divided seeds from **pureline no. 13** into four distinct weight classes: 200 mg, 300 mg, 400 mg, and 500 mg. These seeds were grown, and the average seed weights of the progeny were recorded. The results were revealing:

» Despite selecting seeds based on weight, the progeny from all weight classes exhibited very similar seed weights, with averages around 450 mg.

» Seeds from the 200 mg class produced progeny with an average weight of 475 mg.

» Seeds from the 300 mg class produced progeny with an average weight of 450 mg.

» Seeds from the 400 mg class produced progeny with an average weight of 451 mg.

» Seeds from the 500 mg class produced progeny with an average weight of 458 mg.

This showed that **selection based on weight did not lead to a significant change** in the traits of the progeny. The observed variation in seed size within the parent seeds was due to environmental factors rather than genetic differences. The progeny, all derived from the same pureline, exhibited almost identical seed weights, confirming that the variation within the pureline was not heritable.

» **Genetic Homogeneity of Purelines**: This experiment reinforced the idea that purelines are genetically homogeneous. Since all individuals within a pureline share the same genotype, selecting based on phenotype (such as seed size) will not result in genetic changes in the offspring. In a pureline, the phenotype is largely influenced by the environment, and selecting for traits that vary due to environmental factors does not lead to long-term improvements.

» **Non-Heritable Variation**: The consistency in seed weight across the different classes confirmed that the variation within the pureline was due to **non-heritable** environmental influences. Therefore, selection within a pureline is ineffective because it does not lead to genetic differentiation or improvement.

2. Ineffectiveness of Selection Across Generations

In a follow-up experiment, Johannsen tested whether selection within a pureline could produce changes in traits like seed size over multiple generations. He selected the largest and smallest seeds from a pureline and grew them over **six successive generations**. In each generation, he selected the largest seeds from the large-seeded progeny and the smallest seeds from the small-seeded progeny. However, the results were consistent with his earlier findings:

» After six generations of selection, the **average seed weight** remained nearly constant. The line selected for large seeds stabilized around 690 mg, while the line selected for small seeds stabilized around 680 mg.

» This minimal difference in seed size, despite rigorous selection over multiple generations, demonstrated that **selection within a pureline does not result in genetic changes**. The population was already genetically homogeneous, and selecting for traits like seed size could not alter the genetic makeup of the pureline.

» **Limitations of Selection**: This experiment illustrated the fundamental limitation of selection in genetically uniform populations. In a pureline, where all individuals are homozygous and share the same genotype, selection based on phenotype will not lead to improvements in traits like seed size. Because purelines are already fixed for specific genetic traits, the variation observed in traits such as seed size is due to environmental influences, not genetic differences. Therefore, selecting the largest or smallest seeds within a pureline will not result in meaningful changes in the population over time.

» **Relevance to Breeding Programs**: For breeders working with self-pollinated crops, this finding is crucial. Once a pureline is established, further selection within that pureline will not produce genetic gains. To introduce new variation or improve traits within a crop, breeders must introduce new genetic material through crossing or mutation breeding, rather than relying on selection within an already uniform population.

3. Parent-Offspring Correlation

Johannsen also analyzed the **parent-offspring correlation** for seed size within the pureline to understand the heritability of traits. The parent-offspring correlation measures the degree to which the traits of the offspring resemble those of the parent. If a trait is heritable, the parent-offspring correlation will be greater than zero, indicating that the variation has a genetic basis. Conversely, if the correlation is close to zero, it suggests that the variation is not heritable and is influenced by environmental factors.

» In **line no. 13**, the parent-offspring correlation for seed size was effectively zero (–0.018 ± 0.038). This indicated that the variation observed within the pureline was **non-heritable** and due to environmental influences. Since the parent plants did not pass on their seed size traits to the offspring in a predictable manner, it was clear that seed size variation within the pureline was not genetically controlled.

» In contrast, the parent-offspring correlation in the **original seed lot** (Princess variety) was significantly greater than zero (0.336 ± 0.008). This suggested that the variation in seed size within the original seed lot had a **genetic basis** and was heritable. The greater-than-zero correlation in the original population indicated that the seed size variation in the commercial lot was due to genetic differences between the various purelines.

» **Heritability and Selection**: The zero correlation in purelines further confirmed that the variation within purelines was due to environmental factors. Since the traits were not genetically inherited, selection based on these traits would not lead to genetic changes. In contrast, the significant correlation in the original seed lot suggested that genetic variation existed between different purelines, allowing for effective selection in a mixed population. This evidence reinforced Johannsen's conclusion that **selection within a pureline is ineffective** because the variation is not genetically controlled.

Summary of Johannsen's Findings

Johannsen's experiments provided compelling evidence for the **ineffectiveness of selection within purelines**, based on three key observations:

1. **Seed Weight Classes Within a Pureline**: Despite selecting seeds based on weight, the progeny from different weight classes exhibited minimal differences in seed size. This showed that variation within a pureline is non-heritable and cannot be altered through selection.
2. **Ineffectiveness of Selection Across Generations**: After six generations of selecting for large or small seeds, the average seed weight remained nearly constant, demonstrating that

selection within a genetically uniform population does not result in genetic changes.

3. **Parent-Offspring Correlation**: The zero parent-offspring correlation for seed size within the pureline indicated that the variation was not heritable, further confirming that selection within a pureline is ineffective. In contrast, the significant correlation in the original seed lot indicated that genetic variation existed between purelines, making selection effective in a genetically diverse population.

Implications for Plant Breeding

Johannsen's findings have significant implications for **plant breeding**, especially for self-pollinated crops. His experiments demonstrated that selection within a pureline, which is genetically homogeneous, cannot produce genetic improvements because the variation observed in traits is due to environmental influences rather than genetic differences.

- **Introduction of Genetic Variation**: For breeders to make genetic improvements in a crop, they must introduce new genetic variation. This can be done through **crossbreeding** (introducing genetic material from different varieties) or **mutation breeding** (inducing genetic changes). Once a pureline is established, further selection within that pureline will not lead to improvements.
- **Focus on Environmental Management**: In cases where purelines are used, breeders may need to focus on **environmental management** to optimize the expression of desirable traits. Since traits like seed size can be influenced by environmental factors such as soil fertility or water availability, improving the growing conditions can enhance crop performance, even if the genetic traits remain fixed.

Johannsen's Key Conclusions:

Wilhelm Johannsen's groundbreaking experiments led to two major conclusions that have profoundly influenced the fields of plant breeding and genetics. These conclusions provide fundamental insights into the nature of genetic variation in self-pollinated crops and the limitations of selection within purelines.

1. Genetic Variation in Self-Fertilized Populations

One of Johannsen's most significant findings was that **self-fertilized populations** often consist of **mixtures of several homozygous genotypes**, each with distinct genetic characteristics. In these populations, genetic variation allows for the effective selection of traits. This variation can be exploited to improve crop performance by selecting individuals with desirable traits, such as disease resistance, drought tolerance, or high yield.

- **Genetic Basis of Variation**: In self-pollinated crops, genetic variation between individuals is a key driver of differences in traits. Since self-pollinated crops undergo repeated self-fertilization, they develop distinct **homozygous lines**, which Johannsen referred to as **purelines**. In a genetically diverse population, different purelines possess varying traits, such as differences in height, yield, or resistance to pathogens. This variation arises from the presence of multiple homozygous genotypes within the population.

» **Selection in Genetically Diverse Populations**: Johannsen's work demonstrated that selection within a **genetically diverse population** of self-pollinated crops can be highly effective. For example, if a population contains multiple purelines, each exhibiting different levels of disease resistance, breeders can select the individuals with the highest resistance for further propagation. Over successive generations, the population can be improved by favoring purelines with superior traits.

- **Example in Breeding**: In crops such as wheat or barley, breeders may start with a genetically diverse population and use **phenotypic selection** to identify plants with the most desirable traits, such as increased yield or better resistance to rust diseases. By selecting these individuals and allowing them to self-pollinate, breeders can produce a population enriched with favorable traits. This approach is effective because the variation in the population is **genetic** and can be passed on to future generations.

» **Importance of Genetic Variation for Crop Improvement**: Johannsen's discovery highlighted the importance of **genetic diversity** in crop breeding. Without genetic variation, it would be impossible to make long-term improvements in crop performance. By identifying and selecting for desirable traits in a genetically diverse population, breeders can enhance crop yield, quality, and resistance to environmental stresses. Johannsen's insight has shaped modern plant breeding strategies, which often involve **creating genetic variation** through methods such as crossbreeding, mutagenesis, or the introduction of genes from wild relatives.

2. Pureline Uniformity and Non-Heritable Variation

Johannsen's second major conclusion was that once a **pureline** is established, the **variation within the pureline** is entirely due to **environmental factors** and is not heritable. This means that while purelines offer **stability** and **uniformity**, they cannot be genetically improved through selection because there is no genetic variation within the pureline.

» **Genetic Uniformity of Purelines**: A pureline is the progeny of a single, self-fertilized homozygous plant. Because all individuals within a pureline share the same genotype, they are genetically identical. As a result, any observable variation in traits such as seed size, plant height, or yield within a pureline is not due to genetic differences but to **environmental influences**. For example, differences in soil nutrients, water availability, or sunlight can cause plants within a pureline to exhibit slight variations in their growth or performance, but these differences are not passed on to the next generation.

» **Environmental Influence on Traits**: Johannsen's experiments with seed weight in purelines demonstrated that the variation in seed size within a pureline was **non-heritable** and entirely due to environmental factors. When he selected the largest and smallest seeds within a pureline and planted them over multiple generations, the progeny consistently exhibited very similar seed sizes, regardless of whether they came from large or small seeds. This indicated that the **environment**, rather than genetics, was responsible for the variation in seed size within the pureline.

- **Implications for Selection**: Johannsen's findings led to the conclusion that **selection within a pureline is ineffective**. Since all individuals within a pureline are genetically identical, selecting for traits such as seed size or yield will not result in genetic changes or improvements in future generations. Any observed changes are due to environmental factors and cannot be passed on to progeny. This has important implications for breeders, as it means that once a pureline is established, further selection will not produce genetic improvements.

» **Stability of Purelines**: One of the key advantages of purelines is their **genetic stability**. Because purelines are genetically uniform, they offer **consistent performance** across generations, provided the environmental conditions remain the same. This stability is beneficial for **commercial agriculture**, where farmers value crops that perform predictably in terms of yield, quality, and resistance to pests or diseases.

- **Challenges with Lack of Genetic Variation**: However, the lack of genetic variation within a pureline can also be a disadvantage, particularly in the face of new challenges such as emerging pests, diseases, or changing environmental conditions. Since purelines cannot be genetically improved through selection, breeders must introduce new genetic variation through methods such as **crossbreeding** or **mutation breeding** to address these challenges.

» **Example in Breeding**: In crops like rice or beans, purelines are often used to maintain consistent traits such as grain size or flavor. However, if a new disease emerges that the pureline is susceptible to, breeders cannot rely on selection within the pureline to produce resistant plants. Instead, they must introduce new genetic material from another source, such as a disease-resistant variety or wild relative, to create genetic variation and develop a resistant population.

Implications for Plant Breeding

Johannsen's conclusions have had a profound impact on **plant breeding** and **genetics**. His work clarified the distinction between **genetic** and **environmental** variation and demonstrated the limits of selection in genetically uniform populations like purelines. These insights have shaped modern plant breeding practices, particularly for **self-pollinated crops** such as rice, wheat, and beans.

» **Breeding Strategies for Self-Pollinated Crops**: In self-pollinated crops, breeders must carefully manage genetic variation to ensure long-term improvements. Johannsen's work emphasizes the importance of starting with a genetically diverse population when selecting for traits, as this allows breeders to make meaningful genetic gains. Once purelines are established, breeders must focus on **introducing new variation** through crossbreeding or other methods to address challenges such as disease outbreaks or environmental stresses.

» **Role of Purelines in Commercial Agriculture**: Purelines are valuable for producing **stable, uniform crops**, making them ideal for seed production and commercial farming. However, Johannsen's findings underscore the importance of **diversity** in breeding programs. While purelines provide stability, they can also be vulnerable to changes in the environment or the emergence of new pests and diseases. To ensure the long-term success of crops, breeders must balance the need for uniformity with the introduction of new genetic variation to maintain adaptability.

Origin of Variation in Purelines

Although **purelines** are genetically uniform, they can exhibit genetic variation over time due to a range of external factors and biological processes. This variation does not originate from within the pureline itself but arises from influences that disturb the genetic uniformity. While purelines are stable and homozygous, their genetic purity can be compromised through mechanical processes, hybridization, or mutations. Understanding the origins of these variations is critical for maintaining the **genetic stability** of purelines in breeding programs.

1. Mechanical Mixture

Mechanical mixture refers to the unintended introduction of other genotypes into a pureline population due to the physical processes involved in agriculture. During stages such as **cultivation, harvesting, threshing, and storage**, different genotypes can become mixed with the pureline, thus introducing genetic variability into what should be a genetically uniform population.

» **Sources of Contamination**: In large-scale farming operations, purelines may come into contact with seeds or pollen from other varieties during handling or processing. For example, different genotypes can become mixed during mechanical threshing if machines are not properly cleaned between different lots of crops. Similarly, seeds can mix during packaging, transport, or even in the storage facilities if they are not kept in isolated containers.

» **Impact on Genetic Uniformity**: The accidental introduction of foreign genotypes through mechanical processes compromises the **genetic uniformity** of the pureline. This can lead to undesirable outcomes, such as off-type plants appearing in fields that were expected to consist entirely of pureline individuals. These off-type plants may exhibit different traits such as lower yields, susceptibility to diseases, or variable growth habits, which can undermine the purpose of maintaining purelines.

» **Mitigation Strategies**: To prevent mechanical mixtures, breeders must ensure that strict protocols are followed during the cultivation and processing of purelines. This includes **cleaning equipment** between different lots, using **isolated storage units**, and practicing meticulous handling of seeds. By maintaining isolation during all steps of the breeding and harvesting process, breeders can minimize the risk of contaminating purelines with external genotypes.

» **Challenges in Commercial Seed Production**: In commercial seed production, where large quantities of seeds are processed, the risk of mechanical mixtures is higher. To maintain the purity of pureline seeds, commercial seed producers must implement rigorous **quality control measures**. Regular field inspections and laboratory tests for genetic purity can help ensure that only true-to-type seeds are distributed to farmers.

2. Natural Hybridization

Natural hybridization is another source of genetic variation in purelines, especially in crops that are not fully isolated or that exhibit some degree of cross-pollination. While purelines are typically produced in **self-pollinated crops**, the possibility of cross-pollination with other genotypes exists, introducing new genetic variability into the pureline population.

- **Cross-Pollination in Self-Pollinated Crops**: Even in self-pollinated crops, natural hybridization can occur if pollen from neighboring plants or crops is transferred via wind, insects, or other agents. This is particularly true in species that are not strictly self-pollinating or in cases where **partial cross-pollination** occurs. For example, crops like wheat or rice, which are predominantly self-pollinated, may still exhibit a low level of cross-pollination, leading to the introduction of foreign genetic material.
- **Effect on Genetic Stability**: The introduction of pollen from outside sources can lead to the formation of hybrid plants within a pureline population. These hybrids may exhibit traits that differ from the desired pureline, such as variations in height, seed size, or disease resistance. Over time, the presence of these hybrids can dilute the genetic purity of the pureline and make it difficult for breeders to maintain the uniformity of the population.
- **Isolation and Rouging as Solutions**: One of the most effective ways to prevent natural hybridization is by **isolating the breeding plots**. By growing purelines in fields that are geographically separated from other crops or by erecting physical barriers, breeders can reduce the risk of cross-pollination. Additionally, the practice of **rouging**, which involves identifying and removing off-type or hybrid plants from the population, helps to maintain the genetic integrity of the pureline. Regular field inspections can help detect these off-types early, allowing breeders to take corrective action before they reproduce.
- **Pollinator Management**: For crops that rely on insects or other animals for pollination, managing the **population of pollinators** can also help control hybridization. Limiting the presence of pollinators such as bees in breeding fields or ensuring that only controlled pollination takes place can prevent unwanted cross-pollination from occurring.

3. Mutation

Mutations are spontaneous changes in the DNA sequence of an organism. While mutations occur naturally in all living organisms, they can introduce new genetic variation into a pureline population. Although purelines are genetically uniform, mutations can arise at random, causing changes in traits that may affect the crop's performance or characteristics.

- **Types of Mutations**: Mutations can occur in various forms, including **point mutations** (changes in a single nucleotide), **insertions** or **deletions** of DNA segments, or larger chromosomal rearrangements. Some mutations may be **silent**, meaning they do not affect the phenotype, while others can lead to observable changes in traits such as plant height, leaf color, or disease resistance.
- **Impact of Mutations on Purelines**: In most cases, mutations are **neutral** or **harmful**, resulting in no significant change or a negative impact on plant health and productivity. However, in rare instances, mutations can be beneficial, providing new traits that could be advantageous for breeding purposes. For example, a mutation that confers resistance to a particular pest or disease could be valuable in crop improvement programs.
- **Mutation Breeding**: While spontaneous mutations occur naturally, breeders can also induce mutations through techniques such as **chemical mutagenesis** or **radiation** (e.g., gamma rays). This process, known as **mutation breeding**, is used to generate new genetic variation

that can be selected for desirable traits. Once a beneficial mutation is identified, it can be incorporated into breeding programs to enhance crop performance.

- **Challenges with Mutation in Purelines**: Although beneficial mutations can arise, the random nature of mutations poses a challenge for maintaining the **stability of purelines**. Since purelines are intended to be genetically uniform, any mutation that introduces variation may disrupt the desired consistency of traits. Breeders must carefully monitor pureline populations for the emergence of mutants and decide whether to retain or eliminate them based on their impact on crop performance.
- **Role of Mutations in Long-Term Evolution**: In the context of long-term plant evolution, mutations play a critical role in introducing new genetic material into populations. While breeders often aim to preserve the genetic stability of purelines in the short term, mutations contribute to the overall **diversity** of species, enabling crops to adapt to changing environmental conditions or new pathogens over time.

Characteristics of Purelines

Purelines are central to plant breeding programs, particularly in self-pollinated crops, where breeders aim for genetic uniformity and stability across generations. The concept of purelines, first introduced by Johannsen, has provided a foundation for creating **genetically homogeneous** populations that consistently express desired traits. These purelines are invaluable in crop improvement because they offer consistency in performance, ease of management, and precise trait selection. Below, we explore the primary characteristics of purelines in detail.

1. Genetic Homogeneity

One of the most defining characteristics of purelines is **genetic homogeneity**. All individuals within a pureline are genetically identical, which means they have the same **genotype** and, as a result, exhibit uniform traits. This homogeneity is achieved through multiple generations of self-pollination, which results in a population of homozygous plants.

- **Implications for Breeding**: Genetic homogeneity ensures that traits such as **yield, growth habits**, and **resistance to pests or diseases** are expressed consistently across all individuals in the population. For plant breeders, this uniformity simplifies the process of selecting and maintaining plants with desirable traits. Since there is no genetic variation within the pureline, all offspring will inherit the same traits, making the population predictable and reliable for both seed production and large-scale farming.
- **Example in Agriculture**: In crops such as **rice, wheat**, or **beans**, purelines are used to maintain uniformity in traits like grain size, plant height, and disease resistance. This consistency is particularly important in commercial agriculture, where farmers rely on stable and predictable crop performance to maximize yield and reduce losses due to pests or diseases.
- **Relevance to Crop Improvement**: Purelines are highly valuable in breeding programs that aim to fix certain traits within a population. For example, once a breeder identifies a pureline with excellent resistance to a specific pathogen, they can propagate that line across multiple generations, ensuring that all progeny carry the same level of resistance. This makes purelines

an essential tool for developing crops with **improved resistance** to diseases, enhanced **tolerance to environmental stresses**, or other desired characteristics.

- **Limitations**: While genetic homogeneity offers stability, it also means that the pureline has limited **genetic diversity**, which can be a disadvantage in rapidly changing environments. A pureline that is highly adapted to one set of conditions may become vulnerable if new pests or diseases emerge, or if the climate shifts, because there is no genetic variation to allow for adaptation. This highlights the importance of balancing **genetic uniformity** with the need for diversity in some cases.

2. Environmental Variation

While purelines are genetically uniform, they still exhibit **phenotypic variation** due to environmental factors. This means that the traits expressed by the plants, such as growth rate, yield, or disease symptoms, can vary based on external conditions like soil fertility, water availability, temperature, and light exposure.

- **Non-Heritable Variation**: Johannsen's experiments demonstrated that the variation observed within a pureline is entirely due to environmental factors and is not heritable. In other words, the variation is not passed on to the next generation because it is not caused by genetic differences. For example, two plants from the same pureline might have different heights or yields if one plant grows in nutrient-rich soil and the other in nutrient-poor soil, but these differences are not due to genetics and will not affect the genetic uniformity of the population.
- **Impact on Trait Expression**: Environmental factors can have a significant influence on the expression of traits. A pureline grown in optimal conditions may express its full yield potential, while the same pureline grown in suboptimal conditions may produce lower yields or exhibit poor growth. Understanding the role of environmental factors is crucial for breeders and farmers alike because it helps them optimize growing conditions to get the best performance out of a pureline.
- **Managing Environmental Variation**: Since environmental variation is non-heritable, breeders can focus on **environmental management** to enhance the expression of desired traits in purelines. For example, in crops like **wheat**, where grain size and yield are critical, providing adequate water and nutrients can help ensure that the crop reaches its full potential, even though the plants are genetically identical. Similarly, managing soil conditions, pest control, and other environmental factors can optimize the performance of purelines in the field.
- **Example in Planting Systems**: In large-scale **rice** production, managing water levels in paddy fields can significantly influence the growth and yield of pureline varieties. Similarly, in **bean** farming, ensuring proper irrigation and nutrient application can lead to better pod formation and seed size, even though the genetic makeup of the plants is identical.

3. Stability

One of the key reasons why purelines are highly valued in plant breeding is their **genetic stability**. Once a pureline is established, it remains genetically uniform across generations, provided that it is not exposed to external factors such as cross-pollination, mutation, or mechanical mixing.

- **Consistent Performance Across Generations**: The genetic stability of purelines ensures that they consistently express the same traits over multiple generations. This stability is particularly useful in breeding programs where breeders need to maintain specific traits, such as **disease resistance, drought tolerance**, or **high yield**, without the risk of genetic variation altering the population's performance. Farmers and seed producers benefit from this stability because it allows them to reliably reproduce the same crop year after year, without the need for constant re-selection or re-breeding.
- **Commercial Seed Production**: In commercial seed production, the stability of purelines is crucial for maintaining the **genetic purity** of the seed. When farmers purchase seeds of a pureline variety, they expect all the plants grown from those seeds to be identical and to express the same traits as the parent generation. This consistency in traits is important for large-scale farming operations where uniformity in crop growth and yield is essential for efficient management and harvest.
- **Challenges to Stability**: Despite their genetic stability, purelines can lose their uniformity if they are exposed to external factors such as **cross-pollination**, **mechanical mixing**, or **mutations**. Cross-pollination with other varieties can introduce new genetic material into the pureline, leading to off-type plants in subsequent generations. Mechanical mixing, where seeds from different varieties are inadvertently combined during processing or storage, can also compromise the genetic purity of the pureline. In rare cases, **mutations** may arise spontaneously, introducing genetic variation that can affect the traits of the population.
- **Maintaining Stability**: To maintain the stability of purelines, breeders must take steps to **isolate** pureline populations from other varieties, especially during flowering, to prevent cross-pollination. Similarly, careful handling and storage of seeds are essential to avoid mechanical mixing. Regular **field inspections** and **genetic testing** can help identify and remove any off-type plants or mutations that arise, ensuring that the pureline remains genetically pure and stable.
- **Example in Breeding Programs**: In crops like **wheat** and **barley**, purelines are used to develop stable, uniform varieties that perform consistently across different environments. These purelines are often tested over multiple growing seasons and in different locations to ensure that their performance remains stable under varying environmental conditions. Once a pureline is proven to be stable, it can be released as a commercial variety for farmers to grow.

Progeny Testing:

Progeny testing is a crucial technique in **plant breeding** used to assess the breeding potential of individual plants based on the performance of their offspring. Developed by **Louis de Vilmorin**, this method, also known as the **Vilmorin Isolation Principle**, has been instrumental in advancing plant breeding practices by allowing breeders to make more informed decisions about which plants to select for future breeding cycles. Progeny testing serves two primary functions: determining the **homozygosity or heterozygosity** of plants and assessing the **heritability** of traits. Both functions are essential for improving the accuracy and effectiveness of breeding programs.

1. Determining Homozygosity or Heterozygosity

One of the primary objectives of progeny testing is to determine whether a plant is **homozygous** (genetically uniform) or **heterozygous** (genetically diverse) for specific traits. This information is vital for breeders because the genetic makeup of a plant influences how consistently its traits will be passed on to the next generation.

- » **Homozygosity**: A plant is homozygous for a trait when it carries identical alleles for that particular gene. Homozygous plants are genetically uniform, meaning that they consistently pass the same version of the trait to their offspring. In self-pollinated crops, homozygosity is often achieved through repeated self-pollination over several generations. **Purelines** are an example of homozygous populations, where all individuals are genetically identical, and their traits remain stable over time.
  - **Role in Breeding Programs**: For breeders, identifying homozygous plants is important because these plants will consistently express the desired traits, allowing for more predictable results in future generations. For example, if a plant is homozygous for **disease resistance**, all of its offspring will inherit the resistance trait. This makes homozygous plants valuable for developing stable, uniform crop varieties.
  - **Stability of Traits**: Homozygous plants are favored in breeding programs where **trait stability** and **uniformity** are critical, such as in the production of seeds for commercial agriculture. Once a breeder identifies a homozygous plant with desirable traits, it can be propagated to produce a uniform crop that consistently expresses those traits across different environments and growing conditions.
- » **Heterozygosity**: A plant is heterozygous for a trait when it carries two different alleles for that gene. Heterozygous plants can produce offspring with a wide range of trait expressions because they pass on a mixture of alleles to their progeny. The offspring of heterozygous plants may exhibit considerable variability in traits such as plant height, yield, or disease resistance.
  - **Implications for Breeding**: Heterozygous plants introduce **genetic diversity** into breeding programs, which is beneficial for developing new varieties with improved traits. However, this diversity can also complicate the selection process because the offspring will not be genetically uniform. Breeders must carefully evaluate the progeny of heterozygous plants to identify which individuals express the desired traits.
  - **Example in Cross-Pollinated Crops**: In cross-pollinated crops like maize, breeders often work with heterozygous plants to maintain genetic diversity and introduce new traits. However, to stabilize these traits in future generations, breeders may use progeny testing to identify homozygous individuals that can be used to develop new purelines or hybrid varieties.

2. Assessing Heritability of Traits

Another critical function of progeny testing is determining the **heritability** of traits, which refers to the extent to which a trait is controlled by genetic factors rather than environmental influences. This distinction is important because it allows breeders to focus their selection efforts on traits that can be reliably passed on to future generations.

» **Heritable Traits:** Traits that are determined by the plant's **genotype** are considered heritable. Progeny testing helps breeders evaluate whether the traits for which a plant was selected (such as yield, disease resistance, or drought tolerance) are due to genetics or environmental conditions. By growing the progeny of a plant in different environments, breeders can determine if the trait consistently appears in the offspring, indicating that it is genetically controlled.

- **Importance in Selection:** Knowing the heritability of traits is essential for making effective selection decisions. If a trait is highly heritable, breeders can confidently select plants with that trait, knowing that it will be passed on to future generations. For example, if a plant consistently produces high yields across multiple environments, breeders can conclude that the high yield is genetically controlled and can be selected for further breeding.
- **Quantitative vs. Qualitative Traits:** The heritability of traits can vary depending on whether the trait is **qualitative** (controlled by one or a few major genes) or **quantitative** (controlled by multiple genes). Qualitative traits, such as resistance to a specific disease, tend to have high heritability, making them easier to select. In contrast, quantitative traits, like yield or plant height, may have lower heritability because they are influenced by multiple genes and environmental factors. Progeny testing helps breeders assess the heritability of both qualitative and quantitative traits to guide their selection strategies.

» **Environmental Influence:** Traits that are strongly influenced by the environment are less heritable, meaning that their expression can vary depending on factors such as soil quality, water availability, or temperature. In these cases, selection based solely on the phenotype of the parent plant may not lead to genetic improvements, as the trait may not be passed on to the progeny. Progeny testing allows breeders to distinguish between **genetically determined** traits and those influenced by the environment, ensuring that selection decisions are based on **genetic potential** rather than temporary environmental conditions.

- **Example in Plant Breeding:** For instance, if a breeder selects a plant for its high yield in one environment, progeny testing can reveal whether that high yield is due to genetic factors or to favorable environmental conditions. If the progeny also produce high yields in different environments, the breeder can conclude that the trait is genetically controlled. However, if the progeny exhibit lower yields in other environments, the breeder may determine that the original plant's high yield was due to environmental factors, and selection based on that plant may not lead to genetic improvement.

Application of Progeny Testing in Plant Breeding

Progeny testing is widely used in **self-pollinated** and **cross-pollinated** crops to improve breeding efficiency and ensure the development of high-performing varieties. The method is particularly useful in identifying plants with desirable traits and ensuring that these traits are reliably passed on to future generations.

» **Pureline Selection:** In self-pollinated crops like rice or beans, progeny testing is often used to develop **purelines** with stable, uniform traits. By evaluating the progeny of individual plants over multiple generations, breeders can identify which plants are homozygous for important traits, such as disease resistance or yield. Once these purelines are established, they can be used to create consistent, high-performing crop varieties.

» **Hybrid Breeding**: In cross-pollinated crops like maize, progeny testing helps breeders identify **heterozygous plants** that can be used in hybrid breeding programs. Hybrids often exhibit **heterosis** (hybrid vigor), resulting in improved performance compared to their parent lines. Progeny testing allows breeders to assess the genetic potential of different crosses and select the best parent lines for producing hybrids with desirable traits like high yield, drought tolerance, or pest resistance.

» **Long-Term Genetic Improvement**: Progeny testing plays a critical role in **long-term genetic improvement**. By continuously evaluating the performance of offspring, breeders can track genetic progress and make informed decisions about which plants to retain in the breeding program. This ensures that each generation of plants shows incremental improvements in traits such as yield, quality, and resistance to biotic and abiotic stresses.

B. Pureline Selection

Pureline selection is a widely used and effective method for improving self-pollinated crops. Almost all modern-day varieties of self-pollinated crops are **purelines**, making this method essential for breeding programs aimed at enhancing crop performance. This approach has broad applications in the improvement of various self-pollinated crops by focusing on **homozygous** individuals that exhibit superior traits. Pureline selection works by isolating and propagating the best individual plants within a genetically diverse population, allowing breeders to develop highly uniform and stable varieties.

Pureline Selection and Its Role in Improving Crop Varieties

Pureline selection is an essential method in **plant breeding** that plays a critical role in enhancing the performance of self-pollinated crops. This process allows breeders to isolate and propagate superior plants from genetically diverse populations, leading to more uniform and productive varieties. It is used extensively to improve various types of crop populations, including **local varieties, old pureline varieties**, and **introduced varieties**. Each of these categories presents specific opportunities for improvement through pureline selection.

1. Improvement of Local Varieties

Local varieties, often referred to as **landraces**, have evolved over time in specific geographic regions. These varieties are typically genetically diverse, displaying a wide range of traits due to the natural selection pressures they have experienced in their native environments. While this diversity is valuable, it also means that these varieties may lack uniformity in important traits like **yield**, **maturity**, and **disease resistance**.

» **Isolating Superior Individuals**: Pureline selection helps breeders identify and isolate the **superior individuals** within a genetically diverse landrace population. By selecting plants that exhibit desirable traits such as high yield, resistance to local pests and diseases, and adaptability to specific environmental conditions, breeders can develop **more uniform and productive varieties.** These selected plants, when propagated through self-pollination, give rise to a pureline population that retains the beneficial traits of the parent plants but with greater consistency across generations.

- **Example**: In regions where traditional rice varieties are grown, pureline selection can be used to enhance the performance of local landraces. A breeder may select rice plants with superior **grain quality** or **drought tolerance** from a diverse population, and through several generations of pureline selection, develop a more uniform rice variety that is better suited for commercial cultivation.
- **Advantages for Farmers**: The improvement of local varieties through pureline selection benefits farmers by providing them with varieties that are not only **high-yielding** but also adapted to the specific environmental challenges of their region. This leads to better productivity and increased resilience to local stress factors such as fluctuating weather patterns or endemic diseases.
- **Challenges**: While local varieties often carry valuable traits, their genetic diversity can sometimes mask the full potential of superior individuals. **Pureline selection** helps to isolate these superior plants, but it requires careful evaluation and multiple generations of testing to ensure that the resulting pureline population maintains the desired traits across different environments.

2. Improvement of Old Pureline Varieties

Old pureline varieties are those that were developed in the past through the process of pureline selection but may have lost their genetic vigor over time. These varieties, while once productive and widely grown, may become susceptible to new diseases or environmental challenges as conditions change. Pureline selection provides a way to **rejuvenate** these old varieties by selecting plants that exhibit renewed vigor or improved traits.

- **Rejuvenating Old Varieties**: Over time, old pureline varieties may become less productive due to **genetic drift**, **environmental stress**, or the emergence of new pests and diseases. Pureline selection allows breeders to identify individuals within the old pureline population that show **superior performance** in terms of yield, disease resistance, or adaptability to new conditions. By selecting these individuals and propagating them through self-pollination, breeders can develop a new pureline that retains the best characteristics of the original variety but with improved vigor and resilience.
- **Example**: A classic example is the improvement of old wheat varieties that were once dominant in specific regions but became susceptible to new strains of **rust disease**. By applying pureline selection, breeders can select wheat plants within the old variety that show resistance to the new disease, thus producing a **rejuvenated wheat variety** that retains its desirable characteristics while gaining resistance to current challenges.
- **Maintaining Heritage**: One of the key benefits of improving old pureline varieties is that it allows breeders to maintain the **heritage** and **unique characteristics** of these varieties while enhancing their performance. This is particularly important in crops that are valued for their traditional qualities, such as flavor or appearance, but may need improvement in terms of yield or resistance to stress.

- **Adaptation to Changing Conditions**: In the context of **climate change** and evolving agricultural challenges, old pureline varieties may struggle to perform as well as they once did. Pureline selection provides a method for adapting these varieties to new environmental conditions, ensuring that they continue to provide value to farmers.

3. Improvement of Introduced Varieties

When new crop varieties are introduced from other regions or countries, they often face challenges in adapting to **local environmental conditions**. Pureline selection can be used to adapt these **introduced varieties** by selecting for traits that are better suited to the new growing conditions. This helps breeders create varieties that combine the desirable traits of the introduced variety with the **adaptability** needed for successful cultivation in the new environment.

- **Selecting for Local Adaptation**: Introduced varieties may possess traits that make them highly valuable, such as **high yield potential, quality traits**, or **disease resistance**. However, these varieties may not be well-suited to the specific environmental conditions in the region where they are introduced. Pureline selection allows breeders to select plants from the introduced variety that perform well in the new environment, focusing on traits such as **drought tolerance, pest resistance**, or **cold hardiness**.
- **Example**: Suppose a variety of maize is introduced from a temperate region into a tropical environment. The introduced variety may struggle with the **heat and humidity** of the new environment. Through pureline selection, breeders can select maize plants that show **superior performance** in terms of resistance to tropical pests and tolerance to high temperatures, ultimately developing a maize variety that thrives in the tropical climate while retaining the high yield potential of the original variety.
- **Enhancing Local Performance**: By adapting introduced varieties to local conditions, pureline selection helps farmers take advantage of new genetic material without sacrificing **crop performance**. This is particularly important in areas where **food security** depends on the introduction of new, higher-yielding varieties that can withstand local environmental pressures.
- **Challenges with Introduced Varieties**: One of the challenges with improving introduced varieties is ensuring that the selected traits for local adaptation do not come at the expense of the original desirable traits. For example, a breeder must balance selecting for **disease resistance** with maintaining the **yield potential** of the introduced variety. Pureline selection provides a systematic approach for achieving this balance, allowing breeders to retain the best traits while improving local adaptability.

Procedure of Pureline Selection

Pureline selection is a systematic process that involves identifying and selecting superior individuals from a genetically diverse population to develop uniform and stable crop varieties. This process, spanning several years, focuses on achieving **high performance and consistency** in traits like yield, disease resistance, and grain quality. Here is a comprehensive breakdown of the procedure:

First Year: Initial Selection from an Old Variety or Landrace

The first year marks the **beginning of the pureline selection process**, where breeders work with an old variety or landrace population. Landraces typically show a mix of genetic variations, despite being homozygous in many respects, and offer a **rich source of genetic diversity**. These populations are surveyed for superior traits, and individual plants that exhibit **desirable characteristics** are selected.

- » **Surveying the Population**: The landrace population is visually inspected by breeders to identify genetic variability. Even though these populations are generally self-pollinated and homozygous, they can still display a range of phenotypic traits due to environmental factors or underlying genetic variability. The goal is to identify plants that are **genetically superior** in terms of their ability to produce high yields or resist diseases.
- » **Selection of Individual Plants**: Breeders typically select between **1000 to 2000 individual plants**, depending on the size of the population and the available resources. Selection is based on **visual observation** of key traits:
    - **Plant height**: Uniformity in height is essential for ease of harvesting and managing the crop in the field.
    - **Flowering time**: Plants with consistent and optimal flowering periods are chosen to ensure synchronous pollination and maturity.
    - **Disease resistance**: Plants that show natural resistance to local diseases or pests are particularly valuable, as this reduces the reliance on chemical treatments.
    - **Grain type and quality**: For cereal crops like wheat or rice, grain size, shape, and quality are important indicators of the crop's market potential and nutritional value.
- » **Harvesting Plants Separately**: Each selected plant is **harvested individually** to ensure that their progeny can be evaluated independently in subsequent years. Each plant represents a **potential pureline**, and careful tracking of these plants ensures that only the most promising individuals are selected for further testing.
- » **Focus on Desirable Traits**: During this phase, breeders prioritize traits that are **simply inherited**, meaning they are controlled by a few major genes. These traits are easier to observe and select for in early generations. Traits like plant type, height, and flowering time serve as **early indicators** of a plant's potential for success in the breeding program.

Second Year: Growing and Evaluating Individual Progenies

In the second year, the focus shifts from individual plant selection to the evaluation of their **progenies** (offspring). Each progeny is grown separately, and the goal is to eliminate poor-performing lines while identifying the **top candidates** for further trials.

- » **Growing Progeny Lines Separately**: The seeds harvested from the selected plants are planted as separate **progeny lines**. Each progeny is grown with adequate spacing to allow for independent evaluation. This ensures that each line can be monitored for the traits it was selected for in the first year.
- » **Selection of Top Progenies**: Breeders evaluate the progeny based on the same criteria used

in the first year, focusing on traits like **plant height**, **grain type**, **flowering duration**, and **disease resistance**. Progenies that show poor or inconsistent performance are discarded, and **segregating progenies** (those that show genetic variability) are also eliminated. The top **15-20 progenies** are selected for further testing.

» **Bulking Selected Progenies**: The best-performing progenies are **bulked together** to create a more uniform population. This process helps ensure that only the most promising lines are carried forward, while maintaining genetic uniformity within the selected lines.

Third Year: Unreplicated Yield Trials

Yield is a critical trait for any new crop variety, and the third year focuses on evaluating the **yield potential** of the selected progenies. However, the limited amount of seed produced in the second year often prevents the use of replicated trials, so breeders conduct **unreplicated trials** during this phase.

» **Evaluating Yield Performance**: Each progeny is grown alongside a **check variety** (a standard variety used as a control) to evaluate yield potential. Yield is the primary criterion for selection, although other traits like **disease resistance** and **growth habits** are also considered. Only progenies that show **significant yield improvements** over the check variety are selected for further testing.

» **Focus on Superior Progenies**: The progenies that perform well in terms of yield and other agronomic traits are selected for the next stage of testing. At this stage, breeders aim to identify lines that consistently **outperform the check variety** in terms of yield, quality, and disease resistance.

Fourth Year: Replicated Yield Trials

In the fourth year, the selected progenies undergo **replicated yield trials** to further evaluate their performance. Replicated trials are critical because they allow breeders to **account for environmental variability** and ensure that the selected progenies consistently perform well across different conditions.

» **Replicated Trials for Accuracy**: The progenies are grown in multiple **replications** to provide a more accurate assessment of their performance. Replicated trials help breeders determine whether the superior performance observed in earlier trials is consistent across different environments and growing conditions. The trials are typically conducted in **randomized block designs**, which help minimize the effects of environmental variation.

» **Comparison with Check Variety**: Each progeny is grown alongside the best available **check variety** to ensure that the selected lines outperform standard commercial varieties. Breeders assess not only yield but also other important traits such as **maturity**, **disease resistance**, and **grain quality**. The progenies that consistently outperform the check variety in all important traits are selected for the next stage of testing.

» **Recording Observations**: Detailed data on **yield**, **disease resistance**, **maturity**, **grain quality**, and other important traits are recorded during this phase. These observations provide breeders with a clear understanding of how each progeny performs under different environmental conditions and whether they meet the required standards for commercial cultivation.

Fifth to Eighth Year: Multi-Location Trials and Variety Release

The final stage of pureline selection involves conducting **multi-location trials** to assess the adaptability of the selected progenies to different environments. This is a crucial step in determining the suitability of the new variety for large-scale cultivation.

» **Multi-Location Trials**: The top-performing progenies are tested in **several locations**, often across different regions with varying climates, soil types, and farming systems. Multi-location trials help breeders assess the **stability** and **adaptability** of the new variety. A variety that performs consistently well across different environments is more likely to succeed in commercial cultivation.

» **Assessing Stability and Adaptability**: During the multi-location trials, breeders focus on evaluating the **stability** of key traits like yield, disease resistance, and maturity across different locations. They also assess how well the progenies adapt to **specific environmental challenges** such as drought, high temperatures, or varying soil fertility. Progenies that perform consistently across diverse environments are considered for variety release.

» **Variety Release and Seed Multiplication**: If a progeny consistently performs well across all locations, it is released as a **new variety**. The newly released variety is then entered into **seed multiplication programs** to produce enough seed for distribution to farmers. Seed multiplication ensures that farmers receive **high-quality, certified seeds** of the new variety, allowing them to benefit from improvements in yield, disease resistance, and adaptability.

Advantages of Pureline Selection

Pureline selection is one of the most effective breeding methods for self-pollinated crops and offers several key advantages that contribute to agricultural productivity and crop improvement. Here's a detailed explanation of these benefits:

1. Maximum Improvement Over Original Variety

One of the primary benefits of pureline selection is that it leads to **maximum possible genetic improvement** from the original population. By isolating the **best-performing genotypes**, breeders can enhance important agronomic traits such as **yield, disease resistance**, and **grain quality**. This ensures that only the most desirable traits are passed on to subsequent generations.

» **Genetic Selection**: Pureline selection identifies the most **superior individuals** from a genetically diverse population, often leading to significant advances over traditional or local varieties. For example, a pureline variety might show increased tolerance to drought or resistance to a local pest, making it far more productive and resilient than the original variety.

» **Improved Crop Performance**: Once the best individuals are identified, they are propagated through self-pollination, leading to a uniform population where each plant exhibits the improved traits. This means that pureline selection **enhances the overall performance** of a crop, allowing for increased **yield stability, better quality produce**, and improved **adaptation to local conditions**.

» **Practical Applications**: This method has been successfully used to improve crops such as wheat, barley, and rice, where pureline selection has led to varieties that are far more productive and resistant to diseases compared to older varieties. For instance, modern wheat varieties developed through pureline selection often show superior performance in both yield and quality, making them indispensable in agricultural production.

2. Uniformity

One of the standout features of pureline varieties is their **genetic uniformity**. This uniformity is a significant advantage in modern agriculture, especially for **self-pollinated crops**, where consistency and reliability are key for large-scale production.

» **Consistent Performance**: Pureline varieties are **genetically homogeneous**, meaning that all plants in the population have the same genetic makeup. As a result, they exhibit **consistent traits**, such as uniform growth, flowering, maturity, and yield. This is crucial for farmers because it allows for **predictable performance** in the field, making crop management more efficient.

» **Ease of Management**: Crops with uniform traits are **easier to manage**, particularly when it comes to tasks such as irrigation, fertilization, pest control, and harvesting. Uniformity means that all plants will mature at the same time, making it easier to plan and execute harvesting activities with minimal variability.

» **Consumer Preferences**: Uniformity is also important from a market perspective, as consumers often prefer produce that is **consistent in size, color, and quality**. For example, in crops like rice or wheat, uniform grain size and quality make the processing and marketing of the crop much easier, leading to higher market acceptance and premium prices.

3. Ease of Identification in Seed Certification

Due to their uniformity, pureline varieties are **easily identifiable** during seed certification programs. Seed certification is essential for ensuring that farmers receive **high-quality seeds** that are true-to-type and free from contaminants.

» **Certifying Genetic Purity**: Because pureline varieties are genetically uniform, they are **easy to certify** for genetic purity. Certification ensures that the seeds sold to farmers are of the highest quality, reducing the risk of **seed contamination** or **genetic variability** that could lead to inconsistent performance in the field.

» **Farmer Confidence**: Certified seeds give farmers confidence that they are purchasing a **reliable product**. In crops like wheat, barley, or rice, pureline varieties offer the assurance that the seeds will produce uniform crops with the expected yield and quality traits, thereby reducing the risk of crop failure.

» **Streamlined Certification Process**: The genetic homogeneity of pureline varieties also streamlines the seed certification process, making it easier for regulatory bodies to confirm that the seeds meet the required standards for **quality and performance**.

4. Stable Performance Across Generations

Another major advantage of pureline selection is the **genetic stability** of the resulting varieties. Once a pureline variety is established, it remains **genetically stable** across multiple generations, meaning that it will continue to exhibit the same traits year after year.

- **Reliability for Farmers:** The stability of pureline varieties is particularly beneficial for farmers who rely on **consistent performance** from season to season. Once a variety is established, its traits, such as **yield, disease resistance**, and **quality**, remain stable, allowing farmers to plan their agricultural practices with greater certainty.
- **Seed Saving:** Pureline varieties are often **self-pollinated** crops, which means that farmers can save seeds from one season to the next without losing the genetic traits that make the variety desirable. This helps reduce seed costs and provides a level of independence for farmers who rely on **seed-saving practices**.
- **Minimal Genetic Drift:** In pureline varieties, there is **minimal risk of genetic drift**, as the population is highly homozygous. This ensures that the traits breeders have worked to establish remain fixed and do not degrade over time, making these varieties highly reliable for long-term use.

Disadvantages of Pureline Selection

While pureline selection has numerous advantages, it also has some **notable limitations** that breeders and farmers must consider.

1. No Creation of New Genotypes

A key limitation of pureline selection is that it **does not create new genetic combinations.** The process is based on selecting the best individuals from an existing population, which means that it only **isolates superior genotypes** rather than creating new ones.

- **Lack of Genetic Diversity:** Because pureline selection does not involve cross-pollination or recombination, the resulting varieties lack **genetic diversity.** This can be a disadvantage when breeders are trying to combine multiple traits, such as yield potential, disease resistance, and drought tolerance, into a single variety. Other breeding methods, such as **hybrid breeding,** are more effective for creating **novel genetic combinations** that introduce new traits into a population.
- **Limited Potential for Adaptation:** In situations where new pests, diseases, or environmental stresses arise, the lack of genetic diversity in pureline varieties can limit their ability to adapt. Without the ability to create new genotypes, pureline varieties may struggle to cope with changing environmental conditions, necessitating the use of other breeding methods to introduce **new genetic traits.**

2. Narrow Genetic Base

One of the biggest challenges of pureline selection is the **narrow genetic base** of the resulting varieties. While genetic uniformity is advantageous for ensuring stable performance, it also makes pureline varieties more vulnerable to **environmental changes** and **emerging threats.**

- **Increased Vulnerability to Diseases and Pests:** A narrow genetic base can make pureline varieties more susceptible to **widespread disease outbreaks** or the emergence of new pests. Since all the plants in the population are genetically similar, a pathogen or pest that can overcome the resistance of one plant is likely to affect the entire population, leading to significant crop losses.

» **Limited Adaptability**: Pureline varieties may struggle to adapt to **changing environmental conditions**, such as shifts in climate, soil fertility, or water availability. Their narrow genetic base limits their ability to develop new traits that could help them cope with these challenges, making them less resilient than cross-pollinated varieties with more genetic diversity.

3. Time-Consuming and Costly

Pureline selection is a **time-intensive** and **resource-demanding** process. Developing a pureline variety can take several years of selection, testing, and field trials before a variety is ready for release.

» **Lengthy Selection Process**: Pureline selection requires multiple generations of **self-pollination and evaluation**, which can take **5-10 years** or more before a stable, high-performing pureline variety is developed. This extended timeline can be a significant drawback when breeders need to respond quickly to new challenges, such as emerging diseases or changing climate conditions.

» **High Costs of Field Trials**: The process of evaluating and testing pureline varieties requires **extensive field trials**, often conducted across multiple locations to assess adaptability. These trials are costly, requiring significant investments in land, labor, and resources.

4. Limited Genetic Variation

The effectiveness of pureline selection depends on the amount of **genetic variation** present in the original population. If the population has **low genetic diversity**, the potential for improvement is limited, as there may not be enough variation in traits like yield, quality, or disease resistance.

» **Limited Improvement**: In populations with little genetic variation, pureline selection may yield only **marginal improvements**, as there is not enough diversity to select from. This can limit the overall effectiveness of the process, especially when breeders are looking to make **significant genetic gains** in traits like yield or stress tolerance.

» **Dependency on Original Population**: Pureline selection is **dependent on the existing genetic variation** within the original population, which means that if the population is genetically uniform, there may be little scope for improvement. In such cases, breeders may need to introduce new genetic material through hybridization or other methods to create more variation before proceeding with pureline selection.

Pureline selection remains one of the most effective methods for improving **self-pollinated crops**, offering a pathway to develop **uniform, stable varieties** that perform consistently across generations. While it has limitations, such as its inability to create new genotypes and the time required for selection, the method continues to be widely used in breeding programs aimed at enhancing yield, disease resistance, and other important traits. By carefully selecting and evaluating superior plants over multiple generations, breeders can develop pureline varieties that meet the needs of modern agriculture and ensure food security for the future.

B. Mass Selection: A Detailed Exploration

Mass selection is one of the oldest and simplest methods of plant breeding, often used in improving both self-pollinated and cross-pollinated crops. In this method, a large number of plants exhibiting similar phenotypic traits are selected from a genetically diverse population. The seeds from these selected plants are then **bulked together** to form a new population or variety. While this method helps in developing varieties with broader adaptability, the resulting plants remain genetically diverse. Unlike pureline selection, which results in uniform genotypes, mass selection works by enhancing **phenotypic uniformity** without significantly altering the underlying genetic diversity.

Steps Involved in Mass Selection

The mass selection process spans multiple years, with each year focused on refining the selected population to develop a variety that performs well under different conditions. Below is a breakdown of the key steps involved in mass selection:

First Year: Initial Selection Based on Phenotype

In the first year of the mass selection process, breeders focus on identifying individual plants that display **desirable phenotypic traits** from a genetically diverse population. This is a crucial step because it lays the foundation for developing a new variety. The selection of plants is based purely on visual inspection of their **external characteristics**—a method that emphasizes physical traits rather than genetic composition at this stage. Mass selection is especially useful in traditional landraces or local varieties, where a broad genetic base exists, offering numerous candidates for selection.

Selection of Phenotypically Similar Plants

The **primary focus** during this stage is on selecting individual plants that **look promising** based on their observable characteristics, also known as the **phenotype**. This is important because certain **visual traits**, such as plant height, flowering time, and grain quality, can provide early indications of a plant's performance.

- **Phenotypic Inspection**: Breeders **visually inspect** a large base population, which could be sourced from **landraces, local varieties**, or even a **diverse breeding population**. The goal is to identify plants that display the most promising traits for further development. This is done with an understanding that these selected plants will later be bulked together for evaluation.
- **Number of Plants Selected**: The number of plants chosen for this initial selection can vary depending on factors such as the size of the population, available resources, and the specific breeding objectives. In most cases, breeders select between **200 to 2000 plants**. A larger selection pool increases the chances of isolating **genetically superior individuals** for further testing.
- **Visual Criteria for Selection**: At this stage, selection is based solely on external characteristics without analyzing the plant's genetic makeup. Key phenotypic traits used for selection typically include:
    - **Plant Height**: Uniformity in plant height is a desirable trait, especially in crops like wheat or maize, where uniformity can simplify **harvest operations** and ensure **consistent yield**.
    - **Flowering Time**: Synchronous flowering is crucial for maintaining uniformity in **crop maturity**. Selecting plants that flower at the same time ensures that the population matures uniformly, which is advantageous for **harvesting** and **post-harvest processing**.

- **Disease Resistance**: Natural **resistance to local pests and diseases** is a highly valuable trait. Selecting plants that display resilience against common diseases reduces the need for **chemical interventions**, ultimately lowering the cost of production and ensuring healthier crops.
- **Grain Quality**: For grain crops like rice, wheat, or barley, traits related to **grain size, shape**, and **nutritional content** are critical for determining **market potential**. Crops with high-quality grains often command a premium price in the market, making this an important selection criterion.

» **Importance of Visual Selection**: It is important to note that **visual selection** can be subjective and influenced by the environment. However, it allows breeders to quickly assess a large population and select plants that show potential for further evaluation. The selected plants are not yet guaranteed to perform well, but they exhibit traits that merit additional testing.

Bulk Harvesting of Selected Plants

After selecting the **phenotypically similar plants**, the next step is to harvest these plants as a **bulk**. Bulk harvesting involves combining the seeds from all the selected plants into a single collection. This **bulk seed** will form the basis of the population that will be evaluated in subsequent generations.

» **Why Bulk Harvesting?**: Since the initial selection is based on phenotype rather than genotype, bulk harvesting ensures that a diverse genetic base is retained in the selected population. This diversity is crucial for future **selection rounds**, as it provides a pool of genetic variability from which further improvement can be made.

» **Diverse Genetic Composition**: Even though the selected plants look similar based on their external traits, they may still vary genetically. By mixing the seeds from these selected plants, breeders ensure that they maintain a wide range of genetic diversity within the bulked population. This diversity is important for later stages when the plants will be subjected to more rigorous testing.

» **Bulk Seed as the Basis for Next Generation**: The bulk seed collected from the selected plants serves as the starting point for the **next generation**. In the second year, this seed will be used to grow and evaluate individual progenies for yield, disease resistance, and other agronomic traits.

» **Advantages of Bulk Harvesting**:

- It allows breeders to maintain **genetic diversity**, which is important for improving complex traits such as **disease resistance** or **drought tolerance**.
- It simplifies the selection process by ensuring that all selected plants contribute to the next generation, rather than focusing on individual plants too early in the breeding process.
- It creates a bu**ffer against environmental variability, as the bulked population is more l**ikely to contain individuals that can thrive under different environmental conditions.

Traits Used for Selection

Each phenotypic trait selected in the first year serves a specific purpose in ensuring that the **new**

variety will perform well in terms of **yield**, **quality**, and **resistance to stresses**. Below is a deeper exploration of the importance of these traits:

» **Plant Height**: In many crops, **plant height** directly affects yield and ease of harvest. For instance, shorter plants are less prone to **lodging** (falling over due to wind or rain), which can significantly impact yield. Conversely, very tall plants may require more **nutritional resources**, which can detract from grain development. Selecting for optimal height ensures the plants are robust yet easy to harvest.

» **Flowering Time**: **Synchronous flowering** helps ensure uniform maturation, which is important for managing **irrigation**, **fertilizer application**, and **harvesting schedules**. Plants that flower and mature uniformly allow for more efficient farming practices, reducing labor costs and increasing the ease of mechanized harvesting.

» **Disease Resistance**: **Disease resistance** is a critical factor in reducing the reliance on **pesticides** and other chemicals. Breeders prioritize selecting plants that naturally resist local pests or pathogens, such as **rusts in wheat**, **blight in potatoes**, or **bacterial wilt in rice**. By selecting for natural resistance, breeders help farmers reduce input costs and environmental impact while ensuring stable crop performance.

» **Grain Quality**: The **size**, **shape**, and **nutritional content** of the grain are important for **consumer acceptance** and **market price**. For instance, in rice, consumers often prefer long, slender grains, while in wheat, the **protein content** is important for determining its use in **bread making** or **pasta production**. Selecting for these traits early in the process ensures that the new variety will have market appeal.

Second Year: Preliminary Yield Trials in Mass Selection

The second year of the mass selection process is a critical stage where the selected plants from the first year undergo more rigorous testing for **yield potential** and **uniformity**. This phase is designed to **eliminate plants** that do not meet specific criteria while identifying the most **productive** and **uniform** individuals for further evaluation. By conducting **preliminary yield trials**, breeders can refine the selected population, ensuring that only the top-performing plants advance to subsequent generations.

Growing in Yield Trials

After the initial selection in the first year, the **bulk seeds** collected from the selected plants are divided into **smaller lots** and planted for further evaluation. These plants are grown in **preliminary yield trials** to assess their performance in terms of **yield**, **uniformity**, and other agronomic traits. The primary goal at this stage is to identify plants that perform well under real-world growing conditions and **eliminate those that do not**.

» **Yield Trials**: Preliminary yield trials are typically conducted by growing the selected plants in **experimental plots**. These trials often include a **control variety** or **check** that serves as a benchmark for comparison. The control variety is usually a well-known, high-performing variety that provides a **standard** against which the selected plants are measured.

» **Comparison with Check Variety**: The selected plants are compared to the check variety to determine whether they show **superior performance**. The control variety serves as a baseline for evaluating yield, disease resistance, plant height, and other agronomic traits. If the selected plants outperform the control in terms of yield and uniformity, they are considered for further evaluation.

- **Plot Layout**: The layout of the yield trials is designed to minimize **environmental variability** and ensure accurate comparisons. Typically, plots are randomized to reduce the effects of environmental factors such as **soil fertility** or **moisture levels**. This ensures that the observed differences in plant performance are primarily due to genetic factors, not environmental influences.

Elimination of Dissimilar Phenotypes

One of the primary objectives in the second year is to **eliminate plants** that do not conform to the desired phenotype. This step is crucial for ensuring that the final population is as **uniform** as possible, both in terms of yield and other important traits. The elimination process is based on **visual observation** and field data collected during the yield trials.

- **Identification of Inferior Plants**: Breeders carefully monitor the **growth and development** of each plant during the trial period. Plants that deviate from the desired phenotype are **rejected** at this stage. For example, if a plant is **too tall or too short, flowers too early or too late**, or produces **inferior grain quality**, it is removed from the population.
- **Phenotypic Variability**: Plants that exhibit **phenotypic variability** (e.g., differences in height, flowering time, or disease resistance) are also eliminated. This is important because **mass selection** aims to develop a **uniform population** where all plants display similar traits. Eliminating dissimilar phenotypes helps ensure that the resulting variety will be more consistent in its performance.
- **Uniformity as a Key Criterion**: In addition to yield, **uniformity** in plant characteristics such as height, flowering time, and grain quality is critical for the success of a new variety. A **uniform population** is easier to manage, harvest, and process, making it more attractive to farmers and producers. Plants that do not meet the desired criteria for uniformity are discarded during this phase.

Selection of High-Yielding Plants

Once the dissimilar phenotypes have been eliminated, breeders focus on selecting the **highest-yielding plants** from the remaining population. Yield is the most important trait in most breeding programs, as it directly impacts the economic value of the crop. However, **yield** is not the only criterion—other traits like **disease resistance**, **drought tolerance**, and **grain quality** are also considered.

- **Yield as the Primary Criterion**: The primary goal of mass selection is to develop a variety that produces **high and stable yields**. Plots that demonstrate the **highest yield potential** during the preliminary trials are selected for further testing. These high-yielding plants are considered the most promising candidates for advancing to the next stage of the breeding process.
- **Uniformity in Other Traits**: In addition to yield, breeders evaluate the selected plants for **uniformity** in other key traits such as **plant height**, **flowering time**, and **grain size**. Plants

that are uniform in these traits are more likely to perform consistently across different environments, making them more desirable for **commercial production**.

- **Refinement of the Bulked Population**: As the second year progresses, the bulked population becomes more refined, with only the top-performing plants advancing to the next stage. This process ensures that the resulting variety is both high-yielding and uniform in its growth and development. By the end of the second year, breeders have a more **refined population** that is ready for more extensive testing in subsequent years.

Key Considerations During the Second Year

The second year of mass selection is critical for **narrowing down** the selected population and identifying the most promising candidates for further development. Several important considerations must be kept in mind during this phase:

- **Environmental Variability**: Yield trials are often subject to **environmental influences**, such as soil fertility, moisture availability, and temperature fluctuations. Breeders must account for this variability when selecting plants to ensure that their choices are based on **genetic factors**, not environmental conditions.
- **Replication of Trials**: While the second year may not always involve **replicated trials** due to limited seed availability, replication becomes important in later years to ensure the **accuracy and reliability** of the results. Replication helps account for environmental variability and provides a more accurate assessment of each plant's performance.
- **Selection Pressure**: The **intensity of selection** during the second year is higher than in the first year, as the goal is to refine the population and eliminate plants that do not meet the desired criteria. This increased selection pressure helps ensure that only the **best plants** advance to the next stage of evaluation.
- **Data Collection**: Detailed data on **yield, plant height, flowering time**, and other traits must be collected and analyzed to ensure that the selected plants meet the breeding objectives. This data is used to make informed decisions about which plants should be retained for further testing.

Third to Sixth Year: Coordinated Yield Trials at Multiple Locations

In the third year of mass selection, the focus shifts from preliminary yield evaluation to **rigorous testing across multiple locations**. This phase is critical for determining the **stability, adaptability**, and **overall performance** of the selected variety under varying environmental conditions. By conducting yield trials at multiple locations, breeders can assess whether the variety is capable of thriving in different soil types, climates, and farming systems. This step ensures that the selected variety is not only high-yielding but also **resilient** to environmental changes and other stress factors. The process typically spans three to four years, culminating in the development of a new, well-adapted variety.

Initial Evaluation Trials (IET)

The first step in this phase, typically carried out in the third year, is the **Initial Evaluation Trials (IET)**. This is a **critical testing phase**, where the selected variety is tested for its **yield potential** and other agronomic traits against a control variety. The primary goal is to evaluate whether the selected variety exhibits superior performance compared to the control under the given environmental conditions.

» **Duration and Setup**: IETs are conducted for one year, during which the bulked population from the previous year's trials is grown alongside a **control variety**. The trials are usually carried out in different fields or experimental stations, where the environmental conditions can vary slightly. The goal is to assess whether the selected variety performs consistently better than the control variety in terms of **yield, disease resistance**, and **quality** traits.

» **Control Variety as Benchmark**: The **control variety** serves as a benchmark for comparison. It is typically a well-established, high-performing variety that farmers already use. If the selected variety outperforms the control in terms of **yield** or other desirable traits, it is considered for further evaluation in the following stages of testing.

» **Criteria for Advancement**: The **criteria for advancement** to the next phase (Main Yield Trials) include higher yield, improved disease resistance, better plant structure, and overall agronomic performance compared to the control variety. If the variety performs well in the IET, it is promoted to the next phase, where more rigorous testing is done.

Main Yield Trials (MYT)

Once the selected variety passes the Initial Evaluation Trials, it is promoted to **Main Yield Trials (MYT)**. These trials, conducted over **2-3 years**, are essential for testing the variety's performance in a wide range of environmental conditions. The goal of MYT is to determine whether the variety is stable and adaptable to **diverse environments**, including varying **soil types**, **climate zones**, and **farming practices**.

» **Multi-Location Testing**: The variety is grown at multiple locations across different **agro-climatic zones**. This could include testing in areas with different **rainfall patterns, altitude variations**, and **temperature ranges**. Multi-location testing helps breeders assess whether the selected variety can perform well across a wide spectrum of conditions.

» **Importance of Testing in Diverse Environments**: The ability to perform consistently across different environments is a crucial trait for any new variety. By evaluating the variety in diverse locations, breeders can identify **environmental limitations** or specific **adaptive traits** that make the variety suitable for certain regions or cropping systems. For example, a variety that performs well in both **drought-prone** and **high-rainfall** areas would be considered highly adaptable.

» **Evaluation Criteria in MYT**:

- **Yield Stability**: Consistent yield performance is one of the primary criteria in MYT. Varieties that show high **yield stability** across various locations are considered more resilient to environmental fluctuations.

- **Resistance to Pests and Diseases**: Another key evaluation point is the variety's ability to resist pests and diseases. This is crucial because pests and pathogens can vary across regions, and a successful variety must offer resistance to the common pests and diseases in the target locations.

- **Agronomic Traits**: Traits such as **plant height, flowering time**, and **grain quality** are also evaluated. These characteristics must remain stable across locations for the variety to be considered for commercialization.
- **Quality Consistency**: For crops grown for **food production** or **commercial purposes**, the consistency of quality traits such as **grain size, nutritional content**, and **processing suitability** are vital.

Evaluation of Multiple Traits

Throughout the **Main Yield Trials**, breeders closely monitor the performance of the selected variety, recording detailed observations on multiple traits. This phase is critical for ensuring that the variety not only yields well but also exhibits **desirable agronomic and quality traits**. The evaluation of multiple traits during MYT ensures that the final product is a well-rounded variety that meets the needs of farmers and consumers alike.

» **Detailed Trait Observation**: Breeders focus on collecting detailed data on important traits such as:

- **Yield**: As the primary trait under evaluation, yield data is collected at each location to determine whether the variety consistently produces high yields across different environments.
- **Flowering Time and Maturity**: The **flowering time** and **maturity period** of the variety are crucial for ensuring uniform crop development. Varieties that show consistent flowering and maturity times across locations are highly valued, as this trait affects harvesting schedules and management practices.
- **Resistance to Pests and Diseases**: Varieties that exhibit **strong resistance** to common pests and diseases in different regions are given preference. For example, a whcat variety resistant to **rusts** in both **humid** and **dry** climates would be considered superior.
- **Plant Health and Vigor**: Healthy, vigorous plants that exhibit strong growth and resistance to environmental stressors (such as drought or nutrient deficiency) are more likely to succeed in diverse farming systems. Observations on **plant architecture, leaf health**, and **root development** are also recorded.
- **Grain Quality and Market Traits**: For crops grown for food or industrial purposes, traits such as **grain size, shape, color**, and **nutritional content** are important for determining **market potential**. In the case of crops like rice, maize, or wheat, the variety must meet **market standards** for quality to be accepted by farmers and consumers.

Multi-Location Trials and Evaluation Over Time

The **third to sixth year** of mass selection is an intensive period of **evaluation**. The primary goal is to identify a variety that not only performs well in controlled conditions but also demonstrates adaptability and resilience to **real-world agricultural challenges**. The multi-location trials allow breeders to identify any **potential weaknesses** in the variety's performance and address them through further refinement, if necessary.

- **Long-Term Data Collection**: By conducting yield trials over multiple years, breeders can collect **long-term data** on the variety's performance. This ensures that the variety is not only high-yielding in the short term but also maintains its performance over time, making it a reliable choice for farmers.
- **Environmental Variability**: Evaluating a variety over a range of **environmental conditions** (such as **dry spells, high rainfall seasons**, or **disease outbreaks**) ensures that the selected variety is robust enough to handle these challenges. Varieties that can maintain their performance under **adverse conditions** are highly prized.
- **Confirmation of Superior Performance**: Only the most promising varieties that consistently outperform the control variety across multiple locations and years are selected for the final phase of development. These varieties are then considered for **commercial release**.

Seventh Year: Variety Multiplication and Release

By the seventh year, the mass selection process reaches its final and most crucial phase—**variety multiplication and release**. This stage is critical for ensuring that the new variety, which has been rigorously tested over the previous years, is now ready for **large-scale cultivation** and distribution to farmers. This phase not only marks the culmination of years of breeding efforts but also signifies the point at which the variety becomes commercially available. Here is a detailed explanation of the process:

Final Yield Assessments

Before a variety is released, it undergoes **final yield assessments** to confirm its **consistent performance** under different environmental conditions. This ensures that the variety maintains its **superiority** in terms of yield, uniformity, and resistance to stress factors, including **diseases, pests, drought**, or **extreme temperatures**.

- **Consistency Across Locations**: The final assessments are aimed at verifying the variety's ability to perform consistently across the locations where it was tested. This includes examining **yield stability, uniformity in growth**, and **disease resistance** over multiple seasons.
- **Verification of Traits**: In addition to yield, other key traits like **flowering time, maturity duration, grain quality**, and **plant health** are examined to ensure that the variety meets the desired criteria set out at the beginning of the breeding program.
- **End-Use Quality Testing**: Depending on the crop, additional **quality tests** may be performed to verify the variety's suitability for **processing** or **consumption**. For example, wheat varieties may be tested for their **baking quality**, while rice varieties might undergo **cooking tests** to ensure the **grain texture** and **flavor** meet consumer expectations.

If the variety passes these **final assessments** and continues to demonstrate **superior performance** in comparison to existing varieties, it is then approved for **release**.

Seed Multiplication

Once the variety has been approved for release, the next step is **large-scale seed multiplication**. This is a critical phase in ensuring that there is sufficient high-quality seed available for farmers to plant on a commercial scale. Seed multiplication ensures that the **genetic purity** and **quality** of the variety remain intact.

- **Ensuring Genetic Purity**: During seed multiplication, strict measures are taken to ensure that the seeds produced are **genetically pure.** This includes maintaining **isolation** from other varieties to prevent cross-pollination and **constant monitoring** of seed production fields to eliminate any off-type plants.
- **Breeder, Foundation, and Certified Seed**: The process of seed multiplication often follows a three-tier system:
    1. **Breeder Seed**: The highest quality seed, directly obtained from the original variety, is multiplied first.
    2. **Foundation Seed**: Breeder seed is then multiplied to produce foundation seed, which is further used for large-scale multiplication.
    3. **Certified Seed**: Foundation seed is multiplied to produce **certified seed**, which is the seed ultimately distributed to farmers. Certified seed must meet **strict quality standards** in terms of purity, germination rate, and health.
- **Seed Quality Monitoring**: Continuous **quality control** during seed multiplication is essential. This includes monitoring for **disease-free** seeds, ensuring high **germination rates**, and maintaining **vigorous growth** potential. Testing for genetic purity ensures that the variety remains true to type and retains the traits selected during the breeding process.
- **Distribution to Farmers**: Once enough **certified seed** has been produced, it is distributed to farmers. At this stage, farmers can begin large-scale cultivation of the new variety, benefiting from its improved traits such as **higher yield**, **disease resistance**, or **drought tolerance**.

Naming the Variety

An important step in releasing a new variety is **naming it.** The name often reflects the variety's **unique characteristics**, **region of origin**, or **key traits** that distinguish it from other varieties. This name serves as a form of **identity** for the variety, making it easily recognizable by farmers, seed producers, and agricultural researchers.

- **Significance of Naming**: A well-chosen name helps in **marketing** the variety and **communicating** its advantages to farmers. For example, a variety with **high drought tolerance** might be named to reflect its resilience in dry climates, while a variety bred for **disease resistance** might have a name that reflects its toughness against pests and pathogens.
- **Regulatory Approval**: In many countries, the **release and naming** of a new variety require approval from **agricultural authorities** or **seed certification agencies**. These regulatory bodies ensure that the variety has met all the necessary standards before it is made available to farmers.
- **Farmer Awareness**: After the variety is named, efforts are made to **educate farmers** about the variety's benefits, how to cultivate it, and its **suitability** for specific climates or farming systems. This includes **demonstration plots**, **field days**, and **educational materials** that provide practical insights into how to maximize yields with the new variety.

Schematic Representation of Mass Selection

To summarize, the process of **mass selection** from the initial phase to the release of a new variety can be represented in the following steps:

1. **First Year**:
   - Selection of **200-2000 phenotypically similar plants** from a genetically diverse base population.
   - Bulk harvesting of selected plants, ensuring **genetic diversity** in the bulked population.
2. **Second Year:**
   - Division of **bulk seed** into smaller lots and planting them in **preliminary yield trials**.
   - Elimination of **dissimilar phenotypes** and selection of **high-yielding, uniform plants** for further evaluation.
3. **Third to Sixth Year:**
   - Evaluation of the selected variety in **Initial Evaluation Trials (IET)** and **Main Yield Trials (MYT)** at multiple locations.
   - Focus on **yield stability, disease resistance**, and **performance in diverse environmental conditions**.
4. **Seventh Year:**
   - If the variety proves superior, it is **multiplied** and released as a **new variety** with a suitable name.
   - **Seed multiplication** ensures that farmers have access to **genetically pure, high-quality seeds**.
   - The variety is named to reflect its unique traits, and efforts are made to **promote** the new variety among farmers.

Merits of Mass Selection

Mass selection offers several advantages, making it a valuable tool for crop improvement, especially in self-pollinated and cross-pollinated crops.

1. **Wider Adaptability**
   - Since mass selection involves selecting a **large number of plants**, the resulting variety has a broader genetic base compared to pureline selection. This **wider genetic diversity** allows the variety to adapt to a range of environments and farming practices, making it more resilient to changing environmental conditions.
   - **Practical Example**: In regions with variable weather patterns or soil conditions, va**rieties developed through mass selection often outperform pureline variet**ies due to their ability to adapt to different stress factors.

2. **Cost and Time Efficiency**
   - **Extensive yield trials** are often not required in mass selection, which reduces the **time and cost** associated with developing a new variety. While pureline selection requires years of intensive testing and evaluation, mass selection can achieve genetic improvement with fewer resources.
   - This makes it an attractive method for breeders working with **limited budgets** or in regions where agricultural infrastructure is less developed.
3. **Maintenance of Genetic Variability**
   - Mass selection retains a considerable amount of **genetic variability** in the new variety. This is beneficial because another round of mass selection can be conducted after a few years to further improve the variety, ensuring continuous crop improvement.
   - **Example**: In regions where environmental conditions are rapidly changing, the retained genetic variability allows for greater **long-term adaptability** and resilience.
4. **Less Demanding for Breeders**
   - Mass selection is **less demanding** than other methods like pureline selection or hybrid breeding. It requires less technical expertise and fewer resources, allowing breeders to **focus on other breeding programs** simultaneously.

Demerits of Mass Selection

Despite its advantages, the mass selection also has some drawbacks, particularly when compared to pureline selection or other more precise breeding methods.

1. **Lack of Uniformity**
   - Varieties developed through mass selection are **not as uniform** as those developed through pureline selection. This lack of uniformity makes these varieties less preferred by both farmers and consumers who desire consistent traits like **grain quality, maturity**, and **growth habits**.
   - This variability can also make crop management more difficult, as plants may not mature or yield uniformly, leading to challenges in harvesting and processing.
2. **Less Improvement Compared to Pureline Selection**
   - The improvement achieved through mass selection is generally **less significant** than that obtained through pureline selection. In mass selection, some progenies may perform **worse than others**, leading to an overall lower level of genetic improvement.
   - While pureli**ne selection isolates the absolute best-performing plants, mass selectio**n may allow **inferior progenies** to remain in the population.
3. **Challenges in Identifying Homozygosity**
   - In the absence of **progeny testing**, it can be difficult to determine whether the selected plants are **homozygous or heterozygous**. In self-pollinated crops, there is a chance of some degree of **cross-pollination**, which may introduce heterozygosity into the selected

population. This can result in **genetic instability** in subsequent generations.

4. **Difficulty in Seed Certification**
   - Varieties developed through mass selection can be harder to identify in **seed certification programs** compared to pureline varieties. The **genetic diversity** within mass-selected populations can make it challenging to ensure that seeds maintain the desired traits over time.

5. **No Creation of New Genetic Variability**
   - Like pureline selection, mass selection does not generate new genetic variability. It only works with the **existing genetic variation** in a population. Therefore, it is not effective for creating **entirely new traits** or for breeding crops that need to adapt to **novel challenges**, such as emerging pests or diseases.

Mass selection remains a valuable tool for crop improvement, particularly in **self-pollinated and cross-pollinated crops** where genetic diversity is needed for adaptability and resilience. While it does not create new genetic combinations, its simplicity, cost-effectiveness, and ability to retain genetic variability make it a viable breeding method, especially in regions with limited resources. However, the method's lack of precision and uniformity can limit its effectiveness compared to more advanced techniques like **pureline selection** or **hybrid breeding**.

C. Line Breeding:

Line breeding is a specific breeding method that emphasizes the retention and enhancement of **desirable traits** within a population by repeatedly breeding individuals from the same ancestral line. It focuses on maintaining genetic stability while selectively breeding for superior traits, such as **yield**, **disease resistance**, or **quality traits** in crops and **growth rate**, **fertility**, and **production traits** in animals. Unlike crossbreeding, which introduces new genes from different genetic lines, line breeding concentrates the desirable genes within a single lineage, allowing for more predictable and **stable trait inheritance** over generations.

Key Principles of Line Breeding

Line breeding revolves around the principles of **concentrating desirable traits**, minimizing the effects of **inbreeding depression**, and maintaining **genetic purity** across generations. The method is widely used in breeding programs for both crops and livestock, ensuring that the **elite traits** in a population are passed on consistently.

1. Concentrating Desirable Traits

Line breeding's central objective is to **enhance and retain superior traits** within a population. By selecting parents that possess specific **beneficial traits**, breeders can intensify those traits in subsequent generations. This approach works particularly well in **self-pollinated crops** and **livestock breeding**, where the goal is to achieve a uniform population with **stable, high-performing traits**.

» **In Crops**: In self-pollinated crops like **wheat**, **rice**, and **soybeans**, line breeding is often employed to retain traits such as **high yield, disease resistance**, or **drought tolerance**. Line breeding allows breeders to create **pure lines** that consistently express these traits across multiple generations.

» **In Livestock:** In animal breeding, line breeding is used to enhance traits such as **milk production** in dairy cows, **egg-laying capacity** in poultry, or **growth rate** in beef cattle. By focusing on individuals with superior production traits, breeders can create a line of animals that consistently outperform their ancestors in terms of productivity.

Through line breeding, it is possible to **homogenize traits** in a population, ensuring that the offspring express a **higher level of uniformity** compared to random mating populations. This process enables breeders to reliably maintain desirable characteristics across generations, enhancing both the **economic** and **agronomic value** of the crop or animal population.

2. Managing Inbreeding Depression

While line breeding focuses on increasing genetic uniformity, a major risk associated with this approach is **inbreeding depression**. This phenomenon occurs when **closely related individuals** are bred together too frequently, leading to a reduction in **genetic diversity** and the accumulation of harmful **recessive alleles**. This can cause issues like reduced **vigor**, **fertility problems**, or decreased **yield potential**.

» **Balanced Genetic Diversity:** To mitigate the risks of inbreeding depression, line breeding is often carefully managed by introducing **genetic diversity** at strategic intervals. Breeders avoid breeding individuals that are too closely related (e.g., full siblings) to prevent the excessive concentration of **deleterious alleles**.

» **Introducing Outcrosses**: In some cases, line breeders will introduce **occasional outcrosses** from unrelated lines to increase genetic diversity and **refresh the gene pool** without diluting the selected desirable traits. This is particularly common in animal breeding, where maintaining genetic diversity is critical for long-term success.

By balancing the need for **genetic uniformity** with the need to prevent inbreeding depression, breeders can maintain a healthy, stable breeding line that expresses superior traits without suffering the negative consequences of excessive inbreeding.

3. Pedigree Selection

Line breeding relies heavily on **pedigree selection**, where the **lineage** and **ancestry** of individuals are carefully recorded and analyzed. Pedigree selection enables breeders to make **informed decisions** about which individuals to breed together based on their genetic background and the performance of their ancestors.

» **Detailed Record-Keeping**: Pedigree records track the **heritage** of each individual within the breeding program, including the traits inherited from parents, grandparents, and further back. This enables breeders to monitor the inheritance of key traits like **disease resistance**, **fertility**, or **growth rate**, ensuring that these traits remain stable across generations.

» **Selection Based on Lineage**: By analyzing the pedigree of an individual, breeders can identify the **genetic potential** of offspring before they are even produced. This reduces the risk of unintentional inbreeding and ensures that only individuals with **high breeding value** are used in the program.

Pedigree selection is a cornerstone of line breeding because it ensures that breeders can make strategic, data-driven decisions that enhance the overall **genetic quality** of the line. This meticulous record-keeping process also helps breeders maintain genetic diversity while continuing to concentrate desirable traits.

Steps in Line Breeding

Line breeding is a gradual, iterative process that unfolds over multiple generations. Each step in the line breeding process is crucial for refining traits and minimizing genetic problems such as **inbreeding depression**. Here's an expanded explanation of the key stages involved in line breeding:

1. Initial Selection of Parents

The first step in line breeding involves selecting **parent plants** or **animals** that possess superior traits. These parents will serve as the foundation for the line breeding program, and their **genetic material** will be passed on to future generations.

- » **Selecting High-Quality Individuals**: The parent individuals are selected based on their performance in traits such as **yield, resistance to environmental stresses, disease resistance**, or **production traits** in animals. For example, in a crop breeding program, breeders may select individuals that exhibit **superior grain size**, **drought tolerance**, or **early flowering**.
- » **Pedigree Considerations**: The selected parents are often **genetically related**, but not too closely related. This ensures that the **desired traits** are passed on while still maintaining enough genetic diversity to avoid inbreeding depression. Typically, parents that are cousins or more distantly related are used, rather than siblings or direct descendants.

2. Controlled Mating or Pollination

Once the parent individuals have been selected, **controlled pollination** or **controlled mating** is conducted to produce the **first generation of offspring**. This step is critical for ensuring that the **genetic material** from the selected parents is transferred effectively to the next generation.

- » **In Crop Breeding**: Line breeding in crops involves **self-pollination** or **cross-pollination** between selected individuals. Breeders may manually assist the pollination process to ensure that only the desired genetic material is transferred. This is especially important for crops that are prone to **cross-pollination**, such as maize.
- » **In Animal Breeding**: In animals, line breeding is facilitated through **controlled mating** or **artificial insemination**. Breeders ensure that the **elite traits** from the parents are passed on by managing the reproductive process carefully, selecting only the best individuals for reproduction.

3. Selection of Progeny

After the first generation of offspring is produced, breeders evaluate the progeny for **desired traits**. The goal is to select individuals that exhibit superior performance in these traits while maintaining genetic diversity within the population.

- » **Evaluating Key Traits**: In crops, traits such as **yield stability, plant height, flowering time, disease resistance**, and **grain quality** are evaluated. In livestock, breeders assess traits like **fertility, growth rate, milk production**, or **disease resistance**.

» **Ongoing Selection Process**: The selection process continues over multiple generations, with breeders constantly evaluating each successive generation for the presence of **desirable traits**. Only the best individuals from each generation are selected to **carry the line forward**.

4. Maintenance of Line Integrity

As line breeding progresses, it is crucial to **maintain the integrity of the line** by preventing excessive inbreeding and maintaining genetic diversity. This step involves closely monitoring the **pedigree** and introducing **occasional outcrosses** if necessary.

» **Monitoring Pedigrees**: Breeders maintain detailed records of each individual's **ancestry** to avoid excessive inbreeding. Pedigree management helps prevent the accumulation of harmful recessive alleles that can lead to inbreeding depression.

» **Introducing Genetic Diversity**: When necessary, breeders may introduce **outcrosses** to individuals from outside the line to **refresh the gene pool**. This helps maintain a healthy level of genetic diversity while continuing to concentrate the desired traits.

Advantages of Line Breeding

Line breeding offers several advantages that make it a valuable breeding method for **self-pollinated crops** and **livestock**. These advantages include:

1. Intensification of Desirable Traits

» **Enhanced Trait Expression**: Line breeding allows for the concentration and intensification of specific desirable traits, ensuring that these traits are passed down consistently to future generations. This is particularly important in crops for traits such as **high yield, disease resistance**, or **drought tolerance**. In animals, traits like **milk production** and **fertility** are enhanced.

2. Maintenance of Genetic Stability

» **Predictable Performance**: One of the key benefits of line breeding is the **predictability of performance**. Because the genetic makeup of the line is well-documented and stable, breeders can confidently predict how the line will perform across generations. This is crucial for creating **elite lines** in both crop and livestock breeding.

3. Genetic Purity and Uniformity

» **Consistent Phenotypes**: Line breeding results in a population that is **genetically uniform** and produces consistent phenotypes. This is important for **commercial agriculture** and **livestock production**, where uniformity is crucial for **management** and **marketability**.

4. Long-Term Stability

» **Sustained Performance**: Line breeding produces populations that are stable over long periods, ensuring that the line remains productive and genetically intact for **multiple generations**. This long-term stability is essential for breeders seeking to maintain elite lines that meet commercial demands.

Disadvantages and Challenges of Line Breeding

While line breeding offers many benefits, it also has some significant challenges:

1. Risk of Inbreeding Depression

» **Reduced Vigor**: Excessive inbreeding can lead to **inbreeding depression**, resulting in reduced vigor, fertility, and overall performance. Managing the level of genetic diversity is critical to avoiding this issue.

2. Limited Genetic Diversity

» **Susceptibility to New Threats**: Line breeding can limit **genetic diversity**, making the population more susceptible to environmental changes, new diseases, or pests. A lack of genetic variation can limit the population's ability to **adapt** to new challenges.

3. Time-Intensive Process

» **Multiple Generations Needed**: Line breeding requires **multiple generations** to achieve the desired level of trait intensification while managing genetic diversity. This makes the process **slow** compared to other breeding methods.

4. Requires Detailed Record Keeping

» **Complex Pedigree Management**: Effective line breeding relies on **detailed record-keeping** and careful **pedigree management**. This can be resource-intensive, particularly in large breeding programs.

Applications of Line Breeding

Line breeding is commonly used in both **crop** and **animal breeding** programs.

1. Crop Breeding

» In crops like **wheat, barley, soybeans,** and **maize,** line breeding is used to develop **pure lines** for hybrid production or direct release. Traits such as **yield, disease resistance,** and **quality traits** are intensified through line breeding.

2. Livestock Breeding

» In livestock breeding, line breeding is used to improve traits like **fertility, milk production,** and **growth rate.** Line breeding helps breeders maintain **elite lines** of animals that consistently exhibit superior performance.

D. The Bulk Method of Plant Breeding:

The **Bulk method** is a widely used approach in plant breeding, particularly in crops that self-pollinate. Introduced by **Nilsson Ehle in 1908,** the bulk method allows for the isolation of homozygous lines over multiple generations by maintaining a bulk population, often spanning from the **F2 to the F7** or later generations. The main idea behind this method is to allow the natural **segregation and selection** processes to take place within the bulked generations, thus simplifying the task of the breeder. Once the bulking phase is completed, breeders select individual plants for further evaluation and selection.

This method is both **time-efficient** and **cost-effective**, as it allows breeders to advance generations without intensive labor or detailed record-keeping until the desired level of homozygosity is achieved. It is particularly useful in situations where the breeders have to wait for specific environmental conditions or natural selection pressures to act, allowing the environment to influence the population beneficially.

Key Components of the Bulk Method

1. Isolation of Homozygous Lines in the Bulk Method

The **isolation of homozygous lines** is a central objective in the bulk method of plant breeding, as it aims to achieve genetic stability over successive generations. The process is designed to take advantage of **natural selection** and **self-pollination** to gradually reduce genetic heterozygosity. Over several generations, typically from **F2 to F6 or F7**, the population becomes more genetically homogeneous, with individual plants carrying uniform traits. The bulk method is particularly suited for crops that are self-pollinating, where natural selfing can help reach homozygosity with minimal effort.

Bulking Generation: Advancing Towards Homozygosity

During the **bulk phase**, the population is managed without artificial selection. The seeds from each generation are planted together, and the entire population is harvested in bulk. The lack of artificial selection during this phase helps to maintain the population's natural genetic diversity, allowing for **random recombination of genes** through self-pollination. This process results in the gradual elimination of heterozygous individuals over generations.

» **Self-Pollination and Homozygosity**: In self-pollinating species, the self-pollination process naturally reduces heterozygosity. By the time the population reaches the **F6 or F7** generation, most individuals in the population are nearly homozygous, meaning they carry identical alleles for most traits. This increase in homozygosity occurs because, with each generation of self-pollination, the probability of retaining heterozygous alleles decreases by half. By the sixth or seventh generation, the population is genetically stable.

» **No Artificial Selection**: One of the key characteristics of the bulk method is the absence of artificial selection during the bulking phase. By not imposing artificial selection, breeders allow natural processes like **competition** and **survival under field conditions** to influence the genetic composition of the population. This method is particularly useful in environments where natural selection can favor plants with traits like **disease resistance, drought tolerance**, or **pest resistance**.

Selection for Pure Lines: Stabilizing the Population

Once the population reaches the **F6 or F7** generation, the breeder's role shifts to selecting individual plants for further evaluation. By this stage, the population is almost completely homozygous, meaning most plants exhibit consistent traits due to the **fixation of desirable alleles**.

» **Isolating Desirable Traits**: At this stage, breeders select individual plants that show the best combination of traits, such as high yield, disease resistance, or superior quality. These selected plants are then subjected to **further testing** and evaluation, such as preliminary yield trials, disease screening, or drought tolerance assays.

- **Genetically Fixed Traits**: The primary goal of this selection is to isolate **pure lines** where desirable traits are genetically fixed, meaning they will be consistently expressed in subsequent generations. These pure lines represent the most promising candidates for eventual release as new varieties, as their genetic stability ensures reliable performance across different environments and growing seasons.

Advantages of Isolation in Bulk Method

- **Reduced Effort and Resources**: The bulk method provides a **cost-effective and labor-efficient** way to move towards homozygosity, as breeders do not need to engage in extensive selection or pedigree record-keeping during the bulk phase.
- **Natural Selection's Role**: During the bulk phase, natural selection acts as a silent force, helping to eliminate weaker plants and favoring those better suited to the prevailing environmental conditions. For example, if a disease outbreak occurs during the bulking period, plants that are resistant to the disease will naturally survive and contribute more heavily to the next generation's genetic pool. This results in a population that is naturally adapted to resist certain stresses.
- **Flexibility**: The bulk method offers breeders flexibility, as they can wait until the later generations to make selections based on the environmental conditions or traits that are important in a given season. For example, if disease resistance is a primary concern, breeders can allow the population to experience a natural disease epidemic before selecting resistant individuals.

Challenges in Isolating Homozygous Lines

- **Time-Intensive**: While the bulk method is efficient in terms of labor and resources, it is a **time-intensive process**. Achieving homozygosity can take several years, particularly if natural selection is an important factor. For breeders looking for quick results, this method may be slower compared to methods that involve artificial selection from the early stages.
- **Large Population Sizes**: To maintain genetic diversity and ensure that a wide range of traits can be captured in the final selection, breeders must manage **large populations** during the bulking phase. This can be challenging in terms of space, labor, and resources, especially in environments where field conditions are unpredictable or harsh.
- **Limited Selection Pressure**: Since artificial selection is not practiced during the bulk phase, the breeder has **limited control** over which traits are favored during the early generations. While this allows for natural selection to shape the population, it also means that some potentially valuable traits may not be preserved if they are not favored by the environment at the time.

2. Waiting for the Opportunity for Selection in the Bulk Method

One of the significant advantages of the bulk method is its **flexibility** in selection timing, which allows breeders to delay making selections until specific **environmental conditions** favor the expression of certain traits. This approach is particularly useful when breeding for traits such as **disease resistance**, **cold tolerance**, or **drought tolerance**, where the manifestation of these traits depends heavily on environmental factors. Unlike other breeding methods that may require early selection, the bulk method accommodates the unpredictability of environmental conditions, allowing breeders to capitalize on **natural occurrences** such as disease outbreaks or climate fluctuations.

Environmental Influence on Selection

Environmental factors play a crucial role in the expression of many agriculturally important traits. For example, **disease resistance** may only become apparent during a year when a particular pathogen is prevalent. Similarly, **drought tolerance** may only be observable under water-stress conditions. By adopting the bulk method, breeders are able to **postpone selection** until these specific environmental conditions present themselves, giving the population time to advance while allowing the necessary **genetic variation** to be retained.

- **Advancing Generations Without Selection**: During this waiting period, the population advances through generations (from **F2** to **F6-F7** or beyond) without any artificial selection being applied. This allows for the **natural recombination of genes** while maintaining the overall **genetic diversity** within the population. In self-pollinating species, this process also leads to a gradual increase in **homozygosity**, making it easier for breeders to identify true-breeding lines once the environmental conditions for selection are met.
- **Opportunity to Address Multiple Traits**: In breeding programs aimed at improving multiple traits, such as **disease resistance, cold tolerance**, and **yield**, the bulk method offers the advantage of waiting for the right environmental triggers for each trait. If, for example, a breeder is looking for **cold tolerance**, they can wait for a particularly harsh winter before making selections. This **patience-based approach** ensures that breeders do not prematurely discard genotypes that might otherwise perform well under the right environmental stress.

Mass Pedigree Method of Harlan

A noteworthy variant of the bulk method is the **Mass Pedigree Method**, also known as the **Harlan method**, which integrates bulk handling with pedigree selection under ideal conditions. This method is especially useful in breeding programs targeting traits such as **disease resistance**, where breeders often rely on the occurrence of a natural **disease epidemic** to identify resistant plants.

- **Segregating Generations Bulked**: In the mass pedigree method, segregating generations (such as **F2** to **F5**) are carried as bulk populations until the environmental conditions required for selection are present. For instance, in breeding programs targeting **rust resistance in wheat**, breeders might bulk the population until a natural rust epidemic occurs. Once the desired conditions appear, individual plants exhibiting resistance are selected from the bulk population, and these selected plants are subsequently handled in a manner similar to the **pedigree method**.
- **Selective Pressure Through Natural Events**: The mass pedigree method relies heavily on **natural selective pressure**. In cases where disease resistance is the desired trait, waiting for a **natural outbreak** of the disease allows breeders to observe which plants survive and thrive under these specific conditions. This reduces the need for **artificial inoculation** or **simulated environmental stress**, making the breeding process more aligned with real-world agricultural challenges.

Advantages of Waiting for Selection Opportunities

- **Increased Efficiency in Trait Selection**: By waiting for the right environmental conditions, breeders can make more informed and accurate selections. For example, waiting for a **disease outbreak** allows breeders to identify plants that naturally exhibit resistance, ensuring that only those with the desired trait are selected for further breeding. This reduces the likelihood of **false positives**, where a plant might appear resistant under controlled conditions but fail in the field.
- **Natural Selection's Role**: While breeders are waiting for the ideal selection conditions, **natural selection** continues to act on the population. This means that weak or poorly adapted genotypes may naturally be eliminated, leaving behind a population enriched with plants that are better suited to the prevailing conditions. For example, in regions prone to **drought**, natural selection may favor genotypes with better **water-use efficiency** or **deep root systems**, even before formal selection is applied.
- **Broader Pool of Genotypes**: Since the population is allowed to continue advancing through generations during the bulk phase, breeders have a **broader genetic pool** to choose from when the time for selection arrives. This can be particularly advantageous in breeding programs for **complex traits** like **drought tolerance**, where multiple genes and gene interactions are involved. By waiting until environmental stress conditions manifest, breeders can tap into the full potential of the **genetic diversity** present in the population.
- **Reduced Pressure for Immediate Selection**: The flexibility to wait for the appropriate conditions also reduces the pressure on breeders to make premature selections based on **incomplete data**. In many breeding programs, the expression of certain traits, like **cold hardiness**, may vary from year to year depending on seasonal fluctuations. By using the bulk method, breeders can wait for a particularly cold winter to assess the **true hardiness** of the plants, ensuring that only the most resilient plants are selected for further breeding.

Challenges of Waiting for Selection Opportunities

- **Unpredictable Environmental Conditions**: One of the biggest challenges in waiting for specific environmental conditions is the **unpredictability** of natural events. For example, a breeder working on disease resistance may have to wait several years for a natural outbreak of the target pathogen. Similarly, breeding for drought tolerance requires the occurrence of **consistent drought conditions** during the growing season, which may not happen every year.
- **Longer Time Frames**: Delaying selection until the ideal conditions appear can **prolong the breeding process**. For example, if a breeder is targeting a specific trait like **resistance to a particular pest**, they may need to wait several seasons for an infestation to occur. This extends the **time required** to develop a new variety, making the bulk method more suited to long-term breeding programs rather than those requiring quick results.
- **Environmental Variability**: While the bulk method allows for the selection of traits under natural environmental stress, environmental variability can sometimes introduce **confounding factors**. For instance, the occurrence of a **mild winter** may fail to reveal which genotypes are truly cold-tolerant, leading to incomplete or **inaccurate selections**.

3. Opportunity for Natural Selection in the Bulk Method

One of the most distinctive and advantageous features of the **bulk method** in plant breeding is its allowance for **natural selection** to operate within the population. While many plant breeding methods rely heavily on **artificial selection** by breeders, the bulk method enables natural forces such as **environmental conditions, disease pressure**, and **competition among plants** to play a significant role in shaping the genetic composition of the population. Over time, this process helps favor better-adapted genotypes while **weaker, less-adapted individuals** are naturally eliminated. This process can be particularly beneficial when breeding for complex traits like **yield stability, disease resistance**, or **environmental adaptability**.

Natural Selection Acting on Homozygous Plants

As the population advances through the **F2 to F6 generations**, it remains largely heterozygous. During this period, natural selection has a limited impact on the population because genetic variability is still high, and the plants are segregating for various traits. However, by the **F7 generation** and beyond, the population becomes increasingly homozygous. At this stage, **natural selection** can exert a more pronounced influence on the composition of the population because the plants are more genetically uniform, allowing for better adaptation to environmental pressures.

» **Shift from Heterozygosity to Homozygosity**: Up to the **F6 generation**, the population contains a mix of heterozygous and homozygous individuals. Natural selection during this period may act on some **observable traits** such as **disease resistance, drought tolerance**, or **vigor**, but the influence is more limited due to the genetic variability. By the **F7 generation**, however, most plants are nearly homozygous, meaning that **natural selection** can now act more effectively on the **genetically stable** individuals.

» **Selection of Superior Genotypes**: At this point, it is often assumed that natural selection will favor **higher-yielding genotypes** and **eliminate poor performers**. For example, in a disease-prone environment, natural selection might favor plants that have developed **resistance to a specific pathogen**. Similarly, in drought-prone regions, plants with **deeper root systems** or **better water-use efficiency** may have a competitive advantage, leading to their increased frequency in the population. This gradual improvement of the population by natural selection ensures that the final lines chosen for further breeding or commercialization are **well-adapted to the specific environment** in which they will be grown.

» **Effectiveness in Low-Input Conditions**: The opportunity for natural selection to act on the population is particularly valuable in **low-input agricultural systems** where farmers may not have access to fertilizers, pesticides, or irrigation. In such systems, varieties bred using the bulk method, with natural selection as a key component, are likely to be better suited to **suboptimal growing conditions**, as they are naturally selected for resilience and resource efficiency.

Long-Term Bulks for Enhanced Natural Selection

In some cases, breeders choose to extend the bulk period beyond the **F7 generation**, allowing the population to advance to **F20 or even F30**. This extended period of bulking provides more opportunities for **natural selection** to shape the population and fine-tune the genetic composition over many generations. Long-term bulking is often practiced when breeders are particularly focused on allowing the population to **adapt to local environmental conditions** or when they want to **maximize the influence of natural selection** on traits like **disease resistance**, **stress tolerance**, or **yield stability**.

» **Extended Bulking Period**: In long-term bulks, the population continues to be grown and harvested together for several more generations, allowing natural selection to act continuously. By **F20 or F30**, the genetic composition of the population has shifted significantly, with the **strongest genotypes** (those that are most suited to the local conditions) being more frequent. This approach is often used when breeding for **disease resistance** or **environmental adaptability** because natural selection can effectively eliminate susceptible or weak plants over time.

» **Adapting to Local Conditions**: The **bulk method** is especially useful for breeding crops in environments where **local conditions** are harsh or unpredictable. For example, in regions with high levels of **biotic stress**, such as frequent pest infestations or recurrent disease outbreaks, natural selection can help favor genotypes that are inherently more resistant to these threats. Similarly, in **arid or semi-arid regions**, long-term bulking can help identify plants that have adapted to **drought stress**, resulting in varieties that are more resilient under low water availability.

» **Natural Selection for Complex Traits**: Many traits of interest in crop breeding, such as **yield**, **vigor**, and **resilience to environmental stresses**, are complex and controlled by multiple genes. The bulk method allows for the accumulation of **favorable alleles** over multiple generations, ensuring that the final population contains a higher proportion of **genetically superior individuals**. By allowing natural selection to act over an extended period, breeders can effectively improve complex traits without needing to intervene with artificial selection at every stage.

Advantages of Natural Selection in the Bulk Method

» **Adaptation to Local Conditions**: One of the most significant advantages of allowing natural selection to operate within the bulk method is the ability to breed varieties that are naturally adapted to the **local environment**. For instance, if a crop is grown in a region prone to **fungal diseases**, natural selection will favor plants that have **innate resistance** to these pathogens, ensuring that the final population is better suited to the local conditions.

» **Low Input, High Resilience**: Varieties bred using the bulk method tend to be more **resilient** to environmental stressors, especially when natural selection is allowed to act for multiple generations. This makes these varieties particularly well-suited for **low-input farming systems**, where farmers may not have access to chemical inputs like pesticides or fertilizers. Such varieties can thrive under **minimal management**, as they have been naturally selected for their ability to survive and produce under challenging conditions.

» **Efficient Use of Resources**: By letting **natural selection** operate on the population, breeders can reduce the need for intensive selection efforts in the early generations. This makes the bulk method a **cost-effective** and **resource-efficient** approach to breeding, particularly when compared to methods like the **pedigree method**, which requires careful selection and record-keeping at each generation.

» **Selection for Yield and Stability**: Over time, natural selection may favor **higher-yielding genotypes** within the population, leading to a gradual increase in overall yield potential. In addition, because natural selection often acts on **stability** as well as yield, the final lines produced through the bulk method are likely to perform consistently across a range of conditions, offering **yield stability** over multiple growing seasons.

Challenges of Relying on Natural Selection

» **Unpredictable Selection Pressure**: One of the main challenges of relying on natural selection is its **unpredictability**. For example, if a breeder is targeting **disease resistance**, but there is no disease outbreak during the bulking period, natural selection will not favor the desired trait. Similarly, if the environmental conditions are unusually favorable, natural selection may not exert sufficient pressure to eliminate weaker genotypes, leading to **less effective selection**.

» **Longer Time Frame**: Allowing natural selection to shape the population over multiple generations can be a **slow process**. While the bulk method requires less hands-on selection during the early generations, the extended time frame needed for natural selection to act means that the development of new varieties may take **several years**, especially in long-term bulks.

» **Limited Control Over Selection**: While natural selection is a powerful force, it may not always align with the breeder's objectives. For example, natural selection may favor plants with **vigorous growth** but poor grain quality, or plants that **resist one disease** but are susceptible to another. In such cases, breeders may need to intervene with **artificial selection** to ensure that the desired traits are retained in the population.

Detailed Procedure of Bulk Method

The bulk method follows a structured and systematic approach over several generations to achieve **homozygosity** and to identify superior genotypes for future selection. Below is a step-by-step guide to the bulk method:

1. Hybridization:

Hybridization, defined as the mating or crossing of two genetically distinct plants or lines, is a cornerstone of crop improvement in modern agriculture. Particularly in self-pollinated crops, where natural genetic variability is often exhausted after generations of selection, hybridization creates fresh variability that can be exploited for enhancing yield, disease resistance, and other agronomic traits. Hybridization helps combine complementary traits from two different pure lines, creating a hybrid (F1) that holds genetic material from both parents. The process subsequently leads to **segregating generations** (F2, F3, and beyond) where genetic recombination offers an opportunity to select superior progeny.

Key Elements of Hybridization

Hybridization is employed with various objectives in mind. Each goal focuses on improving one or more traits, ensuring crop productivity and resilience under diverse environmental conditions. Some major aspects of hybridization are:

1. **Creating Genetic Variation**: The most critical goal of hybridization is to introduce genetic

variation, essential for plant breeders to select superior genotypes from the resulting progeny. Crossing two genotypically different plants generates an **F1 population** that combines traits from both parents. From this F1 population, self-pollination leads to the **F2 generation**, where segregation occurs, providing a diverse pool of phenotypes and genotypes for selection.

2. **Aims of Hybridization**:
    - **Transfer of Qualitative Traits**: Hybridization is often aimed at transferring specific qualitative traits (e.g., disease resistance, grain color) controlled by a few genes from one parent to the other. For instance, transferring resistance to rust from a wild wheat relative into a high-yielding commercial variety.
    - **Improvement in Quantitative Traits**: Quantitative traits, such as yield, drought tolerance, and plant height, are polygenic and more complex. Hybridization helps improve these traits by combining superior genes from both parents, aiming for better performance in the F1 and segregating generations.
    - **F1 Hybrid as a Variety**: In some cases, the F1 hybrid itself can be commercially exploited, as it often exhibits **hybrid vigor** (heterosis), which results in superior growth, yield, and other agronomic traits compared to both parents.

Types of Hybridization

Hybridization in self-pollinated crops can take different forms, based on the genetic relationship between the parents:

1. **Inter-varietal or Intraspecific Hybridization**: The most common form of hybridization, it involves crossing parents belonging to the same species but different varieties or strains. The aim is to combine desirable traits from both varieties into a single hybrid. Inter-varietal crosses can be simple (involving two parents) or complex (involving more than two parents).
    - **Simple Crosses**: Crosses between two parents (A × B) that produce an F1 hybrid, which is further advanced through self-pollination or backcrossing.
    - **Complex Crosses**: Involving more than two parents, such as [(A × B) × C], this type of cross is designed to bring together multiple desirable traits into a single hybrid.
2. **Interspecific Hybridization**: This involves crossing individuals from different species within the same genus. For instance, crossing **wheat (Triticum)** with **wild relatives** to incorporate traits like drought tolerance or disease resistance.
3. **Distant Hybridization**: Crosses between different genera are known as distant or wide hybridization. This form of hybridization is more complex and can result in significant evolutionary changes. **Thomas Fairchild (1717)** conducted the first recorded distant hybridization, producing a hybrid between two species of Dianthus. Another example is **Triticale**, a hybrid between **wheat** (Triticum) and **rye** (Secale), which combines the high yield of wheat with the hardiness and disease resistance of rye.

Procedure of Hybridization

The process of hybridization involves multiple steps, from selecting the parents to final evaluation and release of a new variety. Here's a step-by-step breakdown of the procedure:

1. **Selection of Parents**: The first step is the careful selection of parents. One parent is typically a well-adapted, high-yielding variety with desirable agronomic traits, while the other parent provides a complementary trait that is lacking in the first parent, such as disease resistance or drought tolerance. The success of hybridization relies on selecting parents with good **combining ability** to ensure the production of superior F1 hybrids.
2. **Evaluation of Parents**: Before crossing, both parents are evaluated for their **combining ability** to ensure that they will produce a high-performing hybrid. Combining ability refers to how well the traits of two parents blend to create a superior offspring.
3. **Emasculation**: In self-pollinated crops, it is crucial to prevent self-pollination to ensure cross-pollination between the selected parents. This is achieved by emasculation, where the anthers (male reproductive organs) are removed from the flowers of the female parent before they mature and release pollen. This ensures that only the pollen from the chosen male parent will fertilize the female flowers.
4. **Bagging and Pollination**: After emasculation, the flowers are covered with bags to prevent unwanted pollen contamination. Once the flowers are receptive, pollen from the male parent is carefully collected and transferred to the stigma of the emasculated female flowers using a fine brush or other tools.
5. **Tagging**: After pollination, the flowers are tagged to record essential information, such as the names of the parent plants, the date of emasculation and pollination, and other relevant data. This helps maintain the pedigree of the resulting hybrid population.
6. **Harvesting and Storing F1 Seeds**: The crossed plants are harvested, and the F1 seeds are collected. These seeds are stored in optimal conditions to preserve their viability until they are sown in the next generation.
7. **Raising the F1 Generation**: The F1 generation is grown in the field to observe its performance. In self-pollinated crops, the F1 plants are often highly vigorous, displaying hybrid vigor for traits like yield, growth rate, and disease resistance.
8. **Selfing and Advancement**: The F1 generation is allowed to self-pollinate, producing the **F2 generation**, which segregates for the traits of interest. Selection is carried out in the F2 and subsequent generations to identify plants with the desired trait combinations.

Aims of Hybridization

1. **Combination Breeding**: This form of hybridization is used to transfer one or more desirable traits (e.g., disease resistance, improved yield) from one parent to another. It aims to create a new variety that combines the strengths of both parents while minimizing weaknesses.
2. **Transgressive Breeding**: This method focuses on producing **transgressive segregants**, which are offspring that perform better than both parents for specific traits. For example, selecting for high yield and disease resistance may result in a plant that surpasses both parents in these traits due to the **recombination of favorable alleles**.

3. **Hybrid Varieties**: In some cases, the **F1 hybrid** is itself released as a variety due to its superior performance. In crops like maize and rice, F1 hybrids often show significant yield advantages due to heterosis (hybrid vigor), where the hybrid outperforms both parents.

Challenges and Prospects of Hybridization in Self-Pollinated Crops

Although hybridization is more straightforward in cross-pollinated crops, it has been successfully adapted for use in self-pollinated species, particularly to exploit hybrid vigor. However, several challenges remain:

1. **Degree of Heterosis**: In self-pollinated crops, the magnitude of heterosis may not be as high as in cross-pollinated crops. This means that the economic benefit of F1 hybrids in self-pollinated crops must justify the additional costs of producing hybrid seed.
2. **Seed Production Efficiency**: Producing hybrid seed in self-pollinated crops requires **manual pollination** or the use of **male sterility systems**, which can be labor-intensive and costly. Efficient seed production systems are critical for the success of hybrid varieties in self-pollinated crops.
3. **Heterosis in Self-Pollinated Crops**: Despite these challenges, hybrid varieties have been developed in several self-pollinated crops, including rice, barley, tomato, and wheat. The key to their success is the degree of heterosis they express, which can sometimes lead to significant yield increases.

Here's a more detailed expansion on the process and significance of **F1 to F6 generations** within the **Bulk Method** of plant breeding, focusing on key elements like space planting, bulk harvesting, and the role of natural selection and increasing homozygosity:

2. F1 Generation:

The F1 generation is a critical stage in the bulk method of plant breeding. The plants produced from the initial hybridization between the two selected parent lines carry a mix of the genetic material from both parents, with each F1 individual being heterozygous at multiple genetic loci. At this stage, breeders aim to advance the population without focusing on selection. The key goals are to allow the plants to grow, bulk the population, and proceed to the F2 generation.

» **Space Planting of F1**: In the F1 generation, space planting is practiced to ensure that each plant has adequate space to grow and develop without competition. Typically, around 200 or more F1 plants are grown during this phase. The large number of F1 plants ensures that the population has enough genetic diversity to generate a wide range of phenotypes in subsequent generations.

- *Objective of Space Planting*: The space planting technique helps individual plants express their full potential in terms of growth and vigor. This is important for ensuring that the F1 population is robust and capable of yielding healthy progeny in future generations.

» **No Selection in F1**: In this stage, no deliberate selection is performed by the breeder. The purpose of the F1 generation is not to choose specific traits but rather to allow for the recombination of genetic material from the parent lines. Because the F1 individuals are heterozygous, selection at this stage would be premature, as the real genetic variation will only emerge in the segregating F2 generation.

- *Reason for No Selection*: Since all F1 plants are genetically identical hybrids (assuming no mutations), the focus is on advancing the population to the F2 generation, where recombination will generate significant genetic diversity. Thus, breeders focus on bulking the population rather than selecting specific plants.

3. F2 to F6 Generations: Bulk Harvesting and Natural Selection

Once the F1 generation has been allowed to self-pollinate, the F2 generation marks the beginning of segregation, where individual plants begin to show varying traits inherited from both parent lines. This period, from F2 to F6, is crucial for increasing genetic homozygosity and allowing natural selection to act on the population.

» **Bulk Planting and Harvesting**: In the F2 to F6 generations, the plants are grown in bulk, meaning that all plants are treated equally without any deliberate selection. The seeds from the entire population are harvested together and mixed to create a bulk population for the next generation. This method allows the entire genetic diversity created in the F2 generation to be maintained without any bias or artificial interference.

- *Advantages of Bulk Planting*: Bulk planting ensures that the population remains large, preserving genetic diversity. This is crucial because it allows breeders to work with a wide genetic base in later generations when selecting superior plants. By not selecting individual plants too early, breeders can avoid the risk of eliminating potentially valuable genotypes that might express desirable traits in later generations.
- *Importance of Bulk Harvesting*: Bulk harvesting is a labor-efficient method that minimizes the need for detailed individual plant evaluation at an early stage. The main objective is to advance the population through several generations to increase the overall level of homozygosity, as self-pollination naturally leads to the expression of recessive traits and the elimination of heterozygosity.

» **Role of Natural Selection**: Although no artificial selection is performed by the breeder during these generations, natural selection may still influence the population. For example, if a disease outbreak occurs, plants that are naturally resistant to the disease will have a survival advantage. Over time, this can lead to an increase in the frequency of disease-resistant plants in the bulked population. Similarly, plants that are better adapted to the local environment may thrive and contribute more seeds to the bulk.

- *Example of Natural Selection in Bulk Populations*: In a population facing drought conditions, plants with deeper roots or better water retention mechanisms may naturally survive better, contributing more seeds to the bulk. Although the breeder does not actively select these plants, they become more prevalent in the population simply due to their superior survival under the given environmental conditions.

» **Increasing Homozygosity**: One of the primary goals of the bulk method is to allow the population to increase in homozygosity over time. Starting from the highly heterozygous

F2 generation, each subsequent generation experiences increased self-pollination, which leads to the segregation and fixation of genetic traits. By the time the population reaches the F6 generation, it is nearly homozygous, meaning that most individuals carry identical alleles at the majority of loci.

- *Process of Homozygosity*: As self-pollination continues over successive generations, the genetic variability within each plant decreases. The frequency of homozygous plants increases, meaning that the population becomes more genetically stable. This allows for more effective selection in later generations, as each plant will have a more predictable genetic makeup.
- *Significance of F6 Generation*: By the F6 generation, most of the heterozygous loci in the population will have become homozygous. This makes it easier for breeders to select individual plants for desirable traits, as the traits observed are more likely to be genetically fixed. Thus, F6 represents a critical stage where breeders can begin to isolate nearly pure lines for further evaluation and testing.

4. F7 Generation: Selection of Individual Plants

At the F7 generation, the breeding process enters a crucial phase where individual plant selection is emphasized. After several generations of bulk planting, which allowed for natural selection and increasing homozygosity, breeders now have a relatively stable population to work with. The goal of the F7 generation is to identify the best-performing plants from a nearly homozygous population.

» **Space Planting of 50,000 Plants**: In the F7 generation, approximately 50,000 individual plants are space-planted, giving each plant sufficient room to grow and express its full genetic potential. This large-scale planting ensures that breeders have a broad genetic base to select from. Space planting at this stage is critical for observing individual plant characteristics in isolation from their neighbors, which may have influenced their growth in bulk planting.

- *Importance of Space Planting*: By allowing each plant ample space, breeders can more accurately assess the phenotypic traits of each individual. Factors such as plant height, disease resistance, and flowering time can be observed more clearly, and plants are less likely to be influenced by competition for resources, making selection more precise.

» **Selection of 1,000 to 5,000 Plants Based on Phenotype**: Out of the 50,000 plants, breeders will select between 1,000 and 5,000 plants based on their phenotypic characteristics. The selection criteria at this stage focus on traits such as yield potential, disease resistance, quality of grains or fruits, plant architecture, and other agronomically valuable traits.

- *Key Traits for Selection*: Selection at this stage is primarily driven by observable traits, including:
  - **Yield Potential**: Plants that show high yield potential, measured by factors like grain weight and the number of seeds or fruits produced.
  - **Disease and Pest Resistance**: Plants that demonstrate resistance to common diseases or pests are given preference, as this trait is critical for reducing dependency on pesticides.

* **Plant Structure**: Traits such as plant height, branching patterns, and lodging resistance are important for ease of harvest and overall field management.

» **Harvesting Seeds Separately**: Once the plants are selected, the seeds from each individual plant are harvested separately, not bulked. This ensures that the performance of each plant can be tracked and evaluated in subsequent generations. The separation of seeds at this stage allows breeders to maintain the genetic identity of each selected line, which is crucial for progeny testing in the next generation.

5. F8 Generation: Progeny Testing

In the F8 generation, the breeding process moves into progeny testing, where the performance of selected lines is assessed more rigorously. The seeds harvested from individual plants in the F7 generation are now grown and evaluated in single-row or multi-row plots.

» **Single/Multi-Row Progeny Testing**: Each selected plant from the F7 generation is grown in individual single-row or multi-row plots, depending on the available space and resources. This setup allows breeders to evaluate the performance of each progeny group in a controlled environment. By testing each progeny separately, breeders can better assess the heritability of desirable traits and select the most promising lines for further advancement.

- *Traits Evaluated During Progeny Testing*:

  * **Yield**: Yield continues to be a primary selection criterion. Plants that show superior yield potential in single or multi-row tests are advanced for further trials.

  * **Quality**: Grain quality, including size, shape, color, and nutritional value, is evaluated to ensure that the progeny meets market and consumer demands.

  * **Disease and Pest Resistance**: Progeny that consistently resist local diseases and pests are preferred, ensuring that the future variety will have built-in resilience.

» **Elimination of Inferior Lines**: At this stage, progenies that do not meet the required standards are culled from the breeding program. Typically, weak, diseased, or low-yielding progenies are rejected, and only the top-performing 100 to 300 progenies are retained for further trials. This rigorous culling ensures that only the best lines move forward in the selection process.

- *Criteria for Elimination*: Inferior lines may be eliminated based on poor yield, susceptibility to disease, poor plant structure, or any other undesirable trait. The aim is to narrow down the population to the best-performing progenies that consistently exhibit the desired characteristics.

6. F9-F12 Generations: Yield Trials

The F9 to F12 generations mark the beginning of yield trials, which are essential for determining the commercial viability of the selected progenies. These trials are conducted in a more formal and replicated manner to ensure accuracy and reliability of the results.

» **Preliminary Yield Trials in F9**: In the F9 generation, preliminary yield trials are conducted to evaluate the performance of the selected progenies in larger plots. These trials help to assess the yield potential and adaptability of the lines under real-world conditions, including varying soil types, climates, and farming practices.

- *Importance of Preliminary Trials*: Preliminary yield trials provide a snapshot of how the selected lines perform in environments that mimic commercial farming conditions. This is critical for determining which progenies will be advanced to the next stage of testing.

» **Replicated Yield Trials in F10-F12**: From the F10 to F12 generations, replicated yield trials are conducted to ensure that the selected progenies perform consistently across different locations and environments. Replication helps to reduce the impact of environmental variability on the results, providing a more accurate measure of the progeny's true genetic potential.

- *Replicated Trials*: Replicated trials involve growing the same progeny line in multiple plots across different locations. This allows breeders to assess the stability and adaptability of the progeny under various conditions, ensuring that the selected lines are suitable for large-scale commercial cultivation.
- *Key Data Collected*: During these trials, breeders collect detailed data on:
  - Yield per hectare
  - Grain or fruit quality
  - Disease and pest resistance
  - Adaptability to different environments

» **Comparison with Standard Check**: Throughout these trials, the performance of the selected progenies is compared to a standard check variety. The check variety serves as a benchmark, providing a point of reference for evaluating the success of the breeding program. If the progenies outperform the check variety in terms of yield, quality, and resilience, they are advanced to the final stage of evaluation.

7. F13 Generation: Seed Multiplication and Release

The F13 generation represents the final step in the bulk method, where the selected progeny is prepared for release as a new variety.

» **Seed Multiplication**: If the progeny has consistently performed well across all trials, it is released as a new variety. Before the variety is made available to farmers, large-scale seed multiplication is conducted to ensure that enough seeds are produced for commercial cultivation. The seeds are carefully monitored to maintain genetic purity, ensuring that the traits selected during the breeding process remain intact.

- *Significance of Seed Multiplication*: Seed multiplication is a critical step, as it ensures that farmers have access to high-quality seed that performs consistently in the field. The genetic integrity of the seed must be maintained to ensure that the traits selected during the breeding process are passed on to future generations.

» **Variety Release and Distribution to Farmers**: Once the new variety has been tested and proven superior, it is officially released and distributed to farmers. The variety is often named based on its unique characteristics or the region in which it was developed. Farmers can now plant the new variety, benefiting from its improved yield potential, disease resistance, and adaptability.

- *Impact on Agriculture*: The release of a new variety can significantly impact local and regional agriculture by providing farmers with a superior product that enhances productivity, reduces input costs (such as pesticides), and improves profitability. Successful varieties can quickly become popular among farmers, leading to widespread adoption and contributing to food security and economic growth.

Merits of the Bulk Method

The bulk method offers several distinct advantages, particularly in the context of breeding self-pollinated crops. These merits make it a popular choice in certain scenarios, especially for breeders looking to maximize simplicity and natural selection.

1. Simplicity and Cost-Effectiveness

The bulk method is notably simpler and more cost-effective than other methods like pedigree selection. It requires minimal record-keeping, as plants are grown and harvested in bulk, without the need for individual plant tracking in the early generations. This reduces labor, time, and financial resources, making the method ideal for breeding programs with limited funding or personnel.

» **Labor-Intensive Tasks Are Delayed**: In contrast to more meticulous methods where breeders must make detailed selections in every generation, the bulk method defers intense labor and decision-making until the population is closer to homozygosity. This enables breeders to manage large populations without having to monitor every individual plant closely in the early stages.

» **Reduced Costs**: As fewer resources are dedicated to artificial selection, greenhouse maintenance, and record-keeping in the initial years, the bulk method tends to be more affordable. This makes it particularly useful in developing countries or research institutions where financial and labor resources are constrained.

2. Opportunity for Natural Selection

The bulk method creates conditions that allow natural selection to shape the population, especially in long-term bulking scenarios. Natural selection is the process by which plants that are better adapted to the environment survive and reproduce at a higher rate than less adapted plants. By allowing the environment to naturally favor certain genotypes, breeders can take advantage of this "free" selection process.

» **Natural Resistance to Diseases and Pests**: For instance, if a disease outbreak occurs during the bulking period, only those plants with inherent resistance to the disease will survive. This means that subsequent generations will have a higher proportion of plants with natural resistance, reducing the need for artificial selection of resistant individuals.

» **Adaptation to Local Conditions**: The bulk method can be especially beneficial when breeding crops for specific local conditions. Over multiple generations, the population may naturally become better adapted to the unique stresses of a given environment, such as drought or poor soil fertility.

3. Large Populations

A key feature of the bulk method is the cultivation of large populations during the bulking period. This increases the genetic diversity within the population and raises the likelihood of finding **transgressive segregants**—plants that outperform both of their parents in key traits such as yield, resistance to disease, or drought tolerance.

» **Increased Genetic Variation**: Larger populations allow for greater genetic diversity, which is crucial for finding individuals that combine favorable traits in novel ways. These transgressive segregants can lead to the development of superior plant varieties that offer significant advantages over existing ones.

» **Better Chances of Desired Traits**: By growing large populations over several generations, breeders increase the likelihood of encountering rare recombinations of genes that might lead to enhanced productivity or resilience.

4. Effective Selection at Later Stages

In the bulk method, selection is deferred until later generations, typically when the population is nearly homozygous. This makes selection more effective because it occurs at a time when the genetic background of the plants is stable, allowing breeders to identify traits that will be consistently expressed in future generations.

» **Minimized Environmental Influence**: By waiting to select until the population is nearly homozygous, breeders can reduce the influence of environmental factors that might obscure the expression of genetic traits in earlier, more heterozygous generations. This leads to more accurate and reliable selection for desirable characteristics.

5. Suitable for Small Grains

The bulk method is particularly well-suited to small grain crops, such as wheat, barley, and oats, which are typically grown in dense populations. These crops thrive in environments where space is limited, and the bulk method allows for easy management of these densely planted crops over multiple generations.

» **Efficiency in Dense Crops**: For small grain crops, the bulk method takes advantage of the plants' natural growth patterns and enables the cultivation of large, uniform fields with minimal need for individualized attention until the final stages of selection.

Demerits of the Bulk Method

While the bulk method has clear advantages, it also comes with several limitations that may make it less suitable for certain breeding objectives or crop species. These disadvantages are important to consider when deciding whether to use the bulk method in a breeding program.

1. Long Time Frame

One of the most significant drawbacks of the bulk method is the extended time required to develop a new variety. Natural selection only becomes relevant in later generations, often around F10 or even F20, depending on the breeding goals and environmental conditions. This slow pace can be a limitation in cases where rapid varietal improvement is needed.

» **Delayed Results**: In contrast to faster methods like pedigree selection, which allows for selection in early generations, the bulk method requires several years of bulking before significant genetic progress can be made.

» **Not Suitable for Urgent Breeding Programs**: For breeders working in environments that require quick adaptation to changing conditions (e.g., emerging diseases or shifting climates), the bulk method's long time frame may be a significant limitation.

2. Limited Opportunity for Artificial Selection

In short-term bulks, artificial selection has little opportunity to influence the population's genetic composition. Natural selection may not be sufficient to guide the population toward desired traits, particularly when those traits are not heavily influenced by environmental pressures.

» **Breeder's Skill Not Fully Utilized**: Since the bulk method defers most selection decisions until later generations, the breeder's ability to make informed, skillful selections is minimized in the early stages. This may lead to missed opportunities to improve the population through artificial selection in traits that are not strongly influenced by natural conditions, such as specific quality traits or market preferences.

3. Large Number of Progenies

At the end of the bulking period, breeders must handle a large number of progenies, which can be labor-intensive and difficult to manage. While large populations are beneficial for maintaining genetic diversity, they also create logistical challenges during the selection and evaluation phases.

» **Labor-Intensive Handling**: The sheer volume of progenies that must be screened and evaluated can be overwhelming, especially for smaller breeding programs with limited personnel or field space. Managing, evaluating, and selecting from thousands of plants requires a significant investment of time and effort.

» **Logistical Challenges**: The need to grow, harvest, and evaluate such a large population over multiple generations can strain resources, both in terms of labor and field space.

4. Limited Information on Inheritance

The bulk method does not involve detailed record-keeping of individual plants, unlike the pedigree method. As a result, it provides little information about the inheritance of specific traits, making it difficult to track how traits are passed from one generation to the next.

» **Lack of Genetic Insight**: Without detailed pedigree records, breeders lack precise information on the genetic background of the selected plants. This can make it challenging to understand the genetic architecture of important traits, limiting the breeder's ability to make informed decisions.

5. Incompatibility with Off-Season Advancements

The bulk method is less compatible with off-season growing or the use of greenhouse facilities, which can accelerate breeding programs. Since natural selection is a key component of the bulk method, it is important for the population to be exposed to natural field conditions that reflect the target environment.

» **Environmental Mismatch**: Off-season and greenhouse conditions often differ significantly from the actual growing environment, which can lead to selection pressures that do not reflect real-world conditions. This reduces the effectiveness of natural selection and may result in plants that are less well-adapted to the target environment.

F. Backcross method

The **backcross method** is a fundamental breeding technique used to transfer specific traits from one parent to a high-yielding, well-adapted parent while preserving its overall genotype. This process involves repeatedly crossing the hybrid or its segregating progeny with one of its parents, known as the recurrent parent until the progeny closely resembles the recurrent parent except for the introduced trait. Over several backcross generations, the genetic composition of the progeny becomes increasingly similar to the recurrent parent.

The objective of the Backcross Method

The backcross method is primarily used to achieve the following objectives:

1. **Improvement of a High-Yielding Variety**: The primary goal of backcross breeding is to correct one or two specific defects in a high-yielding variety or a well-adapted variety. The goal is to introduce a desirable trait, such as disease resistance or improved grain quality, without altering the high-yielding characteristics of the original variety.
2. **Transfer of a Specific Trait**: Another important objective is to transfer one or a few specific traits from a donor parent to the recurrent parent. These traits, such as pest resistance, drought tolerance, or improved plant architecture, are transferred while maintaining the genotype of the recurrent parent as much as possible. After several backcross generations, the resulting progeny will closely resemble the recurrent parent but will include the desirable trait from the donor parent.

Requirements for Successful Backcross Breeding

Backcross breeding is a powerful technique, but its success depends on several key requirements:

1. **A Suitable Recurrent Parent**: The recurrent parent should be a high-performing variety with desirable traits, such as high yield, good adaptability, and market acceptability. However, this parent should lack one or two key characteristics that can be improved through backcrossing. The recurrent parent is crossed multiple times to the hybrid or progeny to recover its genetic makeup.
2. **A Suitable Donor Parent**: The donor parent is the source of the desirable trait(s) that the recurrent parent lacks. The donor parent should have the trait in question at a high intensity, such as strong disease resistance or tolerance to a particular stress condition.
3. **Heritability of the Trait**: The character being transferred should have high heritability, meaning it is strongly passed from parent to offspring. Ideally, the trait should be governed by one or a few genes, making it easier to transfer through backcrossing. Complex traits controlled by many genes are harder to transfer and stabilize through backcrossing.
4. **Multiple Backcross Generations**: To ensure the recovery of the recurrent parent's genotype,

several backcrosses (typically 6-7 generations) are required. Each backcross generation further reduces the influence of the donor parent's genetic material, until the progeny is nearly identical to the recurrent parent, except for the introduced trait.

Applications of Backcross Breeding

The backcross method has a wide range of applications in plant breeding, particularly for traits that are governed by one or a few genes. It is used in several ways, including:

1. Intervarietal Transfer of Simply Inherited Traits

Backcrossing is highly effective for transferring simply inherited traits, particularly those controlled by one or two genes, such as disease resistance, pest resistance, or herbicide tolerance. By backcrossing the F1 progeny with the recurrent parent, breeders can introduce resistance to diseases like rust, blight, or mildew into a high-yielding variety without altering its other favorable traits.

» **Example**: In wheat breeding, backcrossing has been used to introduce rust resistance from a wild or unrelated variety into a popular high-yielding wheat cultivar.

2. Transfer of Quantitative Traits

Backcrossing can also be used for transferring highly heritable quantitative traits, such as plant height, seed size, and earliness. These traits are usually governed by several genes but can still be successfully transferred through backcrossing if they show high heritability.

» **Example**: In maize, backcrossing has been used to transfer dwarfism and early maturity traits from one variety into another high-yielding but late-maturing variety.

3. Interspecific Transfer of Traits

Backcross breeding can facilitate the transfer of traits from related species into cultivated crops, particularly for simply inherited traits such as disease resistance. Interspecific hybridization followed by backcrossing can introduce traits from wild relatives into domesticated crops.

» **Example**: Resistance to potato late blight has been successfully transferred from wild potato species into cultivated potato varieties using interspecific backcrossing.

4. Cytoplasmic Transfer (e.g., Male Sterility)

The backcross method can be used to transfer cytoplasmic traits, such as male sterility, from wild or unrelated species to cultivated varieties. Male sterility is a valuable trait in hybrid seed production because it eliminates the need for manual emasculation.

» **Example**: In hybrid rice production, cytoplasmic male sterility from wild rice has been transferred to high-yielding rice varieties through backcrossing, enabling large-scale hybrid seed production.

5. Transgressive Segregation

Transgressive segregation refers to the appearance of progeny in the F2 generation that exhibit traits superior to both parents. Backcrossing can promote transgressive segregation by maintaining some degree of heterozygosity. By backcrossing the F1 to the recurrent parent and allowing for further recombination, superior genotypes can emerge.

» **Example**: In maize breeding, backcrossing has been used to combine yield traits from two parents, leading to the production of offspring that outperform both parents in terms of grain yield.

6. Production of Isogenic Lines

Isogenic lines are lines that are genetically identical except for one gene. Backcrossing is used to create isogenic lines by transferring a specific gene into the recurrent parent while keeping the rest of the genotype unchanged.

» **Example**: In tomato breeding, isogenic lines have been developed with different disease resistance genes, allowing breeders to study the effects of individual resistance genes in a uniform genetic background.

7. Germplasm Conversion

Valuable germplasm that may not be directly usable in a breeding program due to undesirable traits can be "converted" through backcrossing. The useful genes from the germplasm can be transferred into an adapted variety, creating "converted lines" that have the desirable traits of the recurrent parent but with the valuable gene(s) from the donor germplasm.

» **Example**: In sorghum breeding, germplasm conversion has been used to introduce drought tolerance from exotic germplasm into locally adapted varieties through backcrossing.

The procedure of the Backcross Method

The **Backcross Method** is a precise and systematic process aimed at transferring specific desirable traits, such as disease resistance or stress tolerance, from a donor parent into an established, high-yielding recurrent parent. The goal is to retain the primary genetic background of the recurrent parent while introducing the new trait without altering its essential characteristics. Here's a detailed look at each step involved in the backcross breeding method:

1. Initial Cross

Objective: To combine the genetic materials of the recurrent parent (the well-adapted, high-yielding variety) with a donor parent that has the desirable trait lacking in the recurrent parent.

» **Selection of Recurrent Parent**: The recurrent parent is typically a well-established, high-yielding variety that performs well under specific conditions but lacks a particular trait (e.g., disease resistance, drought tolerance, or improved grain quality).

» **Selection of Donor Parent**: The donor parent provides the trait that is absent in the recurrent parent. It must possess the trait in a high degree of intensity, such as high resistance to a specific disease, stress tolerance, or quality attributes.

» **Crossing the Parents**: The recurrent parent is crossed with the donor parent to create the F1 hybrid. The F1 hybrid will inherit genetic traits from both parents, but the primary focus is on the desired trait from the donor parent.

2. F1 Generation

Objective: To create a hybrid (F1) that contains a mix of genes from both parents, including the desirable trait from the donor parent.

- **Growing the F1 Plants**: The F1 generation is grown, consisting of plants that have inherited half of their genetic material from the recurrent parent and half from the donor parent.
- **First Backcross**: The F1 plants are crossed back to the recurrent parent. This is the start of the repeated backcrossing process. Since the recurrent parent already possesses most of the desirable agronomic traits, the goal is to gradually eliminate the donor parent's genetic contribution, except for the specific trait being transferred.

3. Repeated Backcrossing

Objective: To recover the genetic background of the recurrent parent while retaining the desired trait from the donor parent.

- **Backcrossing Process**: The progeny from the first backcross are grown, and plants that display the desired trait (e.g., disease resistance) from the donor parent are selected. These selected plants are then backcrossed again to the recurrent parent. This process is repeated for 6-7 generations.
- **Selection of Plants**: In each backcross generation, only plants that exhibit the desired trait from the donor parent are selected. The rest of the genetic material, which comes from the recurrent parent, will be restored over successive backcrosses.
- **Restoration of Recurrent Parent's Genotype**: With each backcross, the genetic composition of the progeny becomes increasingly similar to that of the recurrent parent. By the 6th or 7th backcross, over 98% of the recurrent parent's genotype is restored, with the desired trait from the donor parent fully integrated.
- **Selfing Between Backcrosses**: Occasionally, selfing may be performed between backcrosses to ensure that the desired trait is stable and to facilitate the fixation of the trait within the population.

4. Selection in Advanced Generations

Objective: To finalize the recovery of the recurrent parent's genotype while ensuring that the desired trait from the donor parent is firmly integrated into the progeny.

- **F6 to F7 Generations**: By the time the breeding process reaches the F6 or F7 backcross generations, the plants will closely resemble the recurrent parent, with the main addition being the desirable trait from the donor parent. At this stage, breeders perform a more refined selection, focusing on selecting plants that retain the favorable agronomic traits of the recurrent parent alongside the new trait.
- **Homozygosity**: By the F6 or F7 generation, the population has also reached a high degree of homozygosity for the desired trait, meaning that the plants will consistently pass on the trait to their progeny.

» **Additional Selection for Agronomic Traits**: In addition to the specific trait introduced from the donor parent, breeders may also assess and select for other important traits, such as yield, grain quality, flowering time, and overall plant health.

5. Final Selection and Evaluation

Objective: To evaluate the final backcross progeny for overall performance, confirming that the new variety maintains the desirable traits of the recurrent parent while expressing the introduced trait from the donor parent.

» **Performance Testing**: After completing 6-8 backcross generations, the progeny are evaluated for their agronomic performance under a range of conditions. This evaluation includes testing for the target trait (e.g., disease resistance) as well as overall yield, quality, and stress tolerance.

» **Field Trials**: Field trials are conducted in various locations to test the stability and adaptability of the new variety. These trials assess the plants' performance in different environments to ensure that the desired trait performs well under diverse conditions.

» **Variety Release**: Once the progeny has demonstrated consistent performance in field trials, the new variety is considered for release. The variety will be registered, and seed multiplication programs will begin to produce sufficient seeds for distribution to farmers.

Advantages of the Backcross Method

1. Retention of Favorable Traits

The backcross method excels in its ability to maintain the favorable characteristics of the recurrent parent while introducing one or two specific traits from the donor parent. This allows breeders to improve a high-performing variety by adding traits like disease resistance, drought tolerance, or herbicide resistance without disrupting its essential qualities. This makes it an ideal method for refining already well-adapted varieties that are high-yielding, disease-resistant, or have desirable quality traits. For example, a rice variety that is high-yielding but susceptible to blast disease can be improved by introducing a blast-resistant gene through backcrossing, while preserving the yield and other beneficial traits.

2. Efficient for Simple Traits

The backcross method is particularly efficient when working with traits governed by one or a few major genes. These simply inherited traits, such as resistance to a specific disease or herbicide, are easier to track and transfer using backcross breeding. The method is highly effective when breeders aim to transfer a trait like resistance to rust in wheat or herbicide tolerance in maize. The simplicity of the method allows breeders to make significant improvements in a relatively short amount of time, compared to other methods that might be more suitable for complex traits.

3. Fast-Track Improvement

One of the most significant advantages of the backcross method is its ability to provide fast-track improvement of well-established varieties. Since the process primarily focuses on transferring a specific desirable trait, it avoids the lengthy process of developing a completely new variety from scratch. Once the desired trait is successfully transferred, the variety is almost identical to the original recurrent parent, apart from the new trait. This can save breeders years of development time, particularly for traits that can be easily screened and selected, like pest resistance or herbicide tolerance. It is particularly beneficial for improving crops like rice, wheat, and maize, where high-performing varieties are already widely used but may need enhancement for specific traits.

Limitations of the Backcross Method

1. Limited to Single Traits

The backcross method is largely focused on transferring single traits, making it less effective for improving complex traits governed by multiple genes. Traits like yield, drought tolerance, and overall plant vigor are influenced by multiple genes and environmental factors, making them harder to transfer using the backcross method. For example, improving yield through backcross breeding is often not feasible because it is a quantitative trait influenced by many genetic and environmental factors. This limitation means that the backcross method is often used in conjunction with other breeding methods when more complex traits are involved.

2. Linkage Drag

One of the significant challenges in the backcross method is the risk of **linkage drag**, where undesirable genes that are physically close to the target gene on the chromosome are also transferred from the donor parent. For instance, while transferring a disease-resistance gene, other undesirable traits such as reduced yield or poor grain quality may also be introduced into the new variety. This is because genes that are linked to the desired trait are more likely to be passed along during the backcrossing process. Breeders must carry out further selection to eliminate these unwanted traits, which can prolong the breeding process. In some cases, breaking the linkage between the desired and unwanted traits requires advanced molecular techniques, such as marker-assisted selection, to ensure that only the desired gene is retained.

3. Limited Genetic Variation

The repeated backcrossing to the recurrent parent significantly reduces the genetic variation in the progeny, as most of the donor parent's genetic contribution is eliminated through the backcross generations. This narrowing of genetic diversity can limit the potential for future genetic improvements. While the method is effective for transferring specific traits, it does not contribute to broadening the genetic base of the variety. This lack of genetic diversity can make the new variety more vulnerable to future threats, such as new pests, diseases, or changes in environmental conditions. Moreover, if the same recurrent parent is used repeatedly in different breeding programs, the overall diversity of crop varieties in a region can decline, leading to a potential increase in vulnerability to biotic and abiotic stresses.

G. Single Seed Descent Method

The **Single Seed Descent (SSD) method** is a plant breeding technique used primarily for advancing segregating generations rapidly and efficiently. It was developed as a modification of the bulk method, allowing breeders to fast-track the development of homozygous lines without the need for early selection. This method is commonly used in self-pollinated crops, where it is essential to increase homozygosity and stabilize the genetic material for future selection and evaluation.

Procedure of Single Seed Descent Method

The SSD method involves growing each generation from a single seed of each individual plant in the previous generation, ensuring that all plants in the segregating populations are represented equally. Here's the detailed procedure:

1. **F1 Generation (Hybrid Creation)**

The process begins by creating a cross between two parent plants to generate the F1 hybrid. This hybrid contains the combined genetic material from both parents, with heterozygous loci at multiple genetic sites.

2. **F2 Generation (Single Seed Selection Begins)**

In the F2 generation, single seeds are harvested from each individual plant in the population. These seeds represent the genetic diversity of the F2 population, and each plant is represented equally. The entire population is then advanced by planting these single seeds to grow the F3 generation. This phase marks the beginning of the rapid generation advancement without selection.

3. **F3 to F5 Generations (Rapid Generation Advancement)**

The same procedure is repeated for the F3, F4, and F5 generations. Single seeds are harvested from each plant to grow the next generation. During this stage, the population is allowed to self-pollinate naturally, and no selection is carried out. By the time the plants reach the F5 or F6 generation, the population has attained a high level of homozygosity due to continued selfing. The primary objective here is to speed up the process of homozygosity without interference from environmental factors or early selection pressure.

4. **F6 to F7 Generations (Preliminary Selection Begins)**

After advancing to F6 or F7, the population becomes sufficiently homozygous, and breeders can begin selecting superior plant families or individual plants. By this stage, the population has a much-reduced size because weaker or less viable plants may have been eliminated by natural factors like pests or diseases. Individual plants or progenies that display favorable traits, such as high yield or disease resistance, are selected for further testing and evaluation.

5. **Preliminary Yield Trials (F7 to F8)**

Once the population has achieved significant homozygosity, it is subjected to preliminary yield trials and quality tests. At this point, the breeder can evaluate various lines for agronomic traits, such as yield, disease resistance, maturity, and grain quality. The superior lines are then selected and tested in replicated trials, often over multiple environments, to ensure consistency and stability of the traits across different conditions.

6. **Advanced Yield Trials (F8 and Beyond)**

From the F8 generation onward, selected lines are tested in replicated yield trials. These trials involve growing the selected lines in multiple environments or locations to assess their adaptability, stability, and performance under diverse conditions. These tests are critical for identifying elite lines that can be released as new varieties for commercial production.

Objectives of the Single Seed Descent Method

1. Rapid Advancement of Generations

The SSD method is designed to rapidly advance the generation of crosses. By not focusing on early selection, the breeding program can move quickly through the segregating generations, achieving near-homozygosity in a shorter time frame.

2. High Plant Density Growth

F2 and subsequent generations are grown at very high plant densities. This ensures that breeders can handle large populations while maintaining the genetic diversity of the population. Each F2 plant is equally represented in subsequent generations, and the population remains genetically diverse until the selection phase begins.

3. Equal Representation of F2 Plants

The SSD method ensures that each F2 plant is represented equally in the final population. This is a crucial feature, as it allows all genetic material from the original F2 population to be passed forward, preserving genetic diversity during the early generations.

4. Use of Off-Season Nursery or Greenhouse

The SSD method takes advantage of off-season nursery or greenhouse facilities to advance generations faster. Growing plants in greenhouses or other controlled environments ensures that the breeding cycle continues year-round, maximizing the speed of homozygosity and reducing the overall time required to develop a new variety.

5. Minimal Space, Effort, and Labor Requirements

The SSD method is highly efficient in terms of space, labor, and effort. Since only single seeds are harvested and advanced each generation, the population size is kept manageable, reducing the resources required. This makes it an ideal method for breeders working with limited space or budget.

6. No Selection in Segregating Generations

One of the main principles of the SSD method is that no selection is performed during the segregating generations (F2 to F6). This avoids the influence of environmental factors on the selection process, allowing the population to reach homozygosity without any artificial pressure that could bias the outcome.

7. Natural Population Size Reduction

In each successive generation, the population size naturally reduces due to factors such as disease, pest infestations, or poor plant vigor. These natural factors help eliminate less viable plants, resulting in a smaller, more robust population by the time selection begins in the F6 or F7 generation.

Advantages of the Single Seed Descent Method

1. Speed

The SSD method is one of the fastest ways to achieve homozygosity in self-pollinated crops. By advancing generations rapidly without selection, breeders can develop highly homozygous lines in fewer generations compared to methods that involve selection at each stage.

2. Simplicity

The method is simple to execute, requiring minimal effort in the early generations. No pedigree records need to be maintained, and the process is straightforward, as each generation involves only the collection of single seeds from each plant and their subsequent planting.

3. Maintains Genetic Diversity

Since the population is not subjected to selection pressure in the early generations, the genetic diversity of the original population is preserved until the homozygosity stage is reached. This allows breeders to access a broad range of genetic combinations for selection once the population is nearly homozygous.

4. Cost-Effective

The SSD method is cost-effective because it requires fewer resources in terms of labor, space, and materials compared to other methods like pedigree breeding. Large populations can be managed easily, and the focus on a single seed per plant reduces the need for extensive trial plots in the early generations.

Disadvantages of the Single Seed Descent Method

1. No Early Selection

A key drawback of the SSD method is the absence of early selection, which can result in the advancement of undesirable traits alongside desirable ones. This means that some poor-performing lines are carried forward until the F6 or F7 generation, which could have been eliminated earlier in other breeding methods.

2. Natural Reduction of Population Size

While the natural reduction of population size due to environmental factors is generally advantageous, it can also result in the loss of some valuable genetic material if plants with desirable traits succumb to pests, diseases, or other stresses before selection begins.

3. Limited Use for Complex Traits

The SSD method is less suited for traits influenced by the environment or for breeding objectives that require early selection, such as yield potential or specific quality traits. In these cases, methods that involve earlier selection might be more effective.

4. Limited Genetic Progress in Early Generations

Because selection is deferred until the later generations, there is limited genetic progress in the early stages of the SSD method. This can delay the identification of superior lines and increase the overall time needed to develop a commercially viable variety.

H. Multiline Varieties

Multiline varieties represent an innovative approach to crop breeding that addresses some of the limitations associated with pureline varieties. Pureline varieties, although well-adapted to specific areas, often exhibit limited adaptability and lack stability due to changing environmental conditions and the emergence of new pathogen races. This is particularly problematic when it comes to disease resistance. For instance, many purelines carry only one or a few major resistance genes, making them vulnerable to evolving pathogens. The continuous emergence of new pathogen races can easily overcome the resistance present in a pureline, as seen in the case of the wheat variety *Kalyan Sona*, which, despite its initial resistance to brown rust (*Puccinia recondita*), eventually became susceptible to new races of the pathogen. To mitigate such issues, **multiline varieties** were introduced. These varieties consist of several purelines that share similar agronomic traits, such as height, flowering time, seed color, and yield potential, but each pureline carries different disease-resistance genes. This diversification within a single variety helps buffer against the spread of disease, as not all plants in the field are susceptible to the same pathogen. The purelines that make up the mixture are carefully selected to ensure they are compatible, meaning they do not negatively affect each other's performance when grown together.

Development of Multiline Varieties

The development of **multiline varieties** is a strategic approach in plant breeding, aimed at enhancing disease resistance in crops while maintaining stable agronomic performance. The concept was first suggested by **Norman Borlaug** in 1954, as a response to the limitations of pureline varieties, which often become susceptible to evolving pathogen races. The idea behind multiline breeding is to combine the genetic strengths of multiple lines, each carrying different resistance genes, into a single crop population. This approach enhances the crop's ability to withstand a range of pathogens, particularly those that tend to develop new races that can overcome resistance in pureline varieties.

The Basic Concept of Multiline Varieties

Multiline varieties consist of several genetically distinct lines (usually **five to ten**), all of which are phenotypically similar in terms of height, flowering time, seed color, and other key agronomic traits. However, each line in the mixture carries **different genes for disease resistance**, making the entire population more resilient to disease pressure. When a pathogen infects a crop field, only the lines lacking resistance to that particular pathogen strain are affected, while the rest of the population remains protected. This diversity in resistance genes helps to reduce the spread of the disease, thereby minimizing the overall crop loss. Unlike traditional pureline varieties, where all plants are genetically identical and thus equally vulnerable to pathogen outbreaks, multiline varieties offer a built-in safeguard against disease epidemics. **If one or more lines in the mixture become susceptible to a new pathogen race, they can be replaced** with new resistant lines without compromising the overall yield potential or quality of the crop.

Borlaug's Approach to Multiline Variety Development

Norman Borlaug proposed the use of **backcross breeding** to introduce disease resistance genes into a well-adapted recurrent parent. The recurrent parent serves as the foundation of the multiline variety, providing a stable genetic background for yield, growth habits, and seed quality. By introducing different resistance genes from various donor parents, breeders can create a set of genetically distinct lines that collectively offer broad-spectrum resistance to a range of pathogens. The overall objective is to develop a crop that performs well agronomically but is also highly resilient to evolving pathogen populations.

Step-by-Step Procedure for Developing Multiline Varieties

1. **Selection of a Recurrent Parent**: The process begins with the selection of a well-adapted, high-yielding recurrent parent. This parent is already optimized for the region's environmental

conditions, such as soil type, climate, and farming practices. The recurrent parent is genetically stable in terms of its yield, growth habits, seed quality, and other desirable agronomic traits.

2. **Identification of Donor Parents**: The next step is to identify donor parents that carry specific **resistance genes** to different pathogens. These donor parents may come from different regions or breeding programs and are chosen based on their resistance to key pathogens that are prevalent in the target environment. Each donor parent carries a unique gene or set of genes that provide resistance to a particular pathogen or group of pathogens.
3. **Backcrossing with Recurrent Parent**: Once the donor parents are identified, the process of **backcrossing** begins. Each donor parent undergoes a separate backcross breeding program with the recurrent parent. The aim is to transfer the resistance gene(s) from the donor parent into the recurrent parent while retaining the recurrent parent's agronomic characteristics. After multiple backcrosses (usually 5 to 7), the resulting lines are almost genetically identical to the recurrent parent but now carry the desired resistance gene(s).
4. **Testing and Selection of Lines**: After the backcrossing process is complete, each line is tested for its **resistance** to the target pathogen, as well as for its agronomic performance. Lines that show both good disease resistance and acceptable performance in terms of yield, growth habit, and other traits are selected for inclusion in the multiline variety.
5. **Formation of the Multiline Variety**: The final multiline variety is formed by **mixing the selected lines**, each of which carries different resistance genes but is otherwise agronomically similar. This ensures that the mixture performs uniformly in terms of growth and yield but is protected against multiple races of the target pathogen.
6. **Continuous Monitoring and Replacement**: The breeding process does not stop after the multiline variety is released. Pathogens are continually evolving, and over time, some lines in the multiline variety may become susceptible to new races of the pathogen. **Breeders must continually monitor the performance** of the multiline variety and replace any susceptible lines with new resistant lines developed through backcrossing. This ensures that the multiline variety remains effective in protecting against disease over the long term.

Advantages of Multiline Varieties

The multiline approach offers several key benefits that make it a valuable strategy in modern plant breeding:

1. **Enhanced Disease Resistance**: By incorporating multiple lines with different resistance genes, multiline varieties are much more resilient to disease pressure. Even if one or two lines become susceptible, the overall population remains protected, minimizing the impact of disease outbreaks.
2. **Stability of Agronomic Traits**: Since all the lines in a multiline variety are based on the same recurrent parent, they share similar agronomic traits, such as height, flowering time, and seed quality. This ensures that the crop performs uniformly in the field, making it easier for farmers to manage and harvest.
3. **Slowdown in Pathogen Evolution**: The presence of multiple resistance genes in the multiline variety reduces the selection pressure on pathogens to evolve new virulent races. This helps

to **delay the breakdown of resistance**, prolonging the effectiveness of the variety.

4. **Flexibility in Line Replacement**: If a particular line becomes susceptible to a pathogen, it can be easily replaced with a new resistant line, without needing to redevelop the entire variety. This flexibility makes multiline varieties more adaptable to changing pathogen populations.
5. **Long-Term Sustainability**: Multiline varieties offer a more sustainable solution to disease management compared to varieties that rely on single-gene resistance. By combining multiple resistance genes, multiline varieties help ensure long-term productivity and food security.

Demerits of Multiline Varieties

1. **Frequent Seed Replacement**: One of the major challenges with multiline varieties is that farmers must frequently replace seeds to maintain the effectiveness of the disease resistance. Multiline varieties rely on several distinct lines, each carrying a different resistance gene. Over time, as new pathogen races evolve, some of these lines may become susceptible to the disease, requiring the replacement of these lines with newly developed resistant lines. This process necessitates that farmers continuously update their seed stocks, adding to their operational costs and logistics. Moreover, the frequency of seed replacement can vary depending on the speed at which new pathogen races emerge, which may place additional financial and managerial burdens on farmers.
2. **Inconsistent Disease Control**: While multiline varieties are designed to protect crops from multiple pathogen races, there is still a risk that **new pathogen races could evolve** that overcome all the resistance genes present in the variety. Pathogens are known to evolve rapidly, and in cases where a particularly aggressive race emerges, it can nullify the collective resistance provided by the multiple lines within the multiline variety. In such cases, the entire variety could become susceptible to the disease, leading to substantial crop losses. This issue highlights the constant challenge in plant breeding programs, which must keep pace with the rapid evolution of pathogens.
3. **Complex Seed Production and Certification**: Producing multiline varieties involves mixing multiple lines with similar agronomic traits but different resistance genes. Maintaining the genetic purity of each line in the mixture can be complex, particularly during seed production and certification. Ensuring that each line in the mixture retains its specific resistance gene without mixing up the lines during the process requires strict quality control. The increased complexity in seed production and certification can lead to higher production costs and delays in getting seeds to market.
4. **Yield Trade-offs**: While multiline varieties provide increased disease resistance, there may be minor **trade-offs in yield** compared to optimized pureline varieties. Since multiline varieties consist of several lines with different resistance genes, not all lines may be optimized for yield under all environmental conditions. Some lines may have slightly lower yield potential, leading to a marginal reduction in overall productivity. However, this yield reduction is typically offset by the increased disease resistance and stability offered by multiline varieties, but it remains a factor that farmers need to consider.
5. **Requirement for Continuous Breeding Programs**: Multiline varieties demand **constant**

attention from breeders. As pathogens evolve, breeders must continuously monitor the performance of the lines within the multiline variety and develop new resistant lines to replace those that have become susceptible. This requires ongoing investment in breeding programs, which may not always be feasible for regions with limited resources or access to advanced breeding technologies. Additionally, maintaining a stock of new resistant lines for replacement adds complexity to the breeding process.

6. **Limited Compatibility Across Regions**: Multiline varieties are developed based on the prevalence of specific pathogen races in particular regions. This means that a multiline variety developed for one region may not be effective in another region where different pathogen races dominate. As a result, breeders need to develop **region-specific multiline varieties**, which adds to the overall effort and resources required for maintaining and distributing these varieties.

Achievements in Multiline Varieties

Despite their challenges, multiline varieties have demonstrated significant success in managing diseases, particularly in crops prone to **rapid pathogen evolution**, such as wheat, barley, and rice. Multiline varieties have been particularly effective in controlling rust diseases, which are notorious for producing new pathogen races that overcome resistance in pureline varieties.

Multiline Varieties in Wheat

In wheat (*Triticum aestivum*), several multiline varieties have been developed to combat **rust diseases**, which are a major concern in many wheat-growing regions. Rust pathogens, including **stem rust, leaf rust**, and **stripe rust**, evolve rapidly, often rendering pureline varieties vulnerable within a few years of their release. To address this issue, breeders have developed multiline wheat varieties that combine resistance genes from multiple sources, significantly enhancing the crop's resilience against rust.

» **KSML 3**: This is one of the most notable multiline wheat varieties developed in India. It consists of **eight distinct lines**, each carrying a unique rust resistance gene. The resistance genes in KSML 3 come from a diverse array of sources, including **Robin, Ghanate, K1, Rend, Gabato, Bluebird, and Tobari**. This diversity in resistance genes helps protect the variety against a broad spectrum of rust pathogen races, making it a valuable tool for managing rust diseases in wheat production.

» **MLKS 11**: Another successful multiline wheat variety, **MLKS 11**, includes resistance genes from **eight different donor sources**, including **E 6254, E 6056, E 5868, Frecor, HS 19, and E 4894**. By incorporating resistance genes from multiple sources, MLKS 11 has proven to be an effective line against rust, helping reduce the impact of disease outbreaks in wheat fields. Additionally, its genetic diversity helps slow down the spread of rust diseases, even if one or two lines in the mixture become susceptible.

» **KML 7406**: This variety includes **nine lines**, each carrying resistance genes from different sources. The combination of resistance genes helps protect the wheat crop from a variety of rust pathogen races. Like other multiline varieties, **KML 7406** was developed through a backcross breeding program, using a recurrent parent and introducing resistance genes from various donor parents.

Impact of Multiline Varieties on Disease Management

The development of multiline varieties has had a profound impact on **disease management** in wheat and other cereal crops. The ability to combine multiple resistance genes into a single population has helped prevent large-scale disease outbreaks that can devastate crops and lead to significant economic losses. In particular, multiline varieties have proven to be a valuable tool in managing diseases like **leaf rust**, **stem rust**, and **stripe rust** in wheat, which are often driven by the emergence of new pathogen races. By planting multiline varieties, farmers can reduce their reliance on chemical **fungicides**, which are not only expensive but also pose environmental risks. Multiline varieties provide a more **sustainable approach** to disease management, reducing the need for frequent fungicide applications while ensuring that crops remain productive even in the face of changing pathogen populations.

Future of Multiline Varieties

As pathogens continue to evolve, the development of multiline varieties will remain an important part of plant breeding strategies. **Molecular breeding** techniques, such as **marker-assisted selection** and **genomic selection**, can further enhance the efficiency of developing multiline varieties by allowing breeders to more precisely track the presence of resistance genes and develop new lines more quickly. Moreover, with the advancement of **gene editing technologies** like **CRISPR**, it may become possible to introduce multiple resistance genes into a single line, potentially reducing the need for multiline mixtures while still maintaining broad-spectrum resistance. These advancements could lead to the development of even more **robust and durable** multiline varieties that can better withstand the challenges posed by rapidly evolving pathogens.

Dirty Multiline Concept

The **Dirty Multiline concept** represents an intriguing approach within the framework of multiline breeding, especially for managing plant diseases in crops. In conventional multiline varieties, multiple purelines are used, each carrying different resistance genes to ensure that if one or more lines become susceptible to a pathogen, the overall crop remains protected. The Dirty Multiline takes this concept further by intentionally including **susceptible lines** within the multiline variety, which seems counterintuitive but offers strategic benefits in disease management.

Rationale Behind the Dirty Multiline Concept

The primary rationale behind the Dirty Multiline concept is to reduce **selection pressure** on the pathogen population. When only resistant lines are present in a crop, pathogens that can overcome the resistance genes are selected for, leading to the rapid evolution of **new, more virulent pathogen races**. This accelerated evolution is driven by high selection pressure because only pathogens capable of infecting the resistant lines can survive and reproduce. By deliberately including a small percentage of susceptible lines in the multiline variety, the pathogen has an available host to infect, which reduces the selection pressure to overcome the resistance genes present in the other lines. The presence of these susceptible plants allows the pathogen population to persist without undergoing the same intense evolutionary pressure to mutate. This strategy **slows down the emergence of new pathogen races**, prolonging the effectiveness of the resistance genes in the other lines.

Key Principles of the Dirty Multiline Concept

1. **Inclusion of Susceptible Lines**: A small percentage (usually 1-2 lines) of susceptible plants are deliberately included in the crop mixture. These lines do not possess the resistance genes that protect the other purelines, making them vulnerable to attack by the pathogen.
2. **Balance Between Resistance and Susceptibility**: The key to the success of the Dirty Multiline strategy lies in finding the right balance. The number of susceptible plants should be low enough that the overall yield and crop performance are not significantly affected by disease but high enough to allow a pathogen population to survive without strong selection pressure for virulence.
3. **Reduced Pathogen Evolutionary Pressure**: By providing the pathogen with a host in the form of susceptible plants, the likelihood of the pathogen evolving to overcome the resistance genes in the other lines is reduced. This slows down the evolutionary arms race between the crop and the pathogen, making resistance genes effective for a longer period.
4. **Stability of Crop Resistance**: One of the biggest challenges in breeding for disease resistance is the **stability of resistance over time**. Pathogens are constantly evolving, and it is common for resistant varieties to become susceptible after a few growing seasons due to the emergence of new pathogen races. The Dirty Multiline approach aims to stabilize crop resistance over the long term by reducing the pathogen's evolutionary pressure to develop new virulent strains.
5. **Infection Management**: Even though a small portion of the crop (the susceptible lines) will become infected, this limited infection can act as a **buffer** to protect the resistant lines from being subjected to more virulent pathogen attacks. The overall spread of the disease is thus controlled, and the impact on the entire crop remains minimal.

Benefits of the Dirty Multiline Strategy

1. **Delaying Pathogen Resistance**: By including susceptible lines, the selection pressure on the pathogen population is significantly reduced. This helps delay the emergence of new pathogen races that can overcome resistance genes, providing a more sustainable approach to disease management.
2. **Prolonging the Life of Resistance Genes**: Resistance genes are a crucial part of breeding programs, but they often become ineffective as pathogens evolve. The Dirty Multiline strategy helps prolong the usefulness of these genes by slowing the pathogen's evolutionary response, ensuring they remain effective for multiple growing seasons.
3. **Greater Yield Stability**: Although a small portion of the crop may be susceptible to disease, the overall yield remains stable because the majority of the crop is still protected by resistance genes. This ensures that farmers do not face catastrophic yield losses due to disease outbreaks.
4. **Flexibility and Adaptability**: The Dirty Multiline strategy allows for greater flexibility in breeding programs. Breeders can continually adjust the composition of the multiline variety by replacing susceptible lines with newer resistant lines as needed. This dynamic approach helps the crop population maintain adaptability to changing environmental and pathogen conditions.

Potential Drawbacks and Challenges

1. **Yield Loss in Susceptible Lines**: One of the most obvious downsides of including susceptible lines is the **yield loss** associated with those lines. Since these plants are not resistant to the pathogen, they are likely to experience infections, leading to reduced productivity in the susceptible portion of the crop. The extent of this loss depends on the percentage of susceptible lines included and the severity of the disease.
2. **Risk of Complete Breakdown**: In some cases, new pathogen races may evolve that can infect both the resistant and susceptible lines. This can result in the breakdown of the entire multiline variety. Though the Dirty Multiline concept is designed to delay this outcome, it is not a foolproof method, and there is always a risk that a pathogen could evolve to overcome all resistance present in the multiline variety.
3. **Complexity in Seed Management**: Farmers and breeders need to carefully manage the seed stocks of multiline varieties, particularly when including susceptible lines. It is crucial to maintain the right proportion of resistant and susceptible lines to ensure the effectiveness of the strategy. This requires constant monitoring and adjustments, adding complexity to the breeding and seed distribution process.
4. **Increased Disease Monitoring**: Since the Dirty Multiline strategy intentionally includes susceptible lines, it requires **closer monitoring** of disease progression in the field. Farmers must be vigilant in managing the disease to ensure that the pathogen does not spread too rapidly, causing significant losses.

Examples and Applications of the Dirty Multiline Concept

The Dirty Multiline concept has been explored in crops like wheat and barley, where diseases such as rusts (*Puccinia* spp.) are significant challenges. In these crops, the continual evolution of new pathogen races has led breeders to search for alternative disease management strategies. By incorporating susceptible lines into the multiline mixtures, breeders aim to slow the evolution of new rust races and prolong the effectiveness of existing resistance genes. In **rice breeding programs**, the inclusion of susceptible lines in certain rice cultivars has been tested as a means to manage **blast disease** caused by *Magnaporthe oryzae*. The idea is that by providing a pathogen host, breeders can reduce the need for continuous chemical treatments while still maintaining acceptable yield levels.

In **maize**, where diseases like **northern corn leaf blight** (caused by *Exserohilum turcicum*) and **southern rust** (caused by *Puccinia polysora*) are major concerns, the inclusion of susceptible hybrids in multilines is being considered to manage the evolutionary dynamics of the pathogens involved. This approach is particularly relevant in areas where disease outbreaks are frequent and pose a constant threat to crop production.

I. Hardy-Weinberg Law and its Application in Plant Breeding

The **Hardy-Weinberg Law** is a central concept in population genetics that describes how allele and genotype frequencies remain constant from one generation to the next in a population that is not affected by external evolutionary forces. It is especially relevant in understanding the genetic structure of cross-pollinated crops and forms a theoretical foundation for predicting genetic variation over generations. The law was independently developed by **G.H. Hardy** in 1908 and **Wilhelm Weinberg** in 1909.

In the context of **plant breeding**, this principle helps breeders understand the distribution of alleles in crop populations and provides insights into how different factors, such as selection, mutation, migration, and genetic drift, can impact genetic variation. The Hardy-Weinberg equilibrium serves as a baseline model for a non-evolving population, where mating is random and no evolutionary pressures are acting on the population. Although true Hardy-Weinberg equilibrium is rarely achieved in real-world agricultural systems, it provides a valuable reference point for assessing genetic changes.

Key Formula and Definitions

The Hardy-Weinberg law assumes a population with two alleles for a given gene, typically denoted as **A** (dominant) and **a** (recessive). The law states that allele and genotype frequencies in such a population will remain constant over time in the absence of evolutionary influences. The frequencies of the genotypes in a diploid population (AA, Aa, aa) are calculated as follows:

» $\mathbf{p^2}$ for the homozygous dominant genotype (**AA**),

» **2pq** for the heterozygous genotype (**Aa**),

» $\mathbf{q^2}$ for the homozygous recessive genotype (**aa**).

Where:

- **p** is the frequency of the dominant allele (A),
- **q** is the frequency of the recessive allele (a),
- The sum of p and q is always 1, meaning $\mathbf{p + q = 1}$.

The equation for genotype frequencies becomes: $\mathbf{p^2\ (AA) + 2pq\ (Aa) + q^2\ (aa) = 1}$

This equation illustrates how the alleles combine to form different genotypes in a population.

Assumptions of the Hardy-Weinberg Law

For the Hardy-Weinberg equilibrium to be maintained, several assumptions must be met:

1. **Random Mating**: Individuals must mate randomly, without preference for specific genotypes.
2. **Large Population Size**: Genetic drift (random changes in allele frequencies) is minimized in large populations.
3. **No Selection**: All individuals have equal reproductive success, and no genotype is favored over another.
4. **No Mutation**: Allele frequencies remain constant because no new alleles are introduced.
5. **No Migration**: No individuals enter or leave the population, ensuring the gene pool remains unchanged.

While these conditions are rarely met in practice, the Hardy-Weinberg law is a useful starting point for understanding population genetics in crops and predicting changes due to various breeding practices.

Application of Hardy-Weinberg Law in Plant Breeding

The Hardy-Weinberg law has several important applications in plant breeding. By establishing a baseline for how alleles are distributed in a population, breeders can compare these predictions to actual results and determine the impact of different genetic and environmental factors on a breeding program.

1. Understanding Genetic Variation

In **cross-pollinated crops**, genetic variation is essential for breeding new varieties with improved traits such as yield, disease resistance, and environmental adaptability. The Hardy-Weinberg law helps plant breeders predict how alleles will combine in subsequent generations and how stable the frequencies of desired traits will be. This is particularly useful when attempting to maintain a population with **heterozygous advantages**, such as **hybrid vigor** (heterosis).

In populations that meet Hardy-Weinberg conditions, breeders can expect **allele frequencies** to remain constant. This knowledge can be useful when aiming to maintain specific allele combinations in cross-pollinated crops like **maize, sorghum, and sunflower**. Breeders can design their crossing strategies to optimize the retention of favorable alleles while avoiding those that may negatively impact plant performance.

2. Selection in Breeding Programs

The Hardy-Weinberg law provides a baseline for understanding how **selection** influences a population. In plant breeding, selection is one of the most powerful tools used to increase the frequency of desirable traits. However, selection pressures, such as selecting for disease resistance or higher yields, will disrupt Hardy-Weinberg equilibrium by favoring certain genotypes over others.

By comparing expected allele frequencies under Hardy-Weinberg equilibrium to the observed frequencies after selection, breeders can estimate the **effectiveness of their selection** strategies. If the observed frequency of desirable traits significantly deviates from Hardy-Weinberg predictions, it indicates that selection is successfully altering the population's genetic makeup. This can be especially important in developing disease-resistant crops, where a specific allele conferring resistance might be selected over multiple generations.

3. Inbreeding and Heterozygosity

In **self-pollinated crops** or populations experiencing **inbreeding**, the Hardy-Weinberg law helps breeders quantify the reduction in **heterozygosity** over successive generations. Inbreeding increases the proportion of homozygous individuals in a population, which can lead to **inbreeding depression**, characterized by reduced vigor and fertility.

The Hardy-Weinberg law allows breeders to calculate the expected levels of heterozygosity and homozygosity at each generation of selfing. This information helps breeders decide when to introduce **outcrossing** to reintroduce genetic diversity and prevent inbreeding depression. In crops like **rice and wheat**, which are prone to inbreeding, maintaining a balance between homozygosity for desired traits and enough heterozygosity for vigor is essential.

Factors Affecting Hardy-Weinberg Equilibrium in Plant Breeding

While the Hardy-Weinberg law assumes a population in equilibrium, several factors can disrupt this balance in real-world plant breeding programs:

1. Selection

Selection is a key factor in plant breeding that directly affects **allele frequencies** in a population. By selecting plants with specific traits, breeders intentionally increase the frequency of favorable alleles. For example, selecting plants with disease resistance or high-yield traits will increase the frequency of the associated alleles in the population. Over time, this selective breeding process disrupts the Hardy-Weinberg equilibrium by favoring certain genotypes. In **hybrid breeding programs**, selection is particularly important for maintaining heterosis, where breeders select plants that maintain genetic diversity and heterozygosity. In crops like **maize** and **sunflower**, heterosis results in improved yield, vigor, and resistance, making selection a valuable tool for increasing the frequency of alleles associated with these traits.

2. Mutation

Mutations introduce new genetic variation by altering the DNA sequence of alleles. Although mutation rates are generally low, they can introduce **new alleles** into a population. Over multiple generations, these new alleles can increase in frequency, especially if they provide a selective advantage. For example, a mutation that confers resistance to a particular pathogen can be highly advantageous in breeding programs. While mutations can add valuable diversity to a gene pool, they also disrupt Hardy-Weinberg equilibrium by changing allele frequencies over time. Plant breeders may take advantage of induced mutations, such as **mutagenesis** through chemical treatments or radiation, to create novel alleles for crop improvement.

3. Migration

Migration occurs when individuals from one population move into another and contribute to the gene pool. In plant breeding, this happens when varieties from different regions are introduced and crossed with local varieties. Migration alters the genetic makeup of the population by introducing new alleles, disrupting the Hardy-Weinberg equilibrium. For example, the introduction of a **high-yielding wheat variety** from a different region might bring new alleles into the local population, leading to changes in gene and genotype frequencies. By managing migration, breeders can enhance the diversity of local populations while ensuring the retention of favorable traits.

4. Genetic Drift

Genetic drift refers to random fluctuations in allele frequencies, particularly in small populations. In plant breeding programs with limited genetic material or small breeding populations, genetic drift can lead to the loss of **allelic diversity** over time. This effect is more pronounced in small populations because random events (such as the death of individuals carrying certain alleles) have a greater impact. Drift can lead to the **fixation** of certain alleles, meaning one allele becomes the only allele at a locus, while other alleles are lost. This reduction in diversity can limit future breeding efforts, making it difficult to introduce new traits.

5. Inbreeding

In **self-pollinated crops** or small populations, **inbreeding** increases the proportion of homozygous individuals over time, reducing heterozygosity. Inbreeding can lead to **inbreeding depression**, where plants become less vigorous and more susceptible to diseases or environmental stresses. The Hardy-Weinberg law helps breeders track changes in heterozygosity and predict when inbreeding depression might occur. In crops like **wheat** and **rice**, where inbreeding is common, breeders use outcrossing or **hybridization** to introduce new genetic variation and counteract the negative effects of inbreeding depression.

J. Inbreeding and Inbreeding Depression in Plant Breeding

Inbreeding refers to the mating between individuals that are genetically related by descent or share a common ancestry. This process is frequently observed in self-pollinated crops, where self-fertilization leads to the production of offspring from the same plant. Inbreeding increases the likelihood of homozygosity, which means the alleles for each gene become identical. While this process is essential for stabilizing certain desired traits in plant breeding, it can also lead to a significant reduction in **fitness**, **vigor**, and **reproductive success**, a phenomenon known as **inbreeding depression**. The extent of inbreeding in a population is typically measured by the **inbreeding coefficient**, a value that quantifies the probability that two alleles at a locus are identical by descent.

Historical Context of Inbreeding

Inbreeding and its consequences have been recognized by humans for centuries. Many societies have implemented cultural or legal prohibitions against marriages between closely related individuals due to the known risks associated with inbreeding. In plants, the recognition of inbreeding depression dates back to **Charles Darwin**, who published the book *"Cross and Self-Fertilization in the Vegetable Kingdom"* in 1876. Darwin observed that self-fertilized progeny in maize exhibited weaker growth and reduced vigor compared to cross-fertilized plants. Subsequent detailed studies were conducted by **East (1908)** and **Shull (1909)**, who examined the effects of inbreeding in maize. They found that self-pollination led to the rapid decline in plant vigor, but cross-pollination between inbred lines could restore vigor, laying the foundation for the modern understanding of **heterosis** or hybrid vigor.

Effects of Inbreeding

Inbreeding has several notable effects, particularly in outcrossing species, where genetic diversity is crucial for maintaining plant fitness. The most common effects include the following:

1. Appearance of Lethal and Sublethal Alleles

Inbreeding increases the probability that recessive lethal or sublethal alleles will become homozygous, leading to the expression of deleterious traits. For example, in crops, inbreeding can lead to the manifestation of **chlorophyll deficiencies**, **rootless seedlings**, or **flower deformities**, which result in non-viable or weak plants. Such individuals often fail to survive, resulting in their elimination from the population. However, the presence of these lethal alleles can severely reduce the overall health of the inbred population.

2. Reduction in Vigor

A general decline in vigor, also known as **inbreeding depression**, is one of the most common effects of inbreeding. This reduction manifests as smaller plant size, weaker stems, slower growth, and diminished biomass production. The overall health of the inbred plants is compromised, making them more susceptible to environmental stressors, diseases, and pests.

3. Reduction in Reproductive Ability

Inbred populations typically experience a decline in their reproductive capacity. As a result, fewer flowers are produced, and there is a reduction in seed set and viability. In some severe cases, the reproductive output may be so low that it becomes challenging to maintain the population. This reduced reproductive success can be detrimental to long-term agricultural productivity, especially in crops where high yields are the primary goal.

4. Separation of the Population into Distinct Lines

Inbreeding leads to increased **homozygosity** within the population, causing it to separate into distinct lines or subpopulations. This separation is due to the random fixation of alleles, where different lines may retain different sets of homozygous alleles. As a result, these lines differ in both genotype and phenotype, leading to increased **variance** in the population. The fixation of alleles and the separation of lines limit genetic diversity, making the population more vulnerable to environmental changes and diseases.

5. Increase in Homozygosity

With each successive generation of inbreeding, the population becomes progressively more homozygous. After about 7-8 generations of selfing, the population becomes nearly 99% homozygous. In a homozygous state, plants within a population exhibit very little genetic variation. While this may be desirable for stabilizing certain traits, it reduces the overall adaptive potential of the population, especially when faced with changing environmental conditions.

6. Reduction in Yield

Inbreeding depression often results in a significant loss of **yield**. This yield reduction is a direct consequence of the overall decline in plant vigor, reproductive success, and resistance to stressors. Inbred lines that survive and can be maintained through selfing generally have much lower yields than their outcrossed counterparts or open-pollinated varieties. The loss of yield is particularly critical in commercial crop production, where maximizing output is a primary concern.

Degrees of Inbreeding Depression

Different plant species exhibit varying degrees of inbreeding depression. These species can be categorized into four broad groups based on the severity of the depression they exhibit:

1. High Inbreeding Depression

Species such as **Lucerne** (alfalfa) and **carrot** show severe inbreeding depression. In these plants, the effects of inbreeding are so pronounced that lethal and sublethal alleles often appear after just a few generations of selfing. Maintaining inbred lines in these species is particularly challenging due to the rapid decline in plant health and fertility.

2. Moderate Inbreeding Depression

Crops like **maize, sorghum** (jowar), and **pearl millet** (bajra) experience moderate levels of inbreeding depression. While the effects of inbreeding are noticeable, breeders can still maintain inbred lines for several generations. These species have become important models for understanding the impact of inbreeding depression and hybrid vigor in plant breeding programs.

3. Low Inbreeding Depression

Certain species, such as **cucurbits** (e.g., cucumber, squash) and **sunflower**, exhibit only a mild degree of inbreeding depression. While some negative effects may be observed, these species can tolerate inbreeding to a greater extent without significant loss of vigor or reproductive success.

4. No Inbreeding Depression

Self-pollinated crops, such as **wheat, rice**, and **barley**, show little or no inbreeding depression. In these species, selfing is a natural process, and the plants have evolved mechanisms to cope with high levels of homozygosity. Consequently, these crops do not experience a significant reduction in fitness or yield due to inbreeding.

Managing Inbreeding in Breeding Programs

While inbreeding is an inevitable consequence of breeding programs aimed at stabilizing certain traits, plant breeders must carefully manage inbreeding to minimize its negative effects. One common strategy to counteract inbreeding depression is the use of **hybridization**. By crossing two genetically distinct inbred lines, breeders can exploit **heterosis** or hybrid vigor, where the resulting offspring exhibit improved growth, yield, and resistance to stress compared to the parent lines. In crops like **maize**, hybrid breeding has become a cornerstone of modern agriculture. The hybrid offspring not only exhibit superior vigor but are also more resilient to environmental fluctuations, diseases, and pests. Hybrid seeds are often produced commercially for farmers, ensuring that they benefit from the enhanced performance of hybrid varieties without the detrimental effects of inbreeding depression.

K. Heterosis:

Heterosis, also known as **hybrid vigor**, is a biological phenomenon observed when the offspring (F1 hybrid) resulting from the crossbreeding of two genetically distinct parent lines exhibit superior traits compared to either of the parents. These traits may include enhanced **yield, growth rate, size, vigor, reproductive success**, or even better resistance to diseases and environmental stresses. The term heterosis was first coined by **George H. Shull** in 1914, who laid the foundation for understanding this phenomenon in a scientific context. Heterosis has since become a cornerstone in modern plant breeding, allowing breeders to create highly productive and resilient crop varieties. The concept of heterosis revolutionized agriculture, particularly with the development of **hybrid maize varieties**, which exhibited significantly higher yields than the open-pollinated varieties that were previously used. This phenomenon has been widely studied and applied not only in maize but also in crops like **rice, sorghum, tomatoes, and sunflowers**, among others, allowing for improved agricultural productivity across the world.

Historical Development of Heterosis

The historical observation of hybrid vigor dates back to **Joseph Koelreuter** in the 18th century, who reported increased vigor in tobacco hybrids. However, it was **Charles Darwin** who made significant contributions to the understanding of heterosis through his work in 1876, where he documented that hybrids between unrelated plant types tend to exhibit enhanced growth, health, and productivity compared to their parents. In his book *"Cross and Self-Fertilization in the Vegetable Kingdom,"* Darwin concluded that cross-pollination between different plant strains resulted in stronger and more vigorous offspring.

Later, in the early 20th century, **George H. Shull** and **Edward M. East** expanded on Darwin's findings by conducting detailed studies on **maize**. They discovered that inbred lines of maize, which were genetically homogeneous, showed reduced vigor and productivity. However, when two inbred lines were crossed, the resulting F1 hybrid exhibited superior traits compared to both parents. This phenomenon of hybrid vigor or heterosis was formalized by Shull and East, setting the stage for the development of hybrid varieties, particularly in maize, which has become a global staple.

The Mechanism Behind Heterosis

The exact mechanism of heterosis is still a topic of extensive study, but several genetic theories have been proposed to explain this phenomenon:

1. **Dominance Hypothesis**: According to this hypothesis, heterosis occurs because the hybrid offspring inherit dominant alleles from both parents, which mask the deleterious effects of recessive alleles present in either parent. This leads to the suppression of harmful traits and the expression of beneficial traits in the hybrid.
2. **Overdominance Hypothesis**: This theory suggests that heterosis results from the interaction between different alleles at a single locus. The hybrid genotype (heterozygous) may outperform either homozygous parent due to the combined effects of both alleles, leading to enhanced vigor.
3. **Epistasis**: Another explanation for heterosis involves epistasis, where interactions between genes at different loci lead to a synergistic effect that enhances certain traits in the hybrid offspring.
4. **Physiological and Metabolic Factors**: Some studies have also suggested that heterosis might be due to improved physiological processes such as better nutrient uptake, more efficient photosynthesis, or enhanced metabolic efficiency in hybrid plants.

Manifestations of Heterosis

Heterosis can manifest in various forms depending on the crop and the traits being targeted. The most common manifestations include:

1. **Increased Yield**: One of the most important benefits of heterosis is a marked increase in yield. Hybrids typically outperform both parents in terms of grain, fruit, seed, leaf, or tuber yield. This increase in productivity is crucial for meeting the global food demand, especially in crops like maize, rice, and wheat.

2. **Improved Reproductive Ability**: Hybrids often show a greater number of flowers, fruits, or seeds compared to their parent lines. This enhanced reproductive ability directly translates into higher productivity and better adaptability to different growing conditions.
3. **Increased Size and General Vigor**: Heterosis is commonly associated with an increase in plant size, healthier growth, and a more robust overall appearance. The hybrids are generally faster growing and exhibit improved resistance to environmental stresses such as drought or poor soil conditions.
4. **Better Quality**: In many cases, hybrids also exhibit improved quality traits. For instance, hybrid onions show better storage and quality, while hybrid tomatoes often have better flavor and shelf life. Improved quality traits are particularly important in crops that are consumed fresh or processed, as better quality increases their market value.
5. **Earlier Flowering and Maturity**: Hybrids often mature faster than their parents, which can be a significant advantage, especially in regions where the growing season is limited by climatic factors. Early maturity allows farmers to harvest crops before adverse weather conditions set in, reducing the risk of crop loss.
6. **Greater Resistance to Diseases and Pests**: Another critical advantage of heterosis is that hybrids often show improved resistance to pests and diseases compared to their parent lines. This is especially valuable in regions where crop losses due to biotic stresses are common, as disease-resistant hybrids can reduce the need for chemical inputs, thereby promoting sustainable agricultural practices.
7. **Greater Adaptability**: Hybrids generally exhibit better adaptability to a wide range of environmental conditions compared to inbred lines. This makes them ideal for cultivation in diverse geographical regions, where they can thrive under varying conditions of soil, temperature, and rainfall.
8. **Faster Growth Rate**: Hybrids often have a faster growth rate, allowing them to reach maturity quicker than either parent. This trait is particularly important in crops where rapid growth is desirable, such as vegetables or short-season cereals.

Estimation of Heterosis

The degree of heterosis is typically measured by comparing the performance of the hybrid (F1) to its parents. Several methods are used to estimate heterosis:

1. **Average Heterosis**: This is the superiority of the F1 hybrid over the mid-parent value, which is the average performance of the two parents. This type of heterosis is often used in breeding programs but may not be as agriculturally significant if the improvement is modest.

Average Heterosis = (F1–MP/ MP) × 100

Where F1 is the mean performance of the hybrid, and MP is the mid-parent value (the average of the two parents).

2. **Heterobeltiosis**: This type of heterosis measures the superiority of the F1 hybrid over the better parent. It is more agriculturally useful as it indicates whether the hybrid has exceeded the performance of the best-performing parent.

Heterobeltiosis = (F1–BP/ BP) × 100

Where BP is the performance of the better parent.

3. **Economic Heterosis:** This type of heterosis compares the F1 hybrid's performance to that of a high-yielding commercial variety. It is particularly important in determining whether a hybrid is economically viable for large-scale agricultural production.

Economic Heterosis = (F1–CV/CV) × 100

Where CV is the performance of the commercial variety.

4. **Negative Heterosis:** In some cases, hybrids may perform worse than their parents for certain traits. This is termed negative heterosis and may be undesirable in agricultural contexts. For example, a hybrid may exhibit **delayed maturity** compared to its parents, which may be a disadvantage in regions with short growing seasons.

Agricultural Implications of Heterosis

Heterosis has been a game-changer in modern agriculture, especially in crops like maize, where hybrid varieties have revolutionized production. The use of hybrid seeds has allowed for a dramatic increase in food production worldwide, contributing to food security. However, the benefits of heterosis extend beyond maize to other crops such as rice, sorghum, cotton, and vegetables, where hybrid varieties have led to higher yields, better quality, and improved resistance to biotic and abiotic stresses.

The development of **F1 hybrids** is not only valuable for increasing yield but also for **sustaining crop production** in the face of **climate change, pests**, and **diseases.** As agricultural challenges continue to evolve, the importance of hybrid breeding and heterosis will only grow, helping farmers adapt to new environmental conditions while maximizing productivity.

Chapter - 4

Breeding Methods of Crops

A. Breeding Methods for Asexually/Clonally Propagated Crops

A significant number of crops, particularly those grown for commercial purposes in both agriculture and horticulture, are propagated asexually. Common examples include sugarcane (*Saccharum officinarum*), potato (*Solanum tuberosum*), sweet potato (*Ipomoea batatas*), banana (*Musa paradisiaca*), ginger (*Zingiber sp.*), turmeric (*Curcuma domestica*), various fruit trees like mango (*Mangifera indica*), citrus fruits (*Citrus spp.*), apples (*Malus domestica*), pears (*Pyrus communis*), peaches (*Prunus persica*), and ornamentals and grasses. In these crops, asexual propagation is used to preserve specific traits or to maintain genetic consistency, which is essential in preventing undesirable recombination and segregation that can arise through sexual reproduction. Many asexually propagated crops have reduced flowering or poor seed sets, limiting their natural reproductive processes. A few varieties may not flower at all, such as certain sugarcane, banana, and potato varieties. However, crops that rely on fruits or require seed formation for proper fruiting do flower and set seeds regularly, despite being propagated vegetatively.

Importance of Asexual Reproduction in Breeding

Asexual reproduction plays a crucial role in modern plant breeding, particularly in the propagation of vegetatively reproduced crops. One of the most significant advantages of asexual propagation is its ability to preserve the exact genetic makeup of the parent plant. Since vegetative propagation involves mitotic cell division, the progeny are genetically identical to the parent, which ensures that valuable traits are maintained across generations without alteration. This is a major distinction from sexual reproduction, where meiosis and fertilization introduce genetic variation through the reshuffling and recombination of alleles.

Preservation of Desirable Traits

In sexual reproduction, the crossing of two individuals results in genetic recombination, which can lead to a mix of both desirable and undesirable traits in the offspring. In contrast, asexual reproduction enables breeders to bypass these uncertainties by replicating only the desired genotype indefinitely. This is particularly beneficial for crops that have been selectively bred for specific traits such as high yield, disease resistance, superior fruit quality, or tolerance to environmental stresses. The ability to clone plants ensures that these traits are passed on unchanged to successive generations, making it easier

DOI: 10.1201/9781003757412-4

to maintain the quality and performance of a crop. For example, in crops like **sugarcane** (*Saccharum officinarum*) and **potato** (*Solanum tuberosum*), which are traditionally propagated vegetatively, asexual reproduction ensures that high-yielding, disease-resistant clones can be perpetuated over many cycles of cultivation. This consistency is essential for maintaining uniformity in commercial crop production, where predictability and stability of traits such as yield, size, and resistance are critical to meeting market demands.

Maintaining Genetic Uniformity

The absence of genetic variation in asexually propagated crops allows breeders to achieve a high degree of uniformity within a population. This is particularly important for commercial farming, where uniformity in crop characteristics such as size, growth rate, and ripening time can facilitate more efficient management, harvesting, and processing. For example, fruit trees like **apples** (*Malus domestica*) and **citrus** (*Citrus spp.*), propagated through vegetative means such as grafting, produce orchards where every tree is genetically identical, resulting in uniformity in fruit size, quality, and ripening times. Moreover, this uniformity helps to ensure that consumers receive a consistent product, which is especially important in markets that value standardization, such as the food and ornamental plant industries. In crops like **banana** (*Musa paradisiaca*), where the commercial variety 'Cavendish' dominates global production, asexual propagation ensures that bananas of the same size, taste, and texture are produced worldwide, maintaining consumer expectations and market stability.

Eliminating Segregation and Recombination

In asexual reproduction, there is no opportunity for segregation and recombination, processes that occur during sexual reproduction and lead to genetic variation. This is advantageous in cases where the plant has reached an optimal genotype, and the breeder wishes to maintain the purity of specific traits. In crops like **cassava** (*Manihot esculenta*) and **ginger** (*Zingiber officinale*), where the goal is to propagate a variety with a proven set of desirable traits, asexual reproduction allows for the exact replication of these traits without the unpredictability associated with sexual reproduction. Asexual propagation also allows breeders to work with **heterozygous plants**, which may harbor complex and beneficial combinations of alleles. Through asexual reproduction, heterozygous genotypes can be perpetuated across generations without the loss of heterozygosity that would occur with self-pollination or cross-breeding. This makes asexual reproduction especially important for polyploid crops, which often exhibit high levels of heterozygosity and can suffer from inbreeding depression if reproduced sexually.

Speed and Efficiency in Breeding Programs

In breeding programs, asexual reproduction can significantly accelerate the process of cultivar development and deployment. While sexual reproduction requires several generations of crossing, selection, and evaluation to stabilize a genotype, asexual reproduction allows for the immediate propagation of a superior genotype once it has been identified. This can save breeders years of effort, as they do not need to stabilize the population through recurrent selection or hybridization cycles. For crops with long life cycles, such as fruit trees like **mango** (*Mangifera indica*) or **pear** (*Pyrus communis*), asexual propagation can be invaluable. Instead of waiting for a tree to mature and produce seeds that may or may not exhibit the desired traits, breeders can graft or clone a tree that has the desired characteristics, thus shortening the time required to produce and evaluate new cultivars.

Economic and Environmental Benefits

From an economic standpoint, the stability and uniformity provided by asexual reproduction are highly advantageous. Inconsistent yields or crop performance due to genetic variability can lead to financial losses for farmers. Asexual propagation ensures a predictable output, which is critical for large-scale commercial agriculture. Additionally, the need for fewer breeding cycles and faster cultivar deployment can reduce the costs associated with developing new varieties, making asexual propagation more cost-effective in certain crops. Environmentally, asexual reproduction can reduce the need for chemical inputs, particularly pesticides and fungicides if the propagated clones have been selected for resistance to pests and diseases. By maintaining these resistance traits across generations, farmers can avoid the need for heavy chemical interventions, leading to more sustainable agricultural practices.

Limitations of Asexual Reproduction

While asexual reproduction offers many advantages, it also presents certain limitations. One of the most significant drawbacks is the **lack of genetic diversity**. Since all individuals within a clone are genetically identical, they are equally susceptible to new diseases or environmental stresses. This has been particularly evident in crops like banana, where the 'Cavendish' variety is highly vulnerable to diseases like **Panama disease** caused by *Fusarium oxysporum*. In addition, **mutations** can accumulate over time in clones, leading to **clonal degeneration.** These mutations may affect traits like yield, disease resistance, or overall plant vigor. Once these degenerative traits manifest, the entire clone population may need to be replaced. Despite these challenges, the role of asexual reproduction in modern plant breeding remains vital. Its ability to preserve and perpetuate desirable traits across generations makes it indispensable for maintaining consistency and quality in many of the world's most important crops.

Characteristics of Asexually Propagated Crops

Asexually propagated crops, also referred to as clonal crops, possess a unique set of characteristics that distinguish them from crops propagated through sexual reproduction. These features are central to their agricultural and horticultural management and play a critical role in their breeding and maintenance. The following characteristics are commonly observed in asexually propagated crops:

1. Perennial Nature

A significant proportion of asexually propagated crops are **perennials**, meaning they persist for many growing seasons. These crops continue to grow year after year without the need for replanting. **Perennial crops** like **sugarcane** (*Saccharum officinarum*), **banana** (*Musa paradisiaca*), and many **fruit trees** like **mango** (*Mangifera indica*) and **citrus** (*Citrus spp.*) are propagated vegetatively to preserve their genotype and ensure consistent yields over extended periods. While many asexually propagated crops are perennials, there are also **annual crops**, such as **potato** (*Solanum tuberosum*), **cassava** (*Manihot esculenta*), and **sweet potato** (*Ipomoea batatas*), which are propagated asexually but are harvested annually. In both cases, the asexual propagation of these crops ensures the preservation of the parent genotype, which is vital for maintaining high-quality and uniform traits from season to season.

2. Low Flowering and Seed Set

Asexually propagated crops often exhibit **low or absent flowering** and **reduced seed set**, a common trait in plants that have evolved to reproduce primarily through vegetative means. This is especially true for crops like **banana, cassava**, and certain varieties of **sugarcane** and **potato**. These crops may produce flowers sporadically, but they do not rely on seeds for propagation due to challenges like poor seed viability, reduced seed formation, or the production of non-viable seeds in polyploid species. In many cases, farmers and breeders avoid sexual reproduction in these crops to prevent the genetic segregation and recombination that occurs with seed production, which can lead to a loss of desirable traits. For instance, **commercially important banana varieties**, such as 'Cavendish,' produce very few or no seeds, making them ideal candidates for vegetative propagation methods like suckers or tissue culture.

3. Heterozygosity and Polyploidy

One of the defining characteristics of many asexually propagated crops is their **high degree of heterozygosity**. These crops often contain a large amount of genetic variation within their genotype, which can result in complex and beneficial traits such as **vigor, yield stability**, and **disease resistance**. For example, crops like **cassava** and **sweet potato** are highly heterozygous, which makes it difficult to maintain their genetic makeup through sexual reproduction. Additionally, many asexually propagated crops are **polyploids**, meaning they have multiple sets of chromosomes. **Sugarcane** and **potato**, for instance, are polyploid crops, with **sugarcane** typically being an octoploid (8x) and **potato** commonly a tetraploid (4x). This high level of **polyploidy** adds to the complexity of their genetic makeup, as the presence of multiple chromosome sets can complicate traditional breeding approaches. As a result, vegetative propagation is preferred for preserving the desirable traits of these polyploid crops.

4. Clonal Nature

The most fundamental characteristic of asexually propagated crops is their **clonal nature**. These crops are propagated from a single parent plant, and as a result, all the offspring are genetically identical to the original plant. This means that **clones** are essentially replicates of the parent plant, with no genetic variation among the members of the population. This genetic uniformity is a key advantage in commercial agriculture, as it ensures that crops grown from the same clone will have **consistent traits** such as yield, size, taste, and resistance to pests or diseases. In clonal crops, any differences observed among individuals are not due to genetic factors but are solely the result of **environmental influences**. For example, variation in plant growth, size, or fruit quality within a clone can often be traced back to differences in soil fertility, water availability, or microclimatic conditions, rather than to genetic differences. This **lack of genetic variation** within a clone makes asexual propagation highly advantageous when it comes to maintaining uniformity and stability in crop traits.

Key Advantages of Clonal Crops

Clonal propagation offers several advantages in maintaining the genetic integrity of high-performing crop varieties. Since each clone is genetically identical, traits like **yield stability, disease resistance**, and **fruit quality** can be reliably replicated across large areas of cultivation. This makes clonal propagation ideal for crops that are cultivated on a large commercial scale, such as **sugarcane, potato**, and **banana**. Additionally, the **perpetuation of heterozygosity** in many asexually propagated crops allows breeders to preserve complex genetic combinations that would be lost through sexual reproduction. In highly heterozygous crops, important traits may be the result of interactions between multiple alleles, making it crucial to maintain the genetic integrity of the original parent plant. By using vegetative propagation, breeders can ensure that these valuable traits are maintained without the risks of genetic segregation and recombination that accompany sexual reproduction.

Challenges Associated with Asexual Propagation

Despite its many advantages, asexual propagation is not without its challenges. The most significant drawback is the **lack of genetic diversity** within clonal populations. Because all the individuals in a clonal population are genetically identical, they are equally vulnerable to **diseases**, **pests**, and **environmental stressors**. For instance, a pathogen that can infect one member of the clone will likely be able to infect all members, as there is no genetic variation to provide resistance. This vulnerability was starkly highlighted during the **Irish Potato Famine** in the 19th century, where a single disease devastated large monocultures of genetically identical potato plants.

Another challenge is **clonal degeneration**, a phenomenon where clones may lose vigor over time due to the accumulation of **somatic mutations** or the buildup of diseases, particularly viral infections. While clones can theoretically be maintained indefinitely through asexual propagation, in practice, they may degrade in performance over time. As a result, breeders and farmers must carefully monitor clonal populations and take steps to rejuvenate them through **tissue culture** or the introduction of new, healthy propagules.

Breeding Methods for Asexually Propagated Crops

Breeding methods for asexually propagated crops primarily focus on clonal selection and hybridization to introduce new traits while maintaining the desired genotype.

Clonal Selection in Asexually Propagated Crops

Clonal selection plays a pivotal role in the improvement of asexually propagated crops, as it focuses on identifying and selecting superior clones from a population based on desirable traits. Unlike sexually propagated crops, where genetic recombination and segregation introduce variability, asexually propagated crops maintain genetic uniformity across generations. Therefore, the emphasis in clonal selection is on evaluating the performance of different clones and ensuring their consistent expression of favorable traits.

Key Features of Clones in Clonal Selection

Clone

A clone is a group of plants derived from a single parent through **asexual reproduction**, meaning the progeny is produced without the union of gametes. This method of propagation ensures that all offspring, referred to as **clonal crops**, have the exact same genetic makeup as the original parent. A large number of agricultural and horticultural crops, especially those propagated through vegetative methods such as cuttings, tubers, and rhizomes, consist of clones. For example, crops like sugarcane, banana, potato, and many fruit trees like mango and citrus are propagated clonally. These crops rely on cloning to maintain uniformity in desirable traits such as yield, disease resistance, and fruit quality.

Characteristics of Clones

Clones exhibit several unique characteristics due to their mode of propagation, which differ significantly from sexually propagated crops. Below are the major traits of clonal crops:

1. Identical Genotype

One of the most defining features of clones is that all individuals within a clone are **genetically identical** to the parent plant. This genetic uniformity arises because asexual reproduction involves only **mitotic cell division**, which ensures the faithful replication of the parent plant's DNA in every new plant. Unlike sexual reproduction, which shuffles genes through meiosis and fertilization, cloning preserves the entire genetic structure across generations. For instance, when a potato tuber is planted, all the plants that arise from it share the same genetic information, and thus, they are genetically identical. Since clones are derived from a single parent through asexual reproduction, all the plants within a clone share an identical genetic makeup. This results in a lack of genetic segregation or recombination, as mitotic division maintains the exact genetic composition of the parent. Consequently, any **phenotypic variation** observed within a clone is primarily due to environmental factors, rather than genetic differences.

In this sense, a clone can be considered a "fixed" genetic line, and its genotype remains **stable indefinitely** unless it is affected by external factors like mutations. This stability allows the traits of the parent plant, such as disease resistance, yield, or fruit quality, to be perpetuated over generations without variation. The uniformity within clones allows for the **reliable reproduction of desired traits**, making it easier for breeders to maintain specific characteristics such as disease resistance, yield, and quality over multiple growing seasons. This is particularly advantageous in crops like **sugarcane** (*Saccharum officinarum*), **potato** (*Solanum tuberosum*), **banana** (*Musa paradisiaca*), and many other fruit and ornamental crops, where consistency in trait expression is crucial for commercial production. However, uniformity also has its drawbacks, particularly in terms of **vulnerability to environmental stresses**. Since all the individuals within a clone share the same genotype, they are equally susceptible to diseases, pests, and adverse environmental conditions. For example, if a clone lacks resistance to a particular pathogen, the entire population of that clone is at risk of being wiped out by the disease.

2. Lack of Genetic Variation

Due to the genetic uniformity within a clone, there is **no genetic variation** among the individual plants. Unlike sexually propagated crops, where new genetic combinations are created through segregation and recombination, cloning preserves the genetic identity without introducing new alleles. Consequently, **phenotypic variation** within a clone is purely the result of **environmental factors**, not genetic differences. For example, differences in soil fertility, moisture levels, or light exposure can lead to variability in plant growth or yield within a cloned population, but these differences do not stem from any genetic variation.

The absence of genetic variation has both advantages and disadvantages. While uniformity is beneficial for maintaining consistent traits, it can also make the entire clonal population **vulnerable to diseases and environmental stresses**, as there is no genetic diversity to provide resilience.

3. Phenotypic Variability Due to Environment (G x E Interaction)

Even though clones are genetically identical, their phenotype can vary due to environmental influences. The phenotype (P) of a plant is the result of its **genotype (G)**, the **environment (E)** in which it grows, and the **genotype by environment interaction (G x E)**. The relationship between these factors can be represented by the following equation:

$$P=\mu+G+E+G\times E$$

Where:

- **P** = Phenotypic expression
- **μ** = Population mean (average trait)
- **G** = Genetic contribution to the phenotype
- **E** = Environmental contribution
- **G × E** = Interaction between genotype and environment

In clonal crops, the **G** component remains constant, while **E** and **G x E** influence phenotypic differences. For example, clones grown in nutrient-rich soils may exhibit greater vigor and higher yields compared to those grown in poorer soils. However, this variability is entirely environmental and does not indicate genetic differences within the population.

4. Immortality of Clones

In theory, clones can be propagated **indefinitely** through asexual reproduction, as their genetic makeup remains unchanged with each generation. This **"immortality"** of clones makes them ideal for long-term cultivation, as superior genotypes can be perpetuated without the risks of genetic segregation that occur with sexual reproduction. For instance, a high-yielding sugarcane variety can be propagated year after year through cuttings, maintaining its performance and genetic integrity. However, in practice, clones are not entirely immortal. Over time, clones may experience **degeneration** due to several factors, including **accumulation of viral infections**, **pest infestations**, and **genetic mutations**. Viral infections, in particular, are a significant threat to the longevity of clones. Many crops propagated vegetatively, such as **potatoes** and **banana**, are susceptible to viral infections that can gradually reduce their vigor and yield. As a result, the performance of the clone may decline over time, necessitating the **selection of new, disease-free clones** or the **use of tissue culture** techniques to eliminate viral pathogens from infected plant material. In addition to disease, **somatic mutations**—spontaneous changes in the DNA of somatic cells—can accumulate over successive generations of a clone. These mutations can lead to changes in the phenotypic traits of the clone, which may be undesirable. Thus, while clones can theoretically be maintained indefinitely, careful monitoring and periodic selection of healthy, vigorous clones are essential for ensuring their long-term productivity.

5. Severe Inbreeding Depression

A key challenge with asexually propagated crops is their **high degree of heterozygosity**. Many clonal crops, such as **bananas, potatoes**, and **cassava**, are highly heterozygous, meaning they have multiple alleles for various traits. This heterozygosity contributes to their **vigor**, **high yield**, and **resilience**.

However, if these crops were propagated through sexual reproduction, the resulting offspring would exhibit **severe inbreeding depression** due to the loss of heterozygosity. **Inbreeding depression** refers to the decline in vigor, fertility, and overall fitness that occurs when genetically similar individuals are crossed, leading to an increase in homozygosity. In clonal crops, any attempt to **self-pollinate** these highly heterozygous plants would result in progeny with reduced genetic diversity, leading to poor performance in terms of yield, disease resistance, and growth rate.

As a result, **asexual propagation** is the only practical method for maintaining the desirable traits of these crops. By avoiding sexual reproduction, breeders can ensure that the **heterozygous genetic makeup** of the parent plant is preserved, preventing the negative effects of inbreeding depression. This is especially important for crops like **bananas**, which are sterile and cannot be propagated through seeds, and **potatoes**, where sexual reproduction leads to wide variability in tuber size, shape, and quality.

Steps in Clonal Selection

The process of clonal selection involves a series of carefully planned steps that enable breeders to evaluate and select the best clones for further propagation. Since all individuals within a clone are genetically identical, the goal is to identify clones that consistently express the desired traits across different environments and growing conditions.

1. First Year: Initial Selection

In the first year, breeders select a **large number of desirable plants** from a mixed, variable population. This initial selection is often based on **observable traits** such as plant vigor, disease resistance, and other characteristics with high heritability. At this stage, the breeder focuses on eliminating obviously weak or undesirable plants. Hundreds or even thousands of plants may be selected for further evaluation.

2. Second Year: Clonal Testing

In the second year, the selected clones are grown separately, typically without replication, due to the limited availability of propagating material. This stage allows breeders to gain a better understanding of the clones' **performance** and **uniformity**. Based on visual observations and basic agronomic assessments, the number of clones is drastically reduced, with only **50 to 100 superior clones** advancing to the next stage.

3. Third Year: Preliminary Yield Trials

The selected clones are evaluated in **replicated preliminary yield trials** alongside standard check varieties for comparison. At this stage, breeders assess the clones for traits like **yield**, **quality**, **disease resistance**, and **adaptability**. Clones that perform well in these trials are further evaluated in **multilocation trials**. Quality testing and disease resistance evaluations are also conducted at this stage to ensure the clones meet the desired standards.

4. Fourth to Sixth Years: Multilocation Trials

From the fourth year onward, **replicated yield trials** are conducted at multiple locations to assess the clones' performance in different environmental conditions. This step is crucial for determining the clones' adaptability and stability across various environments. Traits like **yield**, **quality**, and **resistance to biotic and abiotic stresses** are rigorously evaluated. The best-performing clone, which is superior to the check varieties in one or more characteristics, is identified for release as a new variety.

5. Seventh Year: Release and Multiplication

In the final year, the superior clone is **multiplied** and released as a new variety. The new clone is then propagated on a large scale and made available to farmers for commercial cultivation. Once released, the clone is subjected to further **monitoring** to ensure its long-term performance and resistance to diseases and pests.

Merits of Clonal Selection

1. **Preservation of Genotype**: Clonal selection is highly advantageous for preserving the exact genetic makeup of superior plants. In vegetatively propagated crops, the genotype is maintained without the risk of segregation or recombination that occurs in sexually propagated plants. This allows breeders to perpetuate valuable traits, such as high yield or disease resistance, across generations without genetic changes. This preservation ensures that desirable characteristics remain consistent, which is particularly important for maintaining uniformity in commercial production.
2. **Rapid Introduction of Desirable Traits**: Clonal selection enables the quick propagation of plants with superior traits. Once a superior clone is identified, it can be multiplied rapidly through asexual methods like cuttings, tubers, or tissue culture. This speed is crucial in agriculture, where developing a new variety through sexual reproduction could take years of breeding cycles. For crops like potato or sugarcane, once a high-yielding or disease-resistant clone is found, it can be quickly reproduced to meet the demands of large-scale farming, ensuring quicker access to improved varieties.
3. **Highly Efficient for Traits with High Heritability**: Clonal selection is particularly effective when selecting for traits with high heritability, such as plant height, disease resistance, or fruit quality. These traits tend to be passed from generation to generation with little variation, making selection for these attributes more reliable. Since clonal plants are genetically identical, any phenotypic differences are primarily due to environmental factors, allowing breeders to confidently select clones based on observable traits, knowing that the genetic consistency will remain intact.

Demerits of Clonal Selection

1. **Lack of Genetic Variation**: One of the major disadvantages of clonal selection is the absence of genetic diversity within the crop population. Since all plants within a clone are genetically identical, they share the same vulnerabilities to diseases, pests, and environmental changes. This homogeneity can lead to catastrophic losses if the entire population is affected by a new disease or pest that the clone has no resistance to. Unlike sexually propagated crops, which can adapt over generations through genetic variation, clonal crops lack this adaptability.
2. **Degeneration of Clones**: Over time, clones may degenerate due to the accumulation of viral infections, bacterial diseases, or other biotic stresses. In crops like potato or banana, viral infections can significantly reduce plant vigor and productivity. This degeneration necessitates continuous monitoring, and in some cases, the clones may need to be replaced entirely with healthier, disease-free clones. The cost and effort of constantly maintaining the health of clonal populations can be a major challenge for breeders and farmers alike.

Hybridization in Asexually Propagated Crops

Although asexual propagation is the primary method for multiplying vegetatively propagated crops, hybridization plays a crucial role in introducing new genetic combinations that can lead to the improvement of desirable traits such as disease resistance, enhanced yield, or environmental adaptability. Hybridization in these crops allows breeders to combine the best attributes of two-parent lines while still taking advantage of the stability and uniformity provided by clonal propagation.

1. Crossing:

Crossing is the foundational step in hybridization, where two distinct parent lines with desirable traits are deliberately mated to create new genetic combinations. This process allows breeders to combine the best characteristics of both parents, intending to produce progeny that surpass the performance of either parent. For crops that are propagated asexually, such as **sugarcane** (*Saccharum officinarum*) and **potato** (*Solanum tuberosum*), hybridization plays a critical role in introducing new traits, such as **disease resistance**, **higher yields**, and **environmental tolerance**.

Selection of Parents

The success of the crossing process largely hinges on the careful selection of parent plants. Breeders meticulously choose parents based on the specific traits they want to enhance or improve. One of the key goals is to identify **complementary traits** in the parent plants. For example, one parent might be highly resistant to a prevalent disease but may have a lower yield, while another parent might have excellent yield potential but lack disease resistance. The idea is to create a **synergistic combination** that merges the strengths of both parents, resulting in a hybrid that inherits both **high yield** and **disease resistance**.

Key Factors in Parent Selection:

- » **Disease Resistance**: Selecting a parent with strong resistance to specific diseases, such as rust in sugarcane or late blight in potatoes, ensures that the resulting hybrid will have increased resistance to pathogens.
- » **Yield Potential**: A parent with high productivity ensures that the hybrid will have good yield potential, which is crucial for commercial success.
- » **Adaptability to Environment**: Breeders often select parents based on their ability to thrive under specific environmental conditions, such as drought tolerance, which can be combined with other traits to produce a more versatile hybrid.

This strategy has been particularly successful in crops like **sugarcane**, where one parent with superior disease resistance is often crossed with another that has high yield or enhanced sugar content. Similarly, in **potato**, hybridization has been instrumental in combining resistance to **late blight** with superior **tuber quality**, ensuring a crop that not only resists disease but also produces marketable, high-quality tubers.

Hybrid Vigor (Heterosis)

A significant outcome of hybridization is the phenomenon known as **heterosis** or **hybrid vigor**. This occurs when the hybrid progeny exhibit **superior performance** compared to both of their parents. Hybrid vigor is often seen in traits such as **growth rate, yield**, and **disease resistance**. In asexually propagated crops, this vigor can be particularly valuable because the hybrid can be propagated clonally, ensuring that its superior traits are consistently passed on without the risk of genetic recombination.

Manifestation of Hybrid Vigor:

- **Increased Yield**: The hybrid often produces higher yields compared to either parent. This is especially important in crops like sugarcane, where yield directly correlates with the economic value of the crop.
- **Faster Growth**: Hybrids frequently show accelerated growth rates, which can result in earlier harvest times and better adaptability to short growing seasons or adverse environmental conditions.
- **Improved Resistance**: Hybrids are often more resilient to pests and diseases, thanks to the combination of resistant genes from both parents. This helps in reducing the need for chemical inputs, making the crop more sustainable and cost-effective to cultivate.

In **potato breeding**, for instance, crossing a high-yielding variety with a **late blight-resistant** line can result in hybrids that not only have better resistance to the disease but also produce more and better-quality tubers. The same principle applies to sugarcane, where hybrids that exhibit **resistance to smut or rust diseases** while maintaining high sugar content have been particularly valuable to breeders.

Introduction of New Genes and Overcoming Disease Susceptibility

For many asexually propagated crops, one of the biggest challenges is **disease susceptibility**. Because these crops are typically propagated through vegetative means, they tend to maintain the same genetic weaknesses across generations, making them highly susceptible to certain pathogens. Hybridization provides an effective method to **introduce new genes** and break this cycle of vulnerability.

- **Example in Sugarcane**: Sugarcane, which is highly susceptible to diseases like **rust** and **smut**, has benefitted immensely from hybridization. Disease-resistant varieties are often crossed with high-yielding but disease-susceptible lines to produce hybrids that carry the **resistance genes** while maintaining good yield characteristics. This not only improves the robustness of the crop but also reduces reliance on chemical treatments for disease control.
- **Example in Potato**: Hybridization in potato is commonly used to introduce resistance to **late blight**, a devastating disease that can wipe out entire potato fields. By crossing resistant lines with those that have desirable tuber quality, breeders have been able to create hybrids that significantly reduce the impact of the disease while also ensuring that the harvested crop meets market standards for size, shape, and storage quality.

Through hybridization, breeders can **broaden the genetic base** of crops like sugarcane and potato, introducing **new traits** and improving resilience. These hybrids can then be propagated asexually, ensuring that the desired traits are preserved across multiple generations.

Controlled Conditions for Effective Crossing

The crossing process is typically performed under **controlled conditions** to ensure that the desired traits are effectively combined. These conditions may include **greenhouses**, **growth chambers**, or **isolation fields**, where the potential for **cross-pollination with unintended plants** is minimized. Controlled environments also help in ensuring **optimal growing conditions**, which is particularly important in hybridization efforts where environmental stress could mask the desired genetic traits.

- **Greenhouse Hybridization**: Many hybridization programs, especially in crops like potato, utilize greenhouses to control pollination and ensure that only the selected parents are crossed. In crops like sugarcane, greenhouse crossing helps in preventing accidental pollination, ensuring that the hybrid progeny is a true cross between the intended parents.
- **Field Trials**: Once the initial hybrid progeny is produced, **field trials** are conducted to assess the hybrids' performance in real-world conditions. These trials help breeders determine whether the hybrid exhibits the desired traits under **various environmental stresses**, such as drought, pests, or nutrient deficiencies.

2. Clonal Propagation of Hybrids

After a successful hybridization, the primary goal is to **maintain the advantageous traits** of the hybrid across generations through asexual propagation. This is a crucial step in the breeding of asexually propagated crops, ensuring that the **genetic stability** of the hybrid is preserved and its **desirable characteristics** are consistently passed down to future generations. Asexual propagation allows for the multiplication of superior hybrids without the risks associated with sexual reproduction, such as genetic recombination and segregation.

Selection of Superior Hybrids

The selection of superior hybrids is one of the most critical steps in breeding programs for asexually propagated crops. Once the initial cross between parent lines with complementary traits is made, breeders focus on identifying the best-performing hybrids that combine the most desirable characteristics. The goal is to ensure that these hybrids maintain high performance levels across a range of environmental conditions and stresses, making them suitable for large-scale propagation.

Performance Assessment

The evaluation of hybrids is a detailed process that involves rigorous testing against key performance indicators (KPIs). These KPIs help breeders determine whether a hybrid is worth further investment in terms of time and resources. The performance of hybrids is assessed based on several criteria:

- **Yield Potential**: One of the primary objectives in selecting superior hybrids is to ensure high yield. Yield potential refers to the hybrid's ability to produce an abundant crop under both optimal and stressful growing conditions. For example, in crops like **sugarcane** or **potato**, yield potential includes assessing the mass of the sugarcane stalks or the number of tubers produced per plant. High-yielding hybrids are critical for commercial viability, as they can increase profitability for farmers.
- **Disease Resistance**: A key indicator of hybrid quality is its ability to resist diseases that commonly affect the crop. In many cases, diseases caused by fungi, bacteria, or viruses can severely reduce crop yields and quality. For example, in **potato breeding**, resistance to **late**

blight, caused by *Phytophthora infestans*, is one of the most crucial traits. Hybrids that can withstand or avoid infections are less dependent on chemical controls, reducing the cost of production and environmental impact.

» **Drought Tolerance**: With the increasing frequency of climate variability, including droughts, it is essential for hybrids to demonstrate resilience in low-water environments. Drought-tolerant hybrids can continue to produce high yields even in areas with limited water availability. This trait is particularly important for crops grown in arid or semi-arid regions where water scarcity is a constant challenge.

» **Overall Vigor (Heterosis)**: Hybrid vigor, also known as **heterosis**, refers to the enhanced growth and productivity of hybrids compared to their parents. This is often expressed as faster growth rates, improved nutrient uptake, greater biomass accumulation, and more robust root systems. Vigor is an essential criterion because it generally correlates with higher yields, better stress tolerance, and superior overall performance. **Vigor** in hybrids can also lead to quicker canopy development, which helps suppress weeds and reduces the need for herbicides.

Importance of Superior Hybrids

Selecting superior hybrids is essential for ensuring that future generations of the crop continue to deliver **high performance.** Only hybrids that excel across multiple traits are selected for further propagation. This selection ensures that breeders and farmers are investing in lines that will yield consistent results, regardless of external challenges such as climate variability, pest pressure, or nutrient availability.

The selection of superior hybrids also serves several broader purposes:

1. **Long-Term Stability**: Superior hybrids tend to be more stable across various growing environments, maintaining high yields and quality over multiple seasons. This consistency is critical for crops like sugarcane or potato, which are grown across diverse geographical regions.
2. **Maximizing Resource Efficiency**: By focusing on the best-performing hybrids, breeders ensure that time, labor, and resources are concentrated on the most promising lines. This maximizes the efficiency of the breeding program, allowing for faster development and commercialization of new varieties.
3. **High Commercial Value**: Superior hybrids not only perform well agronomically but also have high market value. For example, potato hybrids with enhanced tuber quality (e.g., improved starch content, size uniformity, and skin texture) are more appealing to both processors and consumers. Similarly, sugarcane hybrids with higher sucrose content can significantly increase profitability for sugar producers.
4. **Resilience to Environmental Stress**: A key aspect of selecting superior hybrids is ensuring that they can cope with **abiotic stresses** like drought, salinity, and extreme temperatures. This is becoming increasingly important as climate change impacts agricultural systems. Hybrids that show resilience in stressful environments can secure stable food production in regions vulnerable to these challenges.
5. **Pest and Disease Management**: Hybrids that possess **broad-spectrum resistance** to multiple diseases or pests reduce the need for chemical inputs, which is both economically beneficial

and environmentally sustainable. For example, a potato hybrid that is resistant to both late blight and **potato cyst nematodes** can help farmers reduce losses from both pests while lowering the reliance on costly chemical treatments.

Advantages of Asexual Propagation:

» **Genetic uniformity**: Asexual propagation ensures that all progeny are genetically identical to the hybrid, maintaining the superior traits without variation.

» **Rapid multiplication**: Asexual propagation allows breeders to quickly multiply the hybrid to meet commercial production needs. For crops like sugarcane or potato, this is especially important for ensuring a steady supply of planting material for farmers.

» **Commercial viability**: The ability to propagate the hybrid on a large scale makes it commercially viable, as the hybrid's desirable traits (such as yield and disease resistance) can be consistently reproduced and marketed to farmers.

In crops like **potatoes**, for example, after hybridization, the superior hybrid progeny that show traits like **late blight resistance** and **high tuber quality** are propagated using tubers. Each tuber is genetically identical to the original hybrid, ensuring that subsequent generations maintain the **same resistance** and **quality traits**. This process allows the hybrid to be multiplied on a large scale for eventual commercial release.

Evaluation and Testing

Even after successful hybridization and propagation, the hybrids must undergo **rigorous evaluation and testing** to ensure that they consistently perform well under various environmental conditions. This stage is critical for determining the **stability** and **adaptability** of the hybrid before it is released as a new variety.

» **Multilocation Trials**: The hybrid clones are tested across different geographical locations to assess their adaptability to **varied environments**. These trials involve growing the hybrid in different **climatic conditions**, **soil types**, and **altitudes** to evaluate its performance across diverse settings.

- **Yield stability**: The hybrid's yield is assessed under different environmental conditions to ensure that it consistently produces high yields, regardless of location.
- **Disease resistance**: The hybrid's resistance to diseases is evaluated across multiple growing seasons. This ensures that the hybrid remains resistant to **pathogen pressure** over time and in different environments.
- **Quality traits**: In addition to yield and disease resistance, the hybrids are evaluated for **quality traits**, such as the sugar content in sugarcane or tuber quality in potato. These traits are crucial for determining the market value of the hybrid.

» **Long-Term Evaluation**: Multiseasonal evaluation is necessary to verify that the hybrid maintains its **superior performance** over time. Seasonal changes in weather patterns, disease outbreaks, and other factors are monitored to ensure that the hybrid is robust enough to withstand these variables.

For instance, in **sugarcane**, a hybrid that shows resistance to **rust** in one season might be exposed to other diseases in subsequent seasons, and its overall disease resistance needs to be confirmed over time. Similarly, **potato hybrids** that are resistant to **late blight** must be tested across multiple environments to ensure that their resistance holds up under diverse conditions.

Commercial Release of Hybrid Varieties

Once the hybrid clones have passed **multilocation trials** and **long-term evaluations**, they are ready for **commercial release**. The final step is to multiply the hybrid on a large scale to meet the demands of farmers and agricultural producers.

- » **Seed Multiplication**: For crops like potatoes, this involves the production of **high-quality seed tubers** that are distributed to farmers. In sugarcane, this might involve multiplying the hybrid through **stem cuttings** or **tissue culture**, depending on the scale of production required.
- » **Distribution to Farmers**: After successful multiplication, the hybrid variety is released for commercial production. Farmers can then use the genetically stable hybrid to ensure **consistent yield** and **disease resistance** in their fields.

Asexual Propagation

Asexual propagation is a crucial technique in the breeding of vegetatively propagated crops, especially after identifying superior hybrids. This method allows for the multiplication of the hybrid without genetic variation, ensuring that the favorable traits observed in the hybrid are passed down to every subsequent generation. In contrast to sexual propagation, where genetic recombination can occur through seed production, asexual propagation guarantees that the genetic identity of the hybrid is maintained.

Methods of Asexual Propagation

Different crops employ various asexual propagation methods, each suited to their biological characteristics. The choice of propagation technique depends on the crop species and the availability of resources. Here are some common methods used in asexual propagation:

1. Cuttings:

Cuttings are one of the most basic and efficient methods of asexual propagation, widely used across various crops, particularly in sugarcane, cassava, and ornamental plants. This method involves the use of vegetative parts such as **stems**, **leaves**, or **roots**, which, when properly planted and nurtured, can develop into a fully grown plant genetically identical to the parent. This technique is highly beneficial for maintaining **uniformity in traits**, such as **yield**, **disease resistance**, and **growth habits**, across generations, as the progeny are clones of the parent plant.

How Cuttings Work

The process of propagation via cuttings relies on the ability of plant tissues to regenerate into whole plants. In most cases, stem cuttings are the preferred vegetative part used in asexual propagation. For crops like sugarcane, the stems are segmented into parts known as setts, with each sett containing at least one node (the point on a stem where a leaf is attached). These nodes are crucial because they are where new shoots and roots emerge.

1. **Preparation of Setts**: The cuttings (setts) are made from healthy, mature plants. In sugarcane, for instance, the cuttings are typically taken from the lower to mid-portion of the stem, which contains the most robust nodes capable of regenerating new plants. Each segment is usually a few inches long and contains one or more nodes to ensure successful growth.
2. **Planting and Root Formation**: Once the setts are prepared, they are planted in the field, either horizontally or vertically, depending on the crop. In the soil, the nodes within the cuttings generate new shoots and roots, which eventually give rise to a fully developed plant. **Rooting hormones** or growth regulators may be applied to accelerate the formation of roots and shoots, especially in crops that are slower to regenerate or more sensitive to environmental conditions.
3. **Growing Conditions**: Proper soil moisture, aeration, and temperature control are critical for the successful establishment of plants from cuttings. The environment must provide adequate **humidity** and **temperature** control to support rapid rooting and shoot development. Often, mulching or covering the planted setts with plastic sheets helps retain moisture and create a microclimate that encourages growth.

Applications of Cuttings in Crop Propagation

The use of cuttings is a valuable tool for the rapid multiplication of crops, particularly those that do not readily produce seeds or have limited flowering, like **sugarcane** or **ornamentals**. Here are specific applications in key crops:

» **Sugarcane (Saccharum officinarum)**: In sugarcane cultivation, cuttings are essential because the crop rarely produces viable seeds. **Setts**, containing one or more nodes, are planted directly in the field. Since sugarcane is a perennial crop, farmers often rely on vegetative propagation to maintain **high yields**, **sugar content**, and **disease resistance** across multiple generations.

» **Cassava (Manihot esculenta)**: Cassava is primarily propagated through stem cuttings, with each cutting carrying nodes that regenerate into a new plant. This method ensures that important traits like **root size**, **starch content**, and **drought tolerance** are passed on uniformly to all new plants. Cassava cuttings are typically planted before the rainy season to ensure optimal moisture for growth.

» **Ornamentals**: In the ornamental industry, stem and leaf cuttings are common methods for rapidly multiplying plants with desirable aesthetic traits, such as flowers or foliage. Crops like **roses**, **geraniums**, and **begonias** are propagated through cuttings to maintain their **color**, **size**, and **growth patterns** in successive generations.

Advantages of Propagation by Cuttings

1. **Genetic Uniformity**: Since cuttings are taken from the parent plant, the progeny produced are **genetically identical**. This allows for the exact replication of desirable traits, such as **disease resistance**, **high yield**, and **adaptability**, ensuring consistency in crop production.

2. **Rapid Multiplication**: The use of cuttings allows for **quick establishment** of large numbers of plants. This method is much faster than seed propagation, especially for crops that have long gestation periods or unreliable seed production, such as sugarcane or banana.
3. **Conservation of Resources**: Propagation through cuttings requires fewer inputs, such as fertilizers and water, compared to seed propagation. Additionally, since the new plants are already at an advanced vegetative stage, they establish themselves more rapidly in the field, reducing the time needed to reach maturity.
4. **Preservation of Superior Traits**: Through cuttings, **superior traits** identified in one generation, such as **disease resistance** or **yield stability**, are maintained without the risk of genetic recombination, which could dilute these traits. This ensures the continuity of desirable qualities over multiple generations.

Challenges of Using Cuttings

While the method offers significant benefits, it is not without its limitations. The most prominent challenges include:

1. **Disease Transmission**: Cuttings can carry **pathogens** or **pests** from the parent plant, leading to the spread of diseases within a crop. This is particularly problematic in perennial crops like sugarcane or banana, where systemic diseases can remain dormant in the plant tissue.
2. **Reduced Genetic Diversity**: Since cuttings replicate the genetic makeup of the parent plant, this method does not introduce genetic diversity. Over time, monocultures of genetically identical plants can become vulnerable to environmental changes, new diseases, or pests that the original clone may not withstand.
3. **Dependence on the Mother Plant**: The quality of the new plants depends entirely on the health and vigor of the mother plant. If the parent plant is stressed or weakened, the cuttings taken from it may not perform optimally.
4. **Limited Availability of Planting Material**: Unlike seeds, which can be mass-produced, the number of cuttings that can be taken from a single plant is limited. This can constrain the scale of propagation efforts, especially in large-scale commercial agriculture.

2. **Tubers**:

Tubers serve as one of the most effective and widely used means of asexual propagation, especially in crops like **potato** (*Solanum tuberosum*), **cassava** (*Manihot esculenta*), and **sweet potato** (*Ipomoea batatas*). As underground storage organs, tubers not only store nutrients but also contain the complete genetic material required to produce new plants. Each tuber is genetically identical to the parent plant, which ensures that all the desirable traits, such as **disease resistance**, **yield**, and **tuber quality**, are retained across generations.

How Tubers Work in Propagation

The propagation of crops using tubers capitalizes on the natural ability of these structures to sprout and form new plants. Tubers are essentially modified stems that grow underground, storing carbohydrates and other nutrients needed for the plant's growth. When planted, these tubers produce shoots from their **buds**, commonly known as **eyes**, which develop into new plants.

1. **Structure and Function of Tubers**:
   - Tubers are **enlarged underground stems** that serve as storage organs. The primary function of tubers is to accumulate **starch**, which can later be utilized by the plant during unfavorable growing conditions.
   - Each tuber contains several **buds**, or "eyes," that have the potential to sprout into new plants. These buds are made up of meristematic tissue capable of rapid growth when environmental conditions are favorable.
   - The stored nutrients in tubers provide the energy needed for the initial stages of growth, allowing the new plant to establish itself before it begins photosynthesis.
2. **Planting and Regeneration**:
   - **Tuber Propagation**: When planting tubers, farmers often cut them into pieces, ensuring that each piece contains at least one bud. This practice maximizes the number of plants that can be grown from a single tuber. Once planted, the **bud sprouts** and forms a new plant, while the nutrients stored in the tuber provide initial sustenance until the plant can photosynthesize.
   - **Growth Cycle**: As the new plant grows, it will eventually form its own tubers, ensuring a continuous cycle of vegetative reproduction. In the case of potatoes, this method ensures that the progeny are **genetically identical** to the parent plant, maintaining the desired traits from one generation to the next.

Applications of Tubers in Potato Breeding

Potato is one of the most important tuber crops globally, and **tuber propagation** plays a pivotal role in maintaining high-quality potato varieties over multiple growing cycles. The reliance on tubers as the primary means of propagation ensures that farmers can consistently grow potatoes with desirable characteristics, such as disease resistance, high yield, and uniform tuber size.

1. **Uniformity in Traits**: Since each new plant that grows from a tuber is a genetic clone of the parent plant, traits such as **tuber quality**, **size**, **disease resistance**, and **flavor** remain consistent across generations. This genetic stability is especially important in commercial potato production, where uniformity in crop traits is crucial for both farmers and consumers.
2. **Disease Resistance and Tuber Quality**: In potato breeding, tubers are essential for retaining important traits, such as **resistance to late blight**, one of the most devastating diseases affecting potatoes. Once a potato variety with resistance to late blight is developed through hybridization, it is propagated through tubers to ensure that the resistance is passed down to future crops. Similarly, tuber quality traits, such as **starch content**, **skin texture**, and **shape**, are preserved through tuber propagation.
3. **Efficient Large-Scale Propagation**: Potato farming relies on tuber propagation for **large-scale cultivation**. By planting tuber pieces, farmers can quickly and efficiently propagate high-quality potato varieties. This method is particularly advantageous because it bypasses

the need for seed production, which can be time-consuming and result in genetic variability. Instead, tuber propagation provides an **efficient** and **reliable** method to scale up the production of superior hybrids across large areas of farmland.

Tubers in Other Asexually Propagated Crops

While potatoes are the most notable crop propagated by tubers, other crops, including **cassava** and **sweet potato**, also rely heavily on this reproduction method.

1. **Cassava (Manihot esculenta):**
    - Cassava, a tropical root crop, is primarily propagated using **stem cuttings**, but tuber-like roots also play a role in regeneration. In cassava, the **roots** store carbohydrates and can sprout new plants under favorable conditions. This method allows cassava farmers to maintain the same high-yielding and disease-resistant varieties across generations.
    - Like potatoes, cassava is prone to diseases like cassava mosaic virus and cassava brown streak virus, but once resistant varieties are identified, they can be multiplied through vegetative propagation using cuttings or tuber-like structures.
2. **Sweet Potato (Ipomoea batatas):**
    - Sweet potato is another tuberous crop that is propagated vegetatively. The **storage roots** of sweet potatoes, like those of potatoes, contain the genetic material to produce new plants. **Tuberous roots** are planted, and new shoots emerge from them to grow into new plants, identical to the parent.
    - Sweet potato breeding also focuses on maintaining traits like **starch content**, **tuber shape**, and **resistance to pests** such as the **sweet potato weevil**. Through tuber propagation, these traits can be maintained consistently in the crop, ensuring high-quality harvests over time.

Advantages of Propagation by Tubers

1. **Genetic Stability**: Tubers provide an **asexual reproduction method** that ensures the progeny are genetically identical to the parent plant. This stability is crucial for maintaining desired traits such as **disease resistance**, **yield**, and **tuber quality** across generations. By using tubers, farmers can be assured that the progeny will retain the same beneficial characteristics that made the parent plant successful.
2. **Increased Propagation Efficiency**: Propagation through tubers allows for **rapid multiplication** of crops. Farmers can divide tubers into multiple pieces, each capable of growing into a new plant. This method is particularly efficient for crops like potatoes, where a single tuber can give rise to several new plants, thereby increasing the **rate of propagation**.
3. **Uniformity in Crop Production**: Since tubers produce **clonal progeny**, the resulting plants are uniform in terms of **growth habit**, **tuber shape**, **size**, and **quality**. This uniformity is particularly important in commercial farming, where consistency in crop appearance and yield is highly valued.

Challenges and Limitations

1. **Disease Transmission**: One of the primary challenges of tuber propagation is the potential for **disease transmission.** Tubers can harbor pathogens such as **bacteria, viruses**, and **fungi**, which can be passed from the parent plant to the progeny. Diseases like **potato blight** and **cassava mosaic virus** can significantly impact yield and crop quality if infected tubers are used for propagation.
2. **Storage and Transportation**: Tubers are bulky and can be difficult to store for long periods without proper conditions. Improper storage can lead to tuber **decay**, reducing the number of viable tubers available for planting in the next season. Additionally, the transportation of tubers over long distances can be costly and logistically challenging due to their weight and size.
3. **Limited Genetic Diversity**: Since tuber propagation creates **genetically identical** clones, the resulting crop may lack the genetic diversity necessary to withstand new diseases or changing environmental conditions. A genetically uniform crop is more vulnerable to outbreaks of diseases or pests that could decimate the entire population.

3. Rhizomes:

Rhizomes are specialized underground stems that serve as an important means of asexual propagation in crops like **ginger** (*Zingiber officinale*) and **turmeric** (*Curcuma longa*). These underground stems grow horizontally beneath the soil and contain **nodes, buds**, and **internodes**, which store energy and nutrients required for the plant's growth. When divided, each section of the rhizome is capable of giving rise to a new plant, making it a highly efficient and reliable method of vegetative reproduction.

How Rhizomes Function in Propagation

Rhizomes play a vital role in vegetative propagation by allowing plants to spread underground and create genetically identical offspring. This method not only conserves the **genetic makeup** of the parent plant but also supports rapid expansion in commercial agriculture. Crops propagated through rhizomes retain important traits such as **yield, disease resistance**, and **flavor**, which are critical for consistent crop production over multiple growing seasons.

1. **Structure and Growth of Rhizomes**:
   - Rhizomes are **underground stems** that grow horizontally, sending out roots and shoots from their nodes. These nodes contain **buds**, which, under favorable conditions, sprout into new plants above the soil.
   - As rhizomes grow and expand horizontally, they produce multiple new plants, creating a **clonal population** that is genetically identical to the parent plant.
   - Rhizomes also serve as **storage organs**, accumulating nutrients and energy in the form of carbohydrates. This stored energy supports the growth of new shoots, particularly in adverse conditions or during the dormancy period.
2. **Division and Planting**:
   - **Rhizome Division**: To propagate a crop like ginger or turmeric, the rhizomes are divided

into sections, each containing at least one bud. These divided sections are then planted in the soil, where they grow into new plants. This method ensures that the new plants are genetically identical to the parent plant, maintaining desirable traits across generations.

- **Planting Technique**: When rhizomes are planted, the **buds** sprout and produce new shoots above ground while the underground portion continues to grow horizontally, producing more rhizomes. This process allows for the rapid multiplication of plants, which is especially useful in commercial agriculture, where uniformity and consistency in crop quality are essential.

Advantages of Rhizome Propagation

The use of rhizomes for asexual propagation offers several key benefits, particularly for crops like ginger and turmeric, where maintaining the **genetic integrity** of the crop is vital.

1. **Genetic Uniformity**: One of the greatest advantages of propagating crops through rhizomes is the preservation of **genetic uniformity**. Since new plants grow directly from the rhizome of the parent plant, they inherit an identical genetic makeup. This ensures that the superior traits of the parent, such as **yield**, **disease resistance**, **flavor**, and **quality**, are retained in all progeny. This uniformity is crucial in commercial agriculture, where consistency in crop performance and quality is paramount.
2. **Efficient Multiplication**: Rhizomes allow for **rapid multiplication** of crops. A single parent plant can give rise to numerous offspring through the division of its rhizomes. This is particularly advantageous for farmers seeking to **scale up** production quickly without the need for seed production or the genetic variability associated with sexual reproduction. Each section of the rhizome is capable of producing an entire new plant, making it an efficient propagation method that maximizes yield per planting unit.
3. **Sustained Crop Performance**: Crops propagated by rhizomes often show **consistent performance** over time, as the traits of the parent plant are faithfully passed down. This is particularly important for **quality traits** like flavor in ginger and turmeric, where the market demands a high level of uniformity in taste, aroma, and size. By using rhizomes, farmers can ensure that their crops continue to meet market standards season after season.
4. **Resilience and Adaptability**: Rhizome-based propagation also offers **resilience** in adverse conditions. Rhizomes store nutrients that allow the plant to survive during periods of **environmental stress** such as drought or low temperatures. Once favorable conditions return, the stored nutrients enable the plant to regenerate quickly, making rhizome-propagated crops well-suited for regions with variable climates.

Applications in Ginger and Turmeric Breeding

Both **ginger** and **turmeric** are economically important crops that are widely propagated through rhizomes. The use of rhizomes ensures the **preservation of specific traits** that are essential for commercial success, such as **yield**, **pest resistance**, and **flavor**.

1. **Ginger Propagation:**

    - In ginger cultivation, **rhizomes** are the main source of propagation. Rhizomes not only store essential nutrients but also ensure the continued growth of the crop year after year. Each ginger rhizome is carefully divided into segments, each containing at least one bud. These segments are planted in the field, where they grow into new ginger plants that are **genetically identical** to the parent.
    - Ginger is known for its **aromatic compounds**, which contribute to its value in culinary and medicinal applications. By using rhizomes for propagation, farmers can guarantee that the **flavor profile** of their ginger remains consistent across multiple harvests.

2. **Turmeric Propagation**:

    - Like ginger, **turmeric** is propagated through its underground **rhizomes**. The rhizomes are divided and planted to produce new plants that inherit the desirable traits of the parent, such as **high curcumin content**, **disease resistance**, and **drought tolerance**.
    - Turmeric is valued for its medicinal properties, particularly its **anti-inflammatory** and **antioxidant** effects, which are largely attributed to the compound **curcumin**. By using rhizomes for propagation, breeders ensure that the **curcumin concentration** in the progeny remains consistent, meeting the demands of both consumers and the pharmaceutical industry.

Rhizomes and Disease Management

While rhizome propagation offers many benefits, it also presents challenges, particularly related to **disease management**. Since rhizomes are underground stems, they are vulnerable to **soil-borne diseases** and pests. If the parent plant is infected, the disease can be transmitted to the progeny through the rhizome. Therefore, strict **sanitation practices** are required during the propagation process to ensure that only healthy rhizomes are used.

1. **Prevention of Disease Spread:**

    - To mitigate the risk of disease, farmers and breeders must ensure that rhizomes are **disease-free** before planting. This involves careful inspection of the rhizomes for signs of rot, fungal infection, or pest damage. Infected rhizomes must be discarded to prevent the spread of disease to healthy plants.
    - In addition, **crop rotation** and the use of **disease-resistant varieties** can help reduce the incidence of soil-borne pathogens. For example, in ginger cultivation, pathogens such as **Pythium** and **Fusarium** can cause significant crop losses if not managed properly. By selecting rhizomes from disease-free parent plants and employing good agricultural practices, farmers can minimize the risk of disease transmission.

2. **Tissue Culture for Disease-Free Rhizomes**:

    - In some cases, advanced propagation techniques like **tissue culture** are used to produce disease-free rhizomes. Tissue culture involves growing small sections of plant tissue in sterile laboratory conditions to produce new plants that are free of pathogens. This method is particularly useful for producing **clean planting material** in crops that are highly susceptible to diseases.

- Once the disease-free plants are established, they can be multiplied through traditional rhizome division, ensuring that the crop remains healthy and productive over time.

Challenges of Rhizome Propagation

While rhizome propagation offers many benefits, it also has some limitations:

1. **Disease Susceptibility**: Rhizomes can carry pathogens, leading to the transmission of diseases from one generation to the next. Soil-borne diseases such as **rhizome rot** can significantly reduce crop yield if not properly managed.
2. **Storage and Handling**: Rhizomes can be difficult to store for long periods, as they are prone to rot if not kept under ideal conditions. Farmers must ensure proper storage techniques to prevent the loss of valuable planting material.
3. **Limited Genetic Diversity**: Because rhizomes propagate plants asexually, there is no genetic variation in the progeny. While this ensures consistency, it also makes the crop vulnerable to new diseases or pests that the clone may not be resistant to. Genetic diversity is important for long-term crop resilience.

4. **Tissue Culture**:

Tissue culture, also known as **micropropagation**, is an advanced and highly efficient method of asexual propagation, enabling rapid multiplication of plants in sterile conditions. This technique is particularly valuable for crops like **bananas**, **orchids**, **potatoes**, and **ornamentals**, where uniformity, large-scale propagation, and disease-free planting materials are critical to agricultural success. Tissue culture is one of the most widely used biotechnological tools in modern plant breeding and propagation, offering immense potential for both commercial agriculturc and conservation efforts.

Principles of Tissue Culture

Tissue culture involves cultivating plant cells, tissues, or organs (known as **explants**) in a **sterile, nutrient-rich medium** under controlled environmental conditions. The medium contains essential nutrients, vitamins, growth regulators, and a carbon source like sucrose to support the growth and development of plant tissues. By using explants, which can be taken from various parts of the plant such as **leaves**, **stems**, **roots**, or even **meristematic tissues**, a large number of genetically identical plants can be produced in a relatively short period.

1. **Explant Selection:**
   - The success of tissue culture largely depends on the choice of the explant, the small piece of plant tissue used to initiate the culture. Common explants include **shoot tips**, **meristematic tissue**, **leaf sections**, **nodal segments**, and **root tips**.
   - The selected explant must be **disease-free** and in optimal physiological condition to ensure efficient regeneration of new plants. Explants are usually obtained from healthy, vigorously growing plants to avoid contamination by pathogens.

2. **Sterilization:**
    - To prevent microbial contamination, all tissue culture procedures are conducted under **sterile conditions**. The explants, culture vessels, and growth media are sterilized using chemical agents like **sodium hypochlorite** or **ethanol**, as well as physical methods such as autoclaving.
    - Maintaining sterility throughout the process is crucial, as contamination by bacteria, fungi, or other microorganisms can lead to the failure of the tissue culture.
3. **Nutrient Media:**
    - The nutrient medium provides the explant with all the necessary elements required for growth, including **macronutrients** (nitrogen, phosphorus, potassium), **micronutrients** (zinc, copper, iron), **vitamins**, and **plant hormones** (cytokinins, auxins, and gibberellins).
    - The composition of the medium can be tailored to meet the specific needs of the crop and the stage of development, whether it is for **callus formation**, **rooting**, or **shoot development**. For instance, a high ratio of cytokinin to auxin promotes shoot growth, while a higher concentration of auxin promotes root development.
4. **Regeneration and Multiplication:**
    - Once the explants are cultured on the nutrient medium, they begin to **regenerate** new tissues through the process of **cell differentiation**. Depending on the species and the growth conditions, the explant may develop into **callus** (a mass of undifferentiated cells) or directly form **shoots** and **roots**.
    - **Multiplication** of shoots is achieved by repeated subculturing, where newly formed shoots are transferred to fresh media to encourage further growth and division. Each subculture cycle can produce hundreds or even thousands of identical plants within a short timeframe.

Advantages of Tissue Culture in Crop Propagation

The tissue culture method offers several distinct advantages over traditional propagation techniques, particularly for high-value crops and those requiring disease-free planting materials.

1. **Production of Disease-Free Plants:**
    - One of the major benefits of tissue culture is the ability to produce **disease-free plants**. Explants taken from **meristematic tissues**, such as the apical meristem, are typically free from viral infections, as viruses often do not penetrate these rapidly dividing cells.
    - By growing plants in sterile conditions, breeders can avoid contamination by soil-borne diseases, fungi, or bacteria, ensuring the **health** and **vigor** of the propagated plants. This is particularly critical for crops like bananas, which are highly susceptible to diseases like **Panama disease** caused by *Fusarium oxysporum*.
2. **Rapid Multiplication:**
    - Tissue culture allows for the **rapid multiplication** of plants, producing hundreds or thousands of genetically identical plants in a fraction of the time required by traditional

propagation methods. For crops like **bananas, potatoes**, and **orchids**, where large-scale planting material is needed, tissue culture is invaluable.

- In traditional methods, vegetative propagation may be slow, constrained by the limited availability of plant material. Tissue culture bypasses these limitations, enabling breeders to meet the high demand for planting materials in commercial agriculture.

3. **Genetic Uniformity:**
   - Since tissue culture involves **asexual reproduction**, the plants produced are **genetically identical** to the parent plant, maintaining the desirable traits such as **yield, disease resistance, quality**, and **flavor** across all progeny. This genetic uniformity is crucial in commercial crop production, where consistency in performance is vital.
   - Uniformity is especially important in ornamental plants like **orchids**, where growers demand specific traits like **flower color** and **size** to meet market requirements. Tissue culture ensures that these traits are preserved in every propagated plant.
4. **Year-Round Production**:
   - Tissue culture allows for the **year-round production** of plants in a controlled environment, regardless of the seasonal growing conditions. This is particularly useful for tropical and subtropical crops, enabling breeders to propagate plants continuously without being limited by climate or weather patterns.
   - Furthermore, the **space efficiency** of tissue culture facilities allows for large-scale propagation in a relatively small area, making it ideal for urban agriculture and other resource-limited settings.
5. **Conservation of Endangered Species:**
   - Tissue culture is widely used in the **conservation** of rare and endangered plant species. By propagating plants in vitro, tissue culture provides a means to preserve **genetic diversity** and prevent the extinction of species that may be difficult to propagate by conventional methods.
   - For example, **orchids**, many of which are endangered in their natural habitats, are successfully propagated using tissue culture. This technique not only contributes to biodiversity conservation but also supports the **commercial horticulture** industry.

Applications of Tissue Culture in Crop Breeding

Tissue culture is extensively applied in both crop breeding and commercial agriculture for the propagation of a wide range of crops, including:

1. **Bananas** (*Musa spp.*): Bananas are one of the most commercially important crops propagated through tissue culture. Traditional propagation methods are slow and inefficient, while tissue culture allows for the mass production of **disease-free planting material**. Tissue culture has played a pivotal role in the banana industry's ability to respond to devastating diseases such as **Panama disease** and **Black Sigatoka**.
2. **Orchids** (*Orchidaceae*): Tissue culture is essential for the **commercial propagation** of orchids,

which are highly valued for their aesthetic qualities. This method ensures the production of **uniform plants** with consistent flower size, shape, and color, meeting the high standards of the ornamental plant industry.

3. **Potatoes** (*Solanum tuberosum*): In potato breeding, tissue culture is used to propagate **disease-free seed potatoes**, reducing the risk of viral and bacterial infections that can decimate crops. This method also allows for the rapid multiplication of new varieties that exhibit desirable traits like **late blight resistance**.
4. **Strawberries** (*Fragaria × ananassa*): Tissue culture is used for the **micropropagation** of strawberries to ensure that high-yielding, disease-resistant varieties are consistently propagated. This technique has been especially useful in producing **virus-free plants** for commercial fruit production.

Challenges in Tissue Culture

Despite its advantages, tissue culture also presents several challenges, particularly in large-scale applications:

1. **Cost and Expertise:**
    - Tissue culture requires specialized knowledge, equipment, and facilities, making it a **cost-intensive** process compared to traditional propagation methods. The initial investment in infrastructure, including sterile growth rooms, autoclaves, and culture media, can be prohibitive for small-scale farmers.
2. **Risk of Somaclonal Variation**:
    - Although tissue culture aims to produce genetically identical plants, there is a risk of **somaclonal variation**, where genetic changes occur in the plants during the culture process. This variation can result in undesirable traits, requiring careful monitoring and selection.
3. **Contamination**:
    - Maintaining sterile conditions is critical for the success of tissue culture. Any contamination by bacteria, fungi, or viruses can lead to the loss of an entire batch of plants, making **quality control** essential throughout the process.

Advantages of Asexual Propagation

» **Genetic Uniformity**: Asexual propagation ensures that each new plant is genetically identical to the parent plant. This is particularly important for maintaining specific traits, such as disease resistance, high yield, or fruit quality, across generations without the risk of recombination that occurs in sexual propagation.

» **Faster Multiplication**: Compared to sexual reproduction, which involves the growth of plants from seeds, asexual propagation is faster and can be done on a large scale. Methods like tissue culture and cuttings allow breeders to quickly propagate superior hybrids and introduce them into commercial production.

» **Preservation of Desirable Traits**: By propagating plants asexually, breeders can ensure that the favorable traits identified in the hybrid, such as resistance to pests or improved tolerance to environmental stresses, are preserved across many generations. This is critical for maintaining the economic value of high-performing hybrids.

» **Disease-Free Plants**: Tissue culture, in particular, offers the advantage of producing plants that are free from pathogens, such as viruses or bacteria. This ensures healthier plants, which are more likely to perform well in the field.

Advantages of Hybridization in Asexually Propagated Crops

1. **Combining Desirable Traits**: Hybridization allows breeders to **combine beneficial traits** from two or more parent lines. This is particularly important for introducing disease resistance, stress tolerance, or other desirable traits into a high-performing clone.
2. **Enhanced Genetic Diversity**: Even though asexual propagation preserves genetic uniformity, hybridization introduces **new genetic combinations**, enhancing the diversity of the breeding pool and allowing for continued improvement over time.
3. **Hybrid Vigor**: Many hybrids display **heterosis**, or hybrid vigor, where the offspring outperform both parents in terms of yield, growth rate, or stress tolerance. This leads to more robust plants that are better suited to variable growing conditions.
4. **Stability Through Asexual Propagation**: Once a desirable hybrid is developed, asexual propagation ensures that its traits are **stably maintained** across generations, without the risk of segregation or recombination that comes with sexual reproduction.
5. **Accelerated Improvement**: The combination of hybridization and clonal propagation can accelerate the development of improved varieties. Breeders can quickly scale up the production of superior hybrids through asexual propagation methods like tissue culture, cuttings, or tubers.

Challenges of Hybridization in Asexually Propagated Crops

1. **Complexity of Crossing**: Hybridization can be more **complex** in asexually propagated crops, especially those that have **low seed set** or limited flowering. For example, in banana (*Musa spp.*) and sweet potato (*Ipomoea batatas*), hybridization may require specialized techniques such as **embryo rescue** or **tissue culture** to successfully generate hybrids.
2. **Heterozygosity**: Many asexually propagated crops are **highly heterozygous**, meaning they contain a wide range of alleles at many genetic loci. This can complicate breeding efforts, as the expression of traits may vary significantly in the progeny.
3. **Evaluation of Hybrids**: After hybridization, extensive **field testing** is required to ensure that the hybrid performs well across different environments and over multiple growing seasons. This can be a time-consuming and resource-intensive process.
4. **Long-Term Maintenance**: Although asexual propagation ensures the genetic stability of hybrids, long-term maintenance of clones can be challenging due to the potential accumulation of **viral infections, mutations**, or other factors that can cause a decline in performance.

Chapter - 5

Plant Genetic Resources

A. Plant Introduction

Plant introduction refers to the process of introducing a genotype or group of genotypes into a new geographical region or environment where they were not previously cultivated. This practice has played a vital role in global agricultural development by enhancing crop diversity, introducing new crops, and improving the performance of existing varieties. Plant introduction involves transferring a new crop species, a wild relative of an existing crop, or a new variety of a crop already grown in the region. For example, the introduction of **IRRI rice varieties**, wild sunflower species from **Russia**, and **oil palm** into Tamil Nadu are classic examples of successful plant introductions that have greatly impacted local agriculture.

There are two major types of plant introduction: **Primary Introduction** and **Secondary Introduction**.

1. Primary Introduction

Primary introduction is a straightforward process where a new crop or variety is directly introduced into a new geographical region and cultivated without undergoing any genetic modification or alteration. This method is employed when the introduced genotype is found to be inherently suited to the new environment, allowing it to thrive without requiring any changes. The goal of primary introduction is to enhance agricultural productivity by introducing high-yielding or resilient varieties that are already proven successful in other regions.

Key Characteristics of Primary Introduction:

» **No Genetic Modification**: The introduced variety is cultivated as it is, without any form of genetic alteration or crossbreeding. The focus is on finding a genotype that can naturally adapt to the new environment.

» **Quick Adaptation**: Since the variety does not require modification, it can be rapidly adopted in the new region, allowing for immediate benefits in terms of yield or other agronomic traits.

» **Wide Adoption**: If the variety proves successful, it can be quickly multiplied and distributed among farmers, making it an efficient method for improving crop production.

DOI: 10.1201/9781003757412-5

Example: Introduction of IRRI Rice Varieties

A significant example of primary introduction is the **International Rice Research Institute (IRRI)** rice varieties such as **IR8, IR20, IR34, and IR50**, which were introduced into various parts of Asia. These varieties were initially developed at IRRI in the Philippines and were known for their high yields and resistance to pests and diseases. Upon introduction to countries like India and other parts of Southeast Asia, these varieties proved to be well-adapted to local conditions and were directly cultivated without further modification.

» **IR8**, known as the "miracle rice," was one of the most successful introductions, playing a pivotal role in the **Green Revolution** in Asia. It had a semi-dwarf stature, which reduced lodging (the bending or breaking of plants), and produced significantly higher yields than traditional varieties.

» **IR20** and other IRRI varieties followed, offering similar advantages in terms of yield and disease resistance.

Example: Oil Palm Varieties from Malaysia

Another example is the introduction of **oil palm (Elaeis guineensis)** varieties from **Malaysia** to countries like India and Indonesia. The Malaysian varieties of oil palm were well-suited to the tropical climates of these regions, and they were directly adopted for large-scale cultivation. Oil palm has since become a major commercial crop in these regions, supporting the edible oil industry.

Advantages of Primary Introduction:

» **Speed of Adoption**: The variety can be quickly tested and adopted if it performs well in the new environment, providing immediate benefits to farmers.

» **Minimal Breeding Efforts**: Since the variety is already suited to the local conditions, it does not require extensive breeding or selection processes, saving time and resources.

» **Maintaining Genetic Integrity**: Primary introduction allows the original genetic composition of the crop to be preserved, ensuring that all its desirable traits are retained.

Limitations of Primary Introduction:

» **Environmental Constraints**: Not all introduced varieties will adapt successfully to the new environment. Factors such as climate, soil type, and pest prevalence may limit the success of primary introductions.

» **Limited Scope for Improvement**: Since the variety is not genetically altered, it may not always possess the exact traits required for optimal performance in the new region, leading to suboptimal results in certain cases.

Overall, primary introduction has been an effective strategy for quickly enhancing crop productivity in many regions, particularly when the introduced varieties are already proven to be successful in similar environments.

2. Secondary Introduction

Secondary introduction is a more complex process than primary introduction. It involves not only introducing a variety to a new environment but also selecting desirable traits from the introduced material or using it in hybridization programs to create new varieties. In this method, breeders take the introduced variety and either refine it further through selection to better suit the new conditions or cross it with local varieties to create hybrids that possess the desired traits.

Key Characteristics of Secondary Introduction:

» **Selection for Local Adaptation**: In secondary introduction, breeders focus on improving the introduced variety's performance in the new environment by selecting plants with favorable traits. This may involve selecting for disease resistance, drought tolerance, yield, or any other important agronomic trait.

» **Hybridization with Local Varieties**: Often, the introduced variety is crossed with local varieties to combine the desirable traits of both parents. This process helps improve the adaptability and performance of the introduced variety in the new region.

» **Genetic Improvement**: Secondary introduction often leads to the development of new, improved varieties that are better suited to the local environment, making this method a crucial tool in crop improvement programs.

Example: Soybean Variety Co1 from Taiwan

One of the successful examples of secondary introduction is the development of **Co1**, a soybean variety introduced from Taiwan. The original variety, **EC 39821**, was introduced into India and subjected to selection in local environments. The selection process led to the development of **Co1**, a variety that was better suited to Indian growing conditions and produced higher yields. This illustrates how secondary introduction allows for the refinement and adaptation of introduced varieties.

Example: Rice Variety Co44

In rice, a notable example of secondary introduction is the development of the variety **Co44**. This variety was developed by crossing **ASD4**, a local variety, with **IR20**, an introduced variety from IRRI. The result was a variety that was better suited to **late planting** conditions in Tamil Nadu, India. The hybridization process allowed breeders to combine the desirable traits of both parents, producing a variety that could thrive in specific local conditions.

» **IR20** brought high yield potential and disease resistance, while **ASD4** contributed its adaptability to late planting. The combination of these traits made Co44 a valuable variety for regions where late planting was necessary due to specific agricultural cycles.

Advantages of Secondary Introduction:

» **Customization for Local Conditions**: Secondary introduction allows breeders to fine-tune the introduced variety to suit specific environmental conditions, making it more adaptable and productive in the new region.

» **Combining Desirable Traits**: By hybridizing the introduced variety with local varieties, breeders can combine the best traits of both parents, leading to the development of superior varieties with multiple desirable traits.

- **Improvement of Local Varieties**: Secondary introduction also serves as a valuable tool for improving local varieties by introducing new traits, such as disease resistance or higher yield, through crossbreeding.

Limitations of Secondary Introduction:

- **Time-Consuming**: The process of selection and hybridization can be time-consuming, requiring several generations to achieve the desired results.
- **Genetic Complexity**: Hybridization may lead to genetic complexities that require careful management to ensure that the new variety possesses the desired combination of traits.

Objectives of Plant Introduction

Plant introduction serves a wide range of purposes that contribute to enhancing agricultural productivity, food security, genetic diversity, and economic value. Below are the key objectives of plant introduction:

1. Introduction of New Crops

One of the most significant objectives of plant introduction is to introduce entirely new plant species into a region where they have not been grown before. These new species often bring economic, agricultural, or industrial value and can lead to the establishment of new sectors within a region's economy.

Key Points:

- **Economic Impact**: By introducing new crops with commercial value, plant introduction can contribute to creating new industries and enhancing the economic output of a region. For example, the introduction of **oil palm** (Elaeis guineensis) into India significantly contributed to domestic oil production, reducing the reliance on imported oils. This led to the establishment of oil palm plantations and processing industries, creating employment opportunities and stimulating rural economies.
- **Diversification**: New crops help diversify agricultural production, reducing dependence on a limited number of crops. This can enhance food security by spreading risk across different crops, making the agricultural sector more resilient to adverse conditions such as pest outbreaks or climate variability.
- **Creation of New Industries**: The introduction of previously unutilized crops can lead to the establishment of industries around these crops. For instance, the introduction of **soybean** into India led to the development of industries related to oil extraction, protein-rich animal feed, and soybean-based food products.

Example:

- **Oil Palm in India**: The introduction of oil palm into India from **Malaysia** and **Indonesia** provided the country with a high-yielding crop that significantly boosted oil production. Oil palm has now become an important cash crop in several Indian states, contributing to domestic oil supply and reducing the dependence on imported edible oils.

2. Increasing Food Production

Another critical objective of plant introduction is to increase food production by introducing high-yielding varieties of staple crops. This objective has been central to many agricultural development programs aimed at addressing food insecurity, especially in regions that are highly dependent on agriculture for sustenance.

Key Points:

» **High-Yielding Varieties**: The introduction of high-yielding crop varieties with improved agronomic traits such as disease resistance, better nutrient use efficiency, and resilience to environmental stresses directly enhances food production. These varieties are particularly important in regions where agricultural productivity is low, and food shortages are common.

» **Addressing Food Security**: In regions facing food insecurity, introducing high-yielding varieties has been a proven strategy to boost food availability and ensure that more people have access to sufficient, nutritious food. For example, the **Green Revolution** introduced high-yielding varieties of rice and wheat, which led to a dramatic increase in food production across Asia and Latin America, helping to avert famines and hunger crises.

» **Meeting Growing Demand**: As the global population continues to grow, the introduction of new and improved crop varieties becomes essential to meet the increasing demand for food. Plant introduction plays a crucial role in this effort by introducing crops that are more productive and adapted to various environments.

Example:

» **Green Revolution in Asia**: The introduction of semi-dwarf, high-yielding rice varieties such as **IR8** and wheat varieties such as **Sonora 64** during the Green Revolution transformed agriculture in Asia. These varieties were more productive and disease-resistant, allowing countries like India to become self-sufficient in staple food production.

3. Germplasm Enrichment

One of the foundational objectives of plant introduction is to enrich the **germplasm**—the genetic material available for breeding and research. By introducing a wide range of varieties, species, or wild relatives, plant introduction provides breeders with diverse genetic resources that can be used for crop improvement programs.

Key Points:

» **Genetic Diversity**: Germplasm enrichment introduces new genetic variations that breeders can use to develop improved varieties with desirable traits such as yield, disease resistance, drought tolerance, and better quality. Genetic diversity is essential for ensuring the long-term sustainability of agriculture by allowing crops to adapt to changing environmental conditions.

» **Crop Improvement**: The introduction of diverse genetic material from various regions enriches the local gene pool, providing breeders with new tools for crop improvement. This is especially important for crops that may have a narrow genetic base due to domestication or extensive breeding.

- **Preservation of Genetic Resources**: Plant introduction also plays a role in the conservation of genetic resources by bringing new varieties or species into germplasm banks. These resources are vital for future breeding programs and ensuring food security.

Example:

- **Sorghum and Groundnut**: The introduction of diverse sorghum and groundnut varieties into India has greatly enriched the germplasm available for these crops. These new genetic resources are now being used to breed varieties that are more resistant to pests, diseases, and environmental stresses.

4. Source of Resistance to Biotic and Abiotic Stresses

Another crucial objective of plant introduction is to provide new sources of resistance to **biotic stresses** (such as pests and diseases) and **abiotic stresses** (such as drought, salinity, and extreme temperatures). By introducing varieties or species that possess natural resistance to these stresses, breeders can develop more resilient crops and require fewer chemical inputs.

Key Points:

- **Biotic Stress Resistance**: Many introduced varieties carry genetic resistance to major pests and diseases. By incorporating these resistant genes into local varieties, breeders can develop crops that are less susceptible to damage, reducing the need for pesticides and improving crop yields.
- **Abiotic Stress Tolerance**: In regions prone to environmental stresses like drought, heat, or salinity, plant introduction can bring in varieties that are better adapted to such conditions. This is critical for maintaining productivity in the face of climate change and ensuring that crops can survive under suboptimal growing conditions.
- **Disease Management**: By introducing varieties with resistance to diseases that are prevalent in the target region, plant introduction can prevent major crop losses. For example, groundnut accessions introduced from the U.S. carry resistance to **rust**, a disease that can devastate groundnut crops.

Example:

- **NCAC Groundnut Accessions**: The introduction of **NCAC groundnut accessions** from the United States into India provided breeders with a valuable source of resistance to rust disease, leading to the development of rust-resistant groundnut varieties.
- **Dasal Rice Variety**: The **Dasal rice variety** was introduced for its ability to tolerate saline conditions, making it an important addition to rice breeding programs in regions with high soil salinity.

5. Aesthetic Value

In addition to its agricultural and economic importance, plant introduction also serves the purpose of enhancing the **aesthetic value** of landscapes through the introduction of ornamental plants. Ornamental plant introductions contribute to the beautification of gardens, parks, and public spaces, providing aesthetic and ecological benefits.

Key Points:

- **Landscaping and Gardening**: Introduced ornamental plants, with their unique colors, shapes, and growth habits, are often used in landscaping and gardening to create visually appealing environments. These plants are selected for their beauty, hardiness, and adaptability to local conditions.
- **Cultural and Social Value**: Ornamental plants enhance the cultural and social aspects of communities by creating spaces that promote well-being and recreational activities. Parks, botanical gardens, and urban green spaces benefit from the introduction of diverse ornamental species.
- **Ecological Benefits**: Many ornamental plants also provide ecological benefits by supporting local biodiversity, improving air quality, and contributing to ecosystem services like pollination.

Example:

- **Introduction of Ornamental Plants**: The introduction of ornamental species such as **roses**, **orchids**, and **tulips** has significantly enriched gardens and landscapes around the world. These plants not only add beauty but also play a role in horticultural industries, contributing to floriculture and the sale of ornamental plants.

Plant Introduction Agencies

Historically, the exchange and introduction of plants were primarily driven by traders, invaders, explorers, and pilgrims. Various crops and ornamental plants were introduced to new regions as these groups moved across the world. For example, **Muslim invaders** brought crops such as **cherries** and **grapes** to India, while **Portuguese traders** introduced significant crops like **maize**, **groundnut**, **potato**, **papaya**, and **cashew nut**. Additionally, the **East India Company** played a pivotal role in the introduction of **tea** to India, which has since become a staple crop in the country.

In modern agriculture and horticulture, the introduction of plant species is managed more systematically by formal institutions and agencies that oversee the introduction, maintenance, and distribution of plant germplasm. These agencies ensure that the introduction process is conducted scientifically, minimizing the risks of introducing pests and diseases while maximizing the benefits of new crop varieties and plant species.

1. National Bureau of Plant Genetic Resources (NBPGR)

The **National Bureau of Plant Genetic Resources (NBPGR)** was established in **1976** and is based in **New Delhi, India**. It is the central authority for **plant introduction**, **germplasm maintenance**, and **germplasm exchange** within India. The NBPGR plays a crucial role in managing the introduction and distribution of germplasm, not just for agricultural crops, but also for horticultural species. It acts as the national gateway for germplasm introduction and export, ensuring compliance with international standards and regulations related to plant quarantine and genetic resources.

Key Functions of NBPGR:

- **Germplasm Introduction and Distribution**: The NBPGR handles the introduction of new germplasm from other countries, including crop varieties and wild relatives, and ensures their availability for research and breeding programs.

- **Maintenance of Germplasm**: It maintains **gene banks** for a wide range of crops, preserving seeds and other germplasm materials for long-term storage. The gene banks serve as repositories for plant diversity, protecting it for future use.
- **Quarantine Services**: NBPGR ensures that all introduced plant materials undergo **quarantine procedures** to prevent the entry of harmful pests and diseases into the country. The bureau issues **phytosanitary certificates** for both imported and exported germplasm.
- **Exploratory Surveys and Collection**: It conducts **exploratory surveys** to collect indigenous and exotic plant genetic resources from different parts of the world, contributing to the enrichment of India's plant genetic pool.
- **Training and Capacity Building**: The NBPGR conducts **training programs** for scientists and plant breeders on the introduction, handling, and management of germplasm.

Example:

- **Introduction of High-Yielding Rice Varieties**: The NBPGR played a crucial role in the introduction of high-yielding rice varieties such as **IR8**, **IR36**, and **IR64** from the **International Rice Research Institute (IRRI)**. These varieties were instrumental in increasing rice production in India during the Green Revolution.

2. International Board of Plant Genetic Resources (IBPGR)

The **International Board of Plant Genetic Resources (IBPGR)**, now known as **Bioversity International**, is a global organization based in **Rome, Italy**, dedicated to the conservation and sustainable use of plant genetic resources. It works in collaboration with various national and international organizations to facilitate the exchange of germplasm between countries, with a focus on **food security**, **biodiversity conservation**, and **climate resilience**.

Key Functions of IBPGR:

- **Global Germplasm Exchange**: IBPGR acts as a central body for the international exchange of plant genetic resources, coordinating with countries to ensure that plant materials are distributed safely and responsibly.
- **Conservation of Plant Biodiversity**: IBPGR supports the **conservation of plant genetic resources** through ex situ (gene banks) and in situ (on-farm) conservation strategies. It encourages the use of plant genetic resources in crop breeding and agricultural research to address challenges such as climate change and food security.
- **Support to National Programs**: IBPGR provides **technical and financial support** to national programs that aim to conserve plant genetic diversity. It also helps countries develop policies for the sustainable use of these resources.
- **Collaboration with Global Institutions**: IBPGR works closely with global institutions like the **Food and Agriculture Organization (FAO)**, **Consultative Group on International**

Agricultural Research (CGIAR) centers, and national research organizations to strengthen global plant genetic resource conservation efforts.

Example:

» **Global Wheat Program**: IBPGR facilitated the introduction of **wheat varieties** with **disease resistance** from **Mexico** into several countries, contributing to enhanced wheat production and the prevention of crop losses due to diseases such as rust.

3. Forest Research Institute (FRI), Dehradun

The **Forest Research Institute (FRI)**, located in **Dehradun, India**, is one of the premier institutions for forestry research and plant introduction related to forest species. Established in 1906, FRI plays a significant role in the **introduction, maintenance**, and **testing of forest tree germplasm**. Its mandate extends to the conservation and management of forest biodiversity, which is vital for sustainable forest management and ecological balance.

Key Functions of FRI:

» **Germplasm Introduction for Forestry Species**: FRI is responsible for introducing and evaluating new tree species that are useful for **timber production, fuelwood, fodder**, and **forest conservation**. These introductions help diversify India's forestry sector and ensure the sustainable use of forest resources.

» **Testing and Evaluation**: Introduced tree species undergo rigorous testing for adaptability, growth rates, wood quality, and resistance to pests and diseases. FRI conducts multi-location trials to evaluate how well these species perform in different ecological zones across India.

» **Conservation of Forest Biodiversity**: FRI works to conserve forest genetic resources by maintaining germplasm banks and seed orchards of key forest species. It also collaborates with other institutions to protect endangered forest species.

» **Afforestation and Reforestation Initiatives**: FRI supports **afforestation** and **reforestation** programs by introducing fast-growing, drought-resistant, and high-yielding tree species that contribute to ecological restoration and the mitigation of climate change impacts.

Example:

» **Introduction of Eucalyptus Species**: FRI introduced several species of **Eucalyptus** trees from **Australia**, which have since been widely planted across India for their fast growth and utility in **timber production** and **pulpwood** for the paper industry.

Procedure for Plant Introduction

The process of plant introduction is a critical and regulated pathway that involves multiple stages, ensuring the safe and efficient transfer of plant germplasm from one region to another. This procedure is essential for maintaining plant health, avoiding the spread of diseases and pests, and ensuring that new plant varieties are well-suited for the receiving environment.

1. Request Submission:

The first step in the plant introduction process is the submission of a formal request for germplasm by a scientist, research institution, or university. The National Bureau of Plant Genetic Resources (NBPGR), headquartered in New Delhi, serves as the central agency for handling plant introductions in India. The requesting entity must specify the type of plant material they are interested in acquiring, providing details on its intended use, such as breeding, conservation, or crop improvement.

The request should include:

» A description of the species or variety required.

» The intended research or breeding goals.

» Information on the environmental conditions of the target region.

» A justification for the introduction, ensuring it aligns with national agricultural priorities or research needs.

In cases where the introduction is to enhance genetic diversity or develop disease-resistant crops, clear objectives need to be stated in the application, as this helps NBPGR prioritize the request. Depending on the request, the material can be sourced either domestically or internationally.

2. IBPGR Coordination:

When the requested germplasm is sourced from another country, the International Board of Plant Genetic Resources (IBPGR), now known as Bioversity International, facilitates international coordination. The IBPGR is responsible for:

» Identifying the source of the requested germplasm from gene banks, research institutions, or other global repositories.

» Ensuring compliance with international agreements on biodiversity, such as the Convention on Biological Diversity (CBD), and regulations on the transfer of plant genetic material.

» Conducting collection and quarantine procedures to ensure the material is free from harmful pathogens or pests.

» Arranging safe packaging and transport of the plant material to the receiving country.

The IBPGR ensures that the introduction complies with the phytosanitary standards set by the International Plant Protection Convention (IPPC). Phytosanitary certificates issued by the exporting country must accompany the germplasm to confirm that the material is free of regulated pests and diseases.

3. NBPGR Distribution:

Once the germplasm reaches the National Bureau of Plant Genetic Resources, NBPGR performs several crucial steps to ensure safe and effective distribution:

» **Unique Identification and Storage**: Upon receiving the germplasm, NBPGR assigns a unique identifier to the material. This identifier helps in tracking and maintaining records of the germplasm for future reference. A portion of the seeds or plant material is stored in

NBPGR's germplasm bank for long-term conservation. This ensures that the material is available for future research or breeding programs.

- **Evaluation and Testing**: The remaining portion of the germplasm is sent to the scientist or institution that made the request. Before the material is released for cultivation or research, it undergoes an evaluation to ensure that it performs well under local conditions. This step may include trials in specific agro-climatic zones, where factors such as growth performance, yield, disease resistance, and adaptability are measured.
- **Collaboration with Other Institutes**: In some cases, NBPGR collaborates with other research institutes to further evaluate and test the introduced germplasm. For example, crop-specific research institutions such as the Indian Agricultural Research Institute (IARI) or regional agricultural universities may be involved in testing the material under specific local conditions.

4. Phytosanitary Restrictions:

A vital aspect of the plant introduction process is the enforcement of phytosanitary restrictions. These are regulations put in place to prevent the introduction of pests, diseases, and invasive species that could harm local ecosystems or agriculture. NBPGR ensures that all introduced plant material is thoroughly checked for compliance with these regulations before it is distributed.

Some plant introductions are subject to specific restrictions to prevent the spread of harmful organisms:

- **Nendran Banana**: Due to the risk of bunchy top disease, Nendran banana from Tamil Nadu cannot be exported to other regions. This restriction is in place to prevent the spread of the disease, which can severely affect banana plantations.
- **Cocoa**: The importation of cocoa from Africa is restricted due to concerns about the introduction of pests and diseases that could threaten local cocoa production.
- **Sunflower**: Similarly, sunflower from Argentina is restricted from entering certain regions due to the potential risks of introducing pathogens that could devastate local sunflower crops.

NBPGR plays a central role in ensuring that phytosanitary certificates are issued and maintained for all plant material, guaranteeing that it meets the necessary health standards.

Additional Steps in Plant Introduction:

- **Quarantine Procedures**: Certain plant materials, especially those being introduced from new regions, undergo strict quarantine processes. This is essential to identify any latent diseases or pests that may not be immediately visible. The plants are observed over several weeks or months to ensure they are free from harmful organisms.
- **Training and Awareness**: NBPGR and other relevant agencies conduct training programs for scientists, breeders, and farmers regarding the safe handling and propagation of introduced plant material. This helps in preventing the unintentional spread of diseases or pests.

Functions of NBPGR

The **National Bureau of Plant Genetic Resources (NBPGR)** plays a crucial role in safeguarding and promoting plant biodiversity in India. It is responsible for ensuring the collection, conservation, evaluation, and sustainable use of plant genetic resources (PGR) for agricultural and horticultural crops. Below are the key functions of NBPGR elaborated in detail:

1. Introduction, Maintenance, and Distribution of Germplasm

The NBPGR plays a pivotal role in the **introduction of plant germplasm** from both domestic and international sources. This germplasm refers to the seeds, tissues, or other plant material that contain the genetic information of various plant species and varieties.

» **Introduction**: NBPGR acts as the main gateway for introducing new crop species, varieties, and wild relatives from different parts of the world. It ensures that these materials are safely imported, properly quarantined, and tested for diseases or pests before distribution. The introduction of high-yielding and disease-resistant varieties, such as semi-dwarf wheat and rice, has been facilitated by the bureau.

» **Maintenance**: Once introduced, the germplasm is conserved in seed banks and gene banks. NBPGR ensures the long-term storage of these genetic resources in controlled environments to maintain their viability and genetic integrity. They employ modern storage technologies like cryopreservation, where seeds are stored at ultra-low temperatures, and seed banks, where seeds are periodically regenerated to ensure their health and vigor.

» **Distribution**: After introducing and maintaining the germplasm, NBPGR distributes plant genetic material to researchers, breeders, and farmers across the country. This process helps in crop improvement programs and facilitates access to genetic diversity that is essential for sustainable agriculture.

2. Information Dissemination

One of the key responsibilities of NBPGR is to **disseminate information** related to the germplasm it conserves, making it available to scientists, breeders, policymakers, and other stakeholders.

» **Publications**: NBPGR publishes regular updates, reports, research papers, and catalogs regarding its germplasm holdings. This information provides insights into the genetic diversity maintained by the bureau and its potential applications in breeding programs.

» **Databases**: The bureau maintains extensive **digital databases** of plant genetic resources, offering an inventory of the available germplasm collections. These databases provide detailed descriptions of various plant species and varieties, including their origin, traits, and conservation status. Such databases are accessible to plant breeders and researchers for developing crop varieties that can meet current agricultural challenges.

» **Knowledge Sharing**: Through workshops, conferences, and publications, NBPGR ensures that knowledge on germplasm management, utilization, and conservation is shared with stakeholders, both nationally and internationally. This dissemination helps in promoting the sustainable use of plant genetic resources in crop improvement.

3. Training and Education

NBPGR provides **training programs** aimed at enhancing the knowledge and skills of scientists, researchers, and agricultural professionals in the field of germplasm conservation and management.

These training sessions help to build capacity within the agricultural community and promote the sustainable use of genetic resources.

» **Specialized Training**: The bureau conducts specialized courses on germplasm conservation techniques, such as seed banking, cryopreservation, and in vitro conservation methods. Trainees learn about the protocols involved in handling plant genetic material, conducting exploratory missions for germplasm collection, and performing evaluations of germplasm for important agronomic traits.

» **Workshops and Seminars**: NBPGR regularly organizes workshops and seminars for researchers and plant breeders. These events focus on emerging challenges in plant genetic resource conservation, including topics like climate change, biodiversity loss, and the potential of genetic diversity in improving food security.

» **Collaboration with Institutions**: In collaboration with other agricultural research institutes, both in India and abroad, NBPGR provides training opportunities and research collaboration for postgraduate students, scientists, and extension workers, promoting the development of knowledge networks in plant breeding and genetic resource management.

4. Exploratory Surveys for Germplasm Collection

NBPGR is actively involved in conducting **exploratory surveys** across various regions of India and beyond to collect new and rare germplasm. This function is crucial for enhancing the genetic diversity available for crop improvement and for conserving wild relatives of cultivated crops.

» **Survey Missions**: NBPGR conducts missions in collaboration with agricultural universities, research institutes, and state departments to explore ecologically diverse areas, including tribal and remote regions, where unique plant species may exist. These exploratory missions target crop wild relatives (CWRs) that could possess valuable genetic traits, such as drought tolerance or disease resistance.

» **Characterization and Documentation**: The collected germplasm is systematically characterized and documented for its agronomic traits, genetic composition, and adaptability to different environmental conditions. This detailed information helps researchers utilize the genetic resources effectively in crop breeding programs.

» **Conservation of Underutilized Species**: In addition to major food crops, NBPGR focuses on collecting and conserving underutilized and indigenous species. These species often have unique nutritional, medicinal, or agronomic value and can play a role in enhancing food security and nutritional diversity.

5. Gene Sanctuary Setup for In Situ Conservation

NBPGR plays an instrumental role in setting up **natural gene sanctuaries** for in situ conservation of wild species. In situ conservation refers to the conservation of plants in their natural habitats, allowing them to evolve naturally while maintaining their genetic diversity.

» **Establishment of Sanctuaries**: NBPGR collaborates with state governments, conservation agencies, and forestry departments to establish gene sanctuaries in regions rich in plant biodiversity. These sanctuaries protect wild relatives of crops that may have important genetic traits, such as resistance to pests, diseases, and environmental stresses.

- **Conservation of Wild Relatives:** The primary objective of these gene sanctuaries is to conserve the wild relatives of cultivated crops. These wild relatives often contain untapped genetic resources that are vital for future breeding programs. By conserving these species in their natural habitats, NBPGR ensures the continued availability of these resources for crop improvement.
- **Monitoring and Management**: NBPGR is involved in the continuous monitoring and management of gene sanctuaries to ensure the genetic diversity of plants is maintained. These sanctuaries also serve as research hubs where scientists can study the evolutionary biology of species and their adaptation mechanisms.

Achievements in Plant Introduction

Plant introduction has played a crucial role in global agricultural development, significantly contributing to crop diversity, food security, and the establishment of new agricultural industries. Through the systematic introduction and adaptation of new crop species and varieties, countries have been able to overcome limitations in local crop production, address food security challenges, and create new economic opportunities. Below are some of the key achievements of plant introduction in agriculture.

1. Introduction of Staple Crops: Maize, Potato, and Tomato

One of the most profound achievements of plant introduction has been the successful transfer of major food crops like **maize (Zea mays), potato (Solanum tuberosum)**, and **tomato (Solanum lycopersicum)** across the world, which have since become staple foods in many regions.

- **Maize**: Originating from Central America, maize was introduced to Europe, Africa, and Asia, where it became a major staple food and fodder crop. Its adaptability to a wide range of climates and its high yield potential allowed it to be integrated into various agricultural systems, transforming diets and economies in these regions. In Africa, for example, maize is now a primary food source for millions of people, contributing significantly to food security.
- **Potato**: The introduction of the potato from the Andean region of South America to Europe in the 16th century revolutionized European agriculture. Its high caloric yield per unit of land made it a crucial food crop, especially in regions where grain crops struggled. The potato's success spread globally, and today it is a staple food in many parts of the world, including Asia and Africa. The introduction of new potato varieties continues to play a key role in addressing food security in developing countries.
- **Tomato**: Originally native to South America, the tomato was introduced to Europe in the 16th century and subsequently spread worldwide. Today, it is one of the most important horticultural crops globally, valued both for its culinary uses and its nutritional content, which includes high levels of vitamins A and C, as well as antioxidants. Tomato breeding programs have since introduced a variety of tomato types adapted to different climates, extending its cultivation across diverse environments.

These introductions not only diversified the diets of populations globally but also provided new opportunities for agricultural development and trade.

2. The Green Revolution: Semi-Dwarf Wheat and Rice Varieties

The **Green Revolution** in the mid-20th century is one of the most significant achievements in plant introduction, particularly the development and widespread adoption of **semi-dwarf wheat** and **rice** varieties. These new varieties, developed through international collaboration, were instrumental in addressing global food shortages and dramatically increasing food production, particularly in Asia and Latin America.

» **Semi-Dwarf Wheat**: Semi-dwarf wheat varieties, such as those developed by **Norman Borlaug** at the International Maize and Wheat Improvement Center (CIMMYT), revolutionized wheat production. These varieties had a shorter plant height, which allowed them to support heavier grain loads without lodging (falling over). As a result, yields increased dramatically, particularly in countries like India and Pakistan, where food shortages were previously common. The introduction of these wheat varieties helped to avert famine in these regions and is considered one of the greatest agricultural achievements of the 20th century.

» **Semi-Dwarf Rice**: Similar success was achieved with the introduction of semi-dwarf rice varieties developed at the **International Rice Research Institute (IRRI)** in the Philippines. Varieties such as **IR8** (also known as "miracle rice") were shorter, had a higher harvest index, and responded better to fertilizer, leading to significant yield increases. These rice varieties were introduced across Asia, where they played a critical role in boosting food production and improving livelihoods. Countries like India, China, and the Philippines saw dramatic increases in rice production, helping to secure food supplies for millions of people.

The introduction of these high-yielding, disease-resistant varieties revolutionized food production and is widely credited with preventing widespread famine in developing countries.

3. Introduction of Soybean and Oil Palm: Growth of the Oilseed Industry

The introduction of **soybean (Glycine max)** and **oil palm (Elaeis guineensis)** has had a profound impact on the global oilseed industry, providing the raw materials for vegetable oil production and livestock feed.

» **Soybean Introduction**: Although originally domesticated in East Asia, soybean was introduced to the United States and other countries in the early 20th century, where it became a major crop. The soybean's high protein content and versatility have made it a valuable crop for food, feed, and industrial applications. Today, soybean is one of the most important oilseed crops globally, and its introduction has contributed to the development of the animal feed industry and the production of vegetable oils, biodiesel, and other soy-based products. The soybean industry is now a multi-billion-dollar sector, with major production in countries such as the United States, Brazil, and Argentina.

» **Oil Palm Introduction**: Oil palm, native to West Africa, was introduced to Southeast Asia in the 19th century, where it quickly became a major plantation crop. The climate in Malaysia and Indonesia proved ideal for oil palm cultivation, and these countries now dominate global palm oil production. Palm oil, derived from the oil palm fruit, is used in a wide range of products, from cooking oil to cosmetics and biofuels. The introduction of oil palm to Southeast Asia not only boosted the regional economy but also transformed global vegetable oil markets. Palm oil is now one of the most widely used vegetable oils in the world.

These introductions have contributed to the growth of the oilseed industry, which is essential for meeting global demands for edible oils, biofuels, and protein-rich animal feed.

4. Rescue and Conservation of Crops through Plant Introduction

Plant introduction has also played a critical role in the **rescue and conservation** of certain crops threatened by diseases or pests. By introducing these crops into new regions where the disease or pest is not prevalent, the species are protected from further losses.

- **Coffee in South America**: Coffee (Coffea arabica) was introduced from Africa to South America in the 18th century. One of the major reasons for this introduction was to protect coffee from the devastating effects of **coffee leaf rust**, a disease that threatened coffee production in its native African regions. By growing coffee in disease-free areas of South America, particularly in Brazil and Colombia, coffee production was safeguarded, and these regions are now some of the largest coffee producers in the world.
- **Rubber (Hevea brasiliensis)**: Rubber was introduced to Southeast Asia from South America to protect it from a leaf disease that severely impacted rubber production in its native region. The introduction of rubber to Malaysia and Indonesia in the late 19th century created a thriving rubber industry in Southeast Asia, which now produces the majority of the world's natural rubber supply. This introduction saved the rubber industry from near extinction and ensured a steady global supply of this critical raw material.

5. Diversification of Crop Portfolios through Ornamental and Specialty Crops

Plant introduction has not only enhanced food production but has also diversified crop portfolios by introducing **ornamental plants** and **specialty crops** that contribute to landscaping, floriculture, and niche markets.

- **Ornamentals**: Many ornamental plants, such as roses, tulips, and chrysanthemums, have been introduced to new regions to enrich the biodiversity of ornamental gardens and the global floriculture industry. These introductions have added aesthetic value to landscapes, public parks, and gardens while creating significant economic opportunities for the ornamental plant industry.
- **Specialty Crops**: Crops like **kiwifruit**, introduced from China to New Zealand in the early 20th century, have become successful specialty crops. Kiwifruit is now one of New Zealand's most important horticultural exports, demonstrating how the introduction of new crops can open up new agricultural markets and industries.

B. Concept of Plant Ideotype

The concept of a plant ideotype was first introduced by **Donald (1968)** about wheat breeding. A plant ideotype refers to an ideal plant model or prototype that is expected to perform exceptionally well under a specific set of environmental conditions. The ideotype is defined based on its ability to maximize yield or performance, often by integrating several desirable traits. These traits are carefully chosen to optimize growth, adaptability, and resilience to environmental challenges. Unlike traditional breeding, which often focuses on single traits (e.g., disease resistance or high yield), the ideotype approach emphasizes the holistic integration of multiple traits that collectively contribute to the plant's overall performance. An ideotype can be developed for different purposes, such as increased yield, resistance to pests or diseases, tolerance to abiotic stresses (e.g., drought, salinity), or improved quality traits like grain or fruit quality. The concept applies to various crops, and the traits included in an ideotype vary depending on the crop species and the target environment.

Characteristics of a Plant Ideotype

A plant ideotype represents a theoretical model of an ideal plant that maximizes productivity under a specific set of environmental and agricultural conditions. It is essentially a blueprint for breeding and genetic improvement, aiming to optimize multiple traits simultaneously to enhance yield and sustainability. The concept of plant ideotypes is rooted in the idea of tailoring plants to meet the needs of particular agricultural practices, cropping systems, and environmental challenges.

1. Efficient Light Interception

The ability of a plant to intercept light efficiently is one of the core elements of photosynthesis and biomass production, making it a fundamental characteristic of a plant ideotype. The efficient capture of sunlight allows the plant to convert light energy into chemical energy, which is then used for growth and development. Efficient light interception involves multiple components, including leaf arrangement, leaf size, canopy structure, and plant height. These factors determine how much light a plant can capture throughout the day and its growing season.

Leaf Orientation and Canopy Structure

Leaf orientation plays a critical role in how much sunlight the plant can capture. Ideally, a plant ideotype should have leaves that are vertically arranged at the top and more horizontal in the lower parts. This arrangement ensures that the upper leaves capture direct sunlight while allowing diffused light to penetrate to the lower leaves. This reduces self-shading and maximizes the photosynthetic efficiency of the entire canopy. Canopy structure is equally important; the plant should have an open but balanced canopy that allows adequate air circulation and light penetration throughout the plant.

Leaf Size and Leaf Distribution

The size and distribution of leaves are also important for efficient light interception. Leaves that are too large can shade the lower parts of the plant, reducing the amount of light reaching lower leaves and leading to inefficient use of light. On the other hand, small leaves may not capture enough light to drive robust photosynthesis. The ideal leaf size should allow for maximum light capture while minimizing shading of other leaves, both on the same plant and on neighboring plants in a crop field. Proper leaf distribution along the stem is also critical for uniform light capture and efficient use of resources.

Plant Height

Plant height is another factor in efficient light interception. Taller plants are better suited for capturing light in competitive environments where neighboring plants or structures may cast shade. However, excessively tall plants are prone to lodging, where the plant falls over due to wind or the weight of the crop. An ideotype should balance height with structural strength, ensuring that the plant is tall enough to capture sufficient light but not so tall that it is at risk of lodging. Additionally, an ideal plant height makes the crop more suitable for mechanical harvesting, which is a key consideration in modern agriculture.

2. Nutrient Use Efficiency

Nutrient use efficiency (NUE) is a critical trait in plant ideotypes, especially as the cost of chemical fertilizers continues to rise, and environmental concerns over their use increase. Nutrient-efficient plants can produce high yields with lower inputs of nitrogen, phosphorus, and other essential nutrients. This trait not only reduces the cost of production but also minimizes the environmental footprint of farming practices by reducing nutrient runoff into water bodies, which can lead to problems like eutrophication.

Root Architecture and Nutrient Uptake

The structure of a plant's root system is a key factor in nutrient use efficiency. An ideotype with deep and well-developed roots is better able to access nutrients deep in the soil, particularly in environments where nutrient availability may be low. In addition to depth, lateral root spread is important for capturing nutrients in the upper layers of soil. The ideal plant will have a root system that is both deep and expansive, allowing it to forage for nutrients in different soil layers.

Internal Nutrient Utilization

Nutrient use efficiency also depends on how well a plant can use the nutrients it absorbs. Plants that can more effectively utilize nitrogen or phosphorus, for example, are able to grow and produce yields with less fertilizer input. Breeding for traits like improved nitrogen uptake or the ability to form symbiotic relationships with nitrogen-fixing bacteria can improve NUE. Similarly, enhancing the plant's ability to solubilize phosphorus, which is often bound in the soil and unavailable to plants, can improve phosphorus use efficiency.

Reducing Fertilizer Dependence

The primary goal of breeding for high NUE is to reduce dependence on chemical fertilizers. Plants that can make the most of available soil nutrients require less fertilizer input, which is both economically beneficial for farmers and environmentally sustainable. By incorporating high NUE into plant ideotypes, breeders can create crops that are more resilient in nutrient-poor soils and contribute to sustainable agricultural practices.

3. Pest and Disease Resistance

One of the most critical characteristics of a plant ideotype is its ability to resist pests and diseases. Pest and disease outbreaks can devastate crops, leading to significant yield losses. Developing plants that are resistant to major pests and diseases not only reduces the need for chemical pesticides but also improves the sustainability of the cropping system. Resistance can be conferred through genetic traits, which allow the plant to naturally fend off or tolerate attacks from pests and pathogens.

Genetic Resistance to Pests

Pest resistance in plants can be achieved through several mechanisms. Some plants produce secondary metabolites that act as natural insect repellents. Others may have physical barriers like thicker cell walls, waxy cuticles, or trichomes (hair-like structures on the leaves) that make it difficult for pests to feed. Breeding for these traits ensures that the plant can protect itself without the need for synthetic pesticides. This is particularly important in areas where access to pesticides is limited or where there is a desire to move toward more organic farming practices.

Disease Resistance through Genetic Breeding

Similar to pest resistance, disease resistance can be achieved through genetic breeding. Some plants carry resistance genes that make them less susceptible to common pathogens like fungi, bacteria, and viruses. For example, in cereal crops like wheat, breeding for resistance to rust diseases caused by fungi such as *Puccinia graminis* has been a key focus of crop improvement programs. By incorporating multiple resistance genes, breeders can create ideotypes that are resilient to a wide range of pathogens, thereby reducing the need for fungicides and other chemical treatments.

Reducing Chemical Inputs

The development of pest- and disease-resistant ideotypes is crucial for reducing chemical inputs in agriculture. By incorporating these traits into plant breeding programs, breeders can help farmers reduce their reliance on synthetic pesticides and fungicides, lowering production costs and mitigating the environmental impact of chemical use. This contributes to more sustainable and eco-friendly farming practices, especially in regions where pesticide use is heavily regulated or costly.

4. Abiotic Stress Tolerance

Abiotic stresses such as drought, salinity, and extreme temperatures are major challenges in global agriculture, particularly as climate change exacerbates the frequency and intensity of these conditions. Breeding plants that can withstand abiotic stresses is essential for ensuring food security in areas prone to these challenges. An ideal plant ideotype will have built-in mechanisms to tolerate or even thrive under adverse environmental conditions.

Drought Tolerance

Drought is a significant abiotic stress that affects crop yields worldwide. A drought-tolerant ideotype will have deep roots that can access water from lower soil layers, as well as a high water-use efficiency, meaning it can produce biomass with less water. Traits such as reduced stomatal opening, waxy leaves to minimize water loss, and the ability to store water in tissues are important for drought tolerance. These traits allow the plant to maintain productivity even under conditions of limited water availability.

Salt Tolerance

Saline soils, which contain high levels of salt, pose a significant challenge for many crops. Salt stress affects a plant's ability to take up water and can lead to toxic levels of sodium and chloride in plant tissues. A salt-tolerant plant ideotype will have traits that allow it to either exclude salts from its tissues or compartmentalize salts in vacuoles, where they cannot interfere with cellular processes. These mechanisms allow the plant to survive and even thrive in saline environments, which are becoming more common due to irrigation practices and rising sea levels.

Heat and Cold Tolerance

Extreme temperatures, whether high or low, can severely impact plant growth and development. Heat-tolerant ideotypes will have traits such as improved expression of heat shock proteins, which protect cellular structures from damage due to high temperatures. Cold-tolerant plants, on the other hand, may have mechanisms to prevent ice formation in tissues or the ability to continue metabolic processes at low temperatures. By incorporating heat and cold tolerance into plant ideotypes, breeders can ensure that crops can withstand temperature extremes, ensuring stable yields in an increasingly variable climate.

5. Growth Habit

The growth habit of a plant, which includes its height, branching patterns, and overall architecture, is critical for determining how well the plant will perform in agricultural settings. An ideal growth habit minimizes lodging (where plants fall over before harvest), maximizes light capture, and facilitates mechanical harvesting.

Compact Growth and Lodging Resistance

Lodging is a major problem in crops like wheat and rice, where tall plants are prone to falling over due to wind or the weight of the grain. Lodging reduces yield and complicates harvesting. A plant ideotype with a compact growth habit and strong stems is less likely to lodge, making it easier to cultivate and harvest. Additionally, a compact growth habit can improve the plant's resistance to wind damage and its ability to support heavy yields without collapsing.

Suitability for Mechanical Harvesting

Modern agriculture increasingly relies on mechanization for planting, managing, and harvesting crops. A plant ideotype that is well-suited for mechanical harvesting will have traits like uniform height, synchronous maturity, and a structure that allows for easy cutting or picking. This reduces labor costs and increases the efficiency of harvest operations. For crops like tomatoes or sugarcane, where mechanical harvesting is essential for large-scale production, growth habit plays a crucial role in determining the commercial viability of the crop.

6. Harvest Index

The harvest index is the ratio of economic yield (e.g., grains, fruits, or tubers) to the total biomass produced by the plant. Breeding for a high harvest index means that more of the plant's resources are directed toward the parts that are harvested, rather than vegetative growth. This improves the overall efficiency of the plant in terms of resource use and yield production.

Maximizing Economic Yield

In crops like wheat, rice, and maize, a high harvest index is achieved by directing more photosynthetic products toward grain production rather than leaves, stems, or roots. This results in a higher grain yield without necessarily increasing the total biomass of the plant. By focusing on increasing the harvest index, breeders can improve the efficiency of resource use and maximize the economic return for farmers.

Reducing Vegetative Growth

While vegetative growth is important for supporting photosynthesis, excessive growth can divert resources away from grain or fruit production. A plant ideotype with an optimal balance between vegetative growth and reproductive output will be more efficient in converting sunlight, water, and nutrients into yield. Breeding for a high harvest index involves selecting for plants that allocate more resources to reproductive tissues, thereby increasing the proportion of the plant's biomass that is economically valuable.

7. Early Maturity

Early maturity is a desirable trait in many plant ideotypes, particularly in regions with short growing seasons or unpredictable weather patterns. Early-maturing plants can be harvested before adverse environmental conditions, such as drought, frost, or disease outbreaks, reduce yields. Additionally, early maturity allows for multiple cropping cycles within a single growing season, increasing overall productivity.

Shorter Growing Cycles and Crop Rotation

Plants with shorter growing cycles allow farmers to practice crop rotation or double-cropping systems, where two or more crops are grown in a single year. This increases the efficiency of land use and improves overall farm productivity. Early-maturing ideotypes are especially important in regions with a limited growing season or where rapid turnover of crops is necessary to meet market demand.

Mitigating the Impact of Climate Change

With climate change leading to more erratic weather patterns, early-maturing crops provide farmers with flexibility in planting and harvesting schedules. By maturing before droughts, floods, or extreme temperatures, these crops are less likely to suffer from yield losses due to environmental stress. This makes early maturity a critical trait for developing resilient crop systems in the face of global climate challenges.

8. Uniformity

Uniformity in plant growth and development is essential for large-scale commercial farming. Uniformity ensures that all plants mature at the same time, which simplifies management practices such as fertilization, irrigation, and pest control. It also facilitates mechanical harvesting and processing, reducing labor costs and improving overall efficiency.

Synchronous Flowering and Maturity

Uniformity in flowering and maturity is particularly important for crops like maize and wheat, where a synchronous harvest ensures that all plants are at the optimal stage for processing. This reduces post-harvest losses due to uneven ripening and simplifies the logistics of harvesting and storage. Additionally, uniformity in crop development can reduce the risk of pest and disease outbreaks, as the entire crop can be managed simultaneously, rather than dealing with different growth stages at the same time.

Importance of Mechanized Agriculture

Uniformity is critical for mechanized agriculture, where crops must be planted, managed, and harvested on a large scale. Inconsistent plant sizes or maturity can lead to inefficiencies in mechanized systems, as machines may not be able to handle variable plant structures. By breeding for uniformity, plant ideotypes can improve the efficiency of mechanized farming operations, reducing labor costs and increasing overall productivity.

Role of Plant Ideotype in Crop Improvement

The role of plant ideotypes in crop improvement has become a cornerstone in modern breeding strategies, shifting focus from merely enhancing individual traits to optimizing the overall architecture and functionality of the plant. This approach is especially valuable in addressing the challenges posed by a growing global population, changing climates, and the increasing need for sustainable agriculture.

1. Yield Maximization

Yield improvement is the primary objective of most crop breeding programs, and plant ideotypes are designed with this in mind. The ideotype approach enhances crop yield by optimizing multiple traits that contribute to the plant's ability to produce more food per unit area. These traits include improved light interception, efficient use of water and nutrients, and overall enhanced plant architecture.

For example, in cereals like wheat and rice, ideotypes often focus on reducing plant height to minimize lodging (falling over) during the grain-filling stage, which helps maintain grain quality and overall yield. Improved tillering, or the production of side shoots, can also lead to more grain heads, thus increasing the grain yield per plant. Furthermore, the concept of **harvest index**, which is the ratio of economic yield (grain or fruit) to total biomass, is a major focus of ideotype breeding. By improving the harvest index, breeders ensure that a higher proportion of the plant's energy is directed toward producing grain or fruit, rather than vegetative growth. In practical terms, **the "Green Revolution"** varieties of wheat and ricc in the 1960s and 1970s were early examples of the ideotype approach. These varieties were semi-dwarf, which increased their resistance to lodging, and they had a higher harvest index, contributing to a significant boost in food production globally.

2. Improved Stress Tolerance

Abiotic stresses such as drought, salinity, and temperature extremes are among the most significant challenges facing global agriculture, especially in light of climate change. Plant ideotypes can be designed specifically to enhance tolerance to these environmental stresses, making crops more resilient and stabilizing yields under difficult conditions.

For instance, deeper and more robust root systems are critical for drought tolerance. An ideotype with a deep root system can access water from deeper soil layers, maintaining growth and productivity even in dry conditions. Similarly, **salinity tolerance** can be improved through ideotypes with traits that enable plants to exclude or sequester salts in specific tissues, preventing salt accumulation in sensitive parts of the plant, such as the roots or shoots. In addition to drought and salinity tolerance, ideotypes are being developed for crops that can withstand **heat stress** and **cold tolerance**, two factors that are increasingly important as global temperatures fluctuate. Plants with **osmotic adjustment** mechanisms, which help maintain cell turgor (firmness) during stress, are being incorporated into ideotypes designed for challenging environments.

3. Sustainability

Sustainability in agriculture is no longer a choice but a necessity, as the environmental impacts of traditional farming practices become more apparent. Plant ideotypes play a pivotal role in promoting sustainable farming practices by enhancing resource use efficiency, including water, nutrients, and sunlight.

Nutrient-use efficiency is one of the critical targets in ideotype breeding. For example, developing plants with better nitrogen uptake can significantly reduce the need for synthetic fertilizers, which are costly for farmers and can lead to environmental issues such as water pollution from runoff. Ideotypes with enhanced **phosphorus-use efficiency** can thrive in low-phosphorus soils, further reducing the need for chemical inputs. In terms of water usage, drought-resistan**t ideotypes** ensure that less water is required to maintain crop productivity, contributing to water conservation in agriculture. This is particularly important in regions with limited water availability or where water is overexploited for irrigation purposes. Sustainable ideotypes thus contribute to lower input costs for farmers, reduced environmental degradation, and more sustainable food production systems.

4. Pest and Disease Management

One of the greatest threats to crop productivity worldwide is damage caused by pests and diseases. Traditional approaches to managing these threats often rely heavily on chemical pesticides, which can have significant environmental and health consequences. Plant ideotypes designed with **genetic resistance to pests and diseases** can reduce or even eliminate the need for chemical inputs, making agriculture more sustainable and safer.

For example, incorporating **resistance genes for fungal diseases**, such as rust or blight, into a plant ideotype ensures that the crop can fend off infection without the need for fungicides. Similarly, developing **insect-resistant ideotypes**, which may involve traits such as thicker cell walls or secondary metabolites that deter pests, can significantly reduce the damage caused by insects. The **push-pull system** in maize, where certain plant ideotypes repel harmful pests while attracting beneficial insects, is a notable example of pest-resistant ideotype breeding. This method enhances natural pest control mechanisms, thus minimizing pesticide use.

5. Adaptability to Mechanization

With labor shortages becoming more acute in many agricultural regions, particularly in developed countries, there is a growing need for crop varieties that are compatible with mechanized farming systems. Plant ideotypes can be tailored to fit the requirements of mechanical planting, cultivation, and harvesting, making farming operations more efficient and reducing labor costs.

For instance, cereals like wheat and rice benefit from ideotypes that exhibit **uniform plant height** and **synchronous maturation**. These traits are essential for mechanical harvesters, which require uniformity to function effectively. Additionally, crops with **stronger stems** and reduced lodging tendencies are better suited for mechanized systems, where crops need to stand upright for efficient harvesting. Ideotypes designed for mechanization also focus on **compact growth habits**, which can improve the efficiency of machinery, particularly in row cropping systems where spacing and uniformity are critical.

6. Nutritional Quality and Consumer Preferences

Modern plant breeding is increasingly focused on the **nutritional quality of crops** as well as their agronomic performance. Plant ideotypes can be designed with traits that enhance nutritional content, such as higher protein levels, increased micronutrient concentrations (e.g., iron or zinc), or improved fiber content.

For example, **biofortified crops** like **Golden Rice**, which contains elevated levels of Vitamin A, have been developed to address nutritional deficiencies in regions where populations rely heavily on rice as a staple food. Similarly, ideotypes with **enhanced flavor profiles** or **improved cooking qualities** can cater to specific consumer preferences, thereby increasing the marketability of new crop varieties. By incorporating traits that cater to both **nutritional needs and consumer demands**, ideotypes help improve public health outcomes and create varieties that are more attractive to both producers and consumers.

7. Climate Resilience

As global climates become increasingly unpredictable, there is a pressing need for crop varieties that can withstand a wide range of environmental conditions. Ideotypes designed for **climate resilience** incorporate traits that enable plants to survive and thrive in extreme temperatures, unpredictable rainfall patterns, and other climate-related challenges.

Heat-tolerant ideotypes are crucial for crops grown in tropical and subtropical regions, where high temperatures can cause severe yield losses. Similarly, **cold-tolerant ideotypes** are important for crops grown in temperate regions, where unexpected frosts can damage or destroy crops.

Flood-tolerant ideotypes are also being developed for areas prone to waterlogging, such as rice fields in monsoon climates. These ideotypes often possess traits like **aerenchyma tissue** (which helps with oxygen exchange in waterlogged conditions) or **early maturation**, which allows crops to avoid the worst of the flooding season.

Applications of Plant Ideotype in Crop Breeding

Applications of plant ideotype in crop breeding have been transformative across major crops, enhancing their yield, adaptability, and resilience to environmental and biotic stresses. The ideotype concept, which involves designing plants with optimal characteristics for specific environments, has played a key role in revolutionizing modern agriculture. Below is a detailed explanation of how plant ideotypes have been applied successfully in crop breeding programs for wheat, rice, maize, and soybean.

1. Wheat: Semi-Dwarf Varieties and the Green Revolution

One of the most famous applications of the ideotype approach in crop breeding was during the **Green Revolution**, particularly in the development of **semi-dwarf wheat** varieties. The Green Revolution was initiated in the mid-20th century and is credited with preventing widespread famine in Asia and Latin America. The breeding efforts were primarily led by Norman Borlaug, whose work focused on improving wheat yields through ideotype breeding.

Key Characteristics of Semi-Dwarf Wheat Ideotypes

- **Reduced Plant Height**: Semi-dwarf wheat varieties were bred to have shorter stems, which reduced the risk of lodging (falling over) during the grain-filling stage. Lodging is a major problem in traditional tall wheat varieties, especially when they are grown under conditions of high fertility and water availability. The shorter plants could support the weight of more grain without falling over, which directly contributed to higher yields.
- **Improved Harvest Index**: The harvest index is the ratio of grain yield to total plant biomass. The semi-dwarf varieties focused on increasing this ratio, meaning more of the plant's resources were directed toward producing grain rather than stems and leaves. This optimized plant architecture for maximum food production per unit area.
- **Early Maturity**: The semi-dwarf wheat varieties also had shorter growing seasons, allowing them to mature earlier. This trait was particularly beneficial in regions with short growing seasons or where multiple cropping cycles in a year were practiced, enabling farmers to grow more than one crop annually.
- **Disease Resistance**: Semi-dwarf wheat ideotypes incorporated resistance to major diseases like rust, which could devastate wheat fields. Disease-resistant varieties reduced the need for chemical inputs, making farming more sustainable and cost-effective.

Impact on Wheat Yields

The ideotype-driven breeding of semi-dwarf wheat had a profound impact on global wheat production. Countries like India and Mexico saw wheat yields double or even triple within a few years of adopting these varieties. This ideotype-based breeding approach became a model for future crop improvement efforts.

Rice: Improved Productivity through Ideotype Breeding

Rice, the staple food for more than half of the world's population, has also benefited significantly from ideotype breeding. Similar to wheat, rice breeding programs have focused on improving yield, resource use efficiency, and stress tolerance through the development of ideal plant types.

Key Characteristics of Rice Ideotypes

- **Shorter Plant Height**: Just as in wheat, reducing the height of rice plants was a key component of ideotype breeding. Shorter plants have better lodging resistance, which is crucial in high-fertility environments or in regions prone to heavy rainfall and wind. This trait was central to the development of high-yielding rice varieties during the Green Revolution, such as IR8, which drastically improved rice production in Asia.
- **Improved Tillering**: Tillering refers to the production of side shoots that contribute to grain-bearing stems. Rice ideotypes were designed to promote more efficient tillering, which increases the number of grain-bearing stems per plant, thereby enhancing yield.
- **Enhanced Nitrogen-Use Efficiency**: Modern rice ideotypes are designed to utilize nitrogen more efficiently. Nitrogen is a crucial nutrient for plant growth, but excessive use of nitrogen fertilizers can lead to environmental problems such as water pollution and greenhouse gas

emissions. Breeding for better nitrogen-use efficiency allows rice plants to produce higher yields with less nitrogen input, making cultivation more environmentally sustainable.

- **Drought and Flood Tolerance**: Ideotype breeding in rice has also focused on improving tolerance to abiotic stresses such as drought and flooding. Varieties with deeper root systems or the ability to withstand temporary waterlogging have been developed to increase the resilience of rice in challenging environments, especially in regions affected by climate change.

Impact on Rice Yields

The introduction of ideotype-bred rice varieties has had a tremendous impact on global food security. The high-yielding semi-dwarf varieties that were developed during the Green Revolution, like **IR8**, helped avert widespread famine in Asia and remain the foundation of many modern rice-breeding programs. Today, rice ideotypes continue to evolve to meet the challenges of sustainable agriculture and climate resilience.

Maize: Modern Hybrid Varieties for Enhanced Yield and Resilience

Maize is one of the most important cereal crops worldwide, and ideotype breeding has significantly improved its yield and adaptability. Modern maize hybrids exhibit ideotype characteristics that enhance their performance under various environmental conditions.

Key Characteristics of Maize Ideotypes

- **Increased Kernel Size**: One of the primary targets in maize ideotype breeding has been increasing kernel size and weight, which directly contributes to higher grain yields. Larger kernels translate into more food per plant, improving the overall productivity of maize crops.
- **Uniformity in Ear Size**: Breeding for uniform ear size ensures that the plants produce consistent yields across a field. Uniformity is particularly important in mechanized farming systems, where uniform plant characteristics facilitate efficient mechanical harvesting and processing.
- **Enhanced Drought Resistance**: Drought resistance has been a major focus in maize breeding, especially in regions like sub-Saharan Africa, where erratic rainfall patterns can devastate crops. Modern maize ideotypes feature traits such as deeper root systems and improved water-use efficiency, which help plants maintain productivity during dry spells.
- **Adaptability to Various Climates**: Maize ideotypes have also been bred for broader adaptability to different climates. Whether in temperate regions or tropical environments, ideotypes are designed to perform well under diverse growing conditions, contributing to the global expansion of maize cultivation.

Impact on Maize Yields

The introduction of ideotype-driven hybrid maize varieties has led to significant increases in maize yields, particularly in regions where abiotic stresses like drought are prevalent. The development of drought-resistant maize hybrids has been a game-changer for smallholder farmers in Africa and Latin America, helping to improve food security in these vulnerable regions.

Soybean: Ideotypes for Drought Tolerance and Nitrogen Fixation

Soybean is a major source of protein and oil globally, and ideotype breeding has played a crucial role in improving its productivity and adaptability, particularly in marginal environments.

Key Characteristics of Soybean Ideotypes

- » **Improved Drought Tolerance**: Given that soybeans are often grown in rainfed environments, drought tolerance has been a key target in ideotype breeding. Traits such as improved root architecture and osmotic regulation have been incorporated into soybean ideotypes to help plants maintain water balance and productivity during dry periods.
- » **Enhanced Root Architecture**: The ideotype concept in soybean breeding often focuses on optimizing root systems for better nutrient uptake, particularly nitrogen and phosphorus. A more efficient root system allows soybeans to thrive in soils with limited nutrient availability, reducing the need for fertilizer inputs.
- » **Efficient Nitrogen Fixation**: Soybeans are legumes and have the unique ability to fix atmospheric nitrogen through symbiosis with soil bacteria. Ideotype breeding has aimed to improve this natural nitrogen fixation process, allowing the plant to produce high yields even in low-nitrogen soils. This reduces the need for nitrogen fertilizers, which benefits both the environment and farmers' budgets.

Impact on Soybean Yields

Ideotype-driven breeding has improved soybean yields by creating varieties that are more resilient to drought and more efficient in nutrient use. The ability to grow soybeans in marginal soils without heavy fertilizer inputs has expanded soybean cultivation into new areas, increasing global production and meeting the rising demand for plant-based protein.

The application of the plant ideotype concept in crop breeding has revolutionized the agricultural landscape, particularly in key crops like wheat, rice, maize, and soybeans. By focusing on optimizing plant architecture and incorporating traits that enhance yield, stress tolerance, and resource efficiency, ideotype breeding has contributed to significant gains in food security and sustainability. As global agriculture continues to face challenges from climate change, population growth, and environmental degradation, the role of ideotype-based breeding will only become more critical in ensuring a resilient and productive food system.

C. Participatory Plant Breeding (PPB)

Participatory Plant Breeding (PPB) is an innovative approach to crop improvement that actively involves farmers, breeders, and other stakeholders in the plant breeding process. Unlike traditional breeding methods, where scientists and breeders make decisions in research institutions or laboratories, PPB decentralizes the breeding process and incorporates the knowledge and preferences of local farmers. This method ensures that the varieties developed are better suited to local environmental conditions, farming practices, and cultural preferences, resulting in higher acceptance and faster adoption rates by the farming community.

Key Components of Participatory Plant Breeding (PPB)

1. Involvement of Farmers and Local Communities

One of the most defining features of Participatory Plant Breeding (PPB) is the active involvement of farmers and local communities throughout the breeding process. Unlike conventional plant breeding, where scientific institutions play a central role, PPB decentralizes the breeding effort by placing farmers at the forefront of decision-making. Farmers contribute their practical knowledge of local ecosystems, weather patterns, and market needs, which are invaluable in developing crop varieties that meet real-world challenges.

Farmers as Collaborators: In PPB, farmers are treated as collaborators rather than end users. They are engaged from the very beginning, participating in the selection of parent plants for hybridization and influencing key decisions throughout the breeding process. Their input helps address practical needs, ensuring that the developed varieties align with local agricultural practices.

Experiential Knowledge: Farmers have experiential knowledge that is often overlooked in scientific breeding. They understand the specific climatic conditions of their regions, the types of soils they work with, and how their crops respond to various biotic (pests, diseases) and abiotic (drought, salinity) stresses. Their day-to-day experience with these variables allows for more targeted and practical breeding solutions that directly address the challenges they face.

Customized Varieties: By actively involving local communities, PPB ensures that the resulting varieties are not only suited to the unique environmental conditions but also to the cultural and socio-economic preferences of the farming community. For example, taste, cooking qualities, or market demands like fruit size or color are given due consideration in the breeding process. This helps avoid the one-size-fits-all approach common in traditional breeding, instead leading to the creation of varieties that resonate with farmers' needs and preferences.

2. Decentralized Trials

A key component of PPB is the decentralization of breeding trials. Instead of limiting field trials to controlled research stations, PPB emphasizes conducting these trials on actual farms. This on-farm experimentation provides more accurate insights into how new crop varieties will perform in real-world agricultural environments.

Real-World Conditions: On-farm trials expose new crop varieties to the actual growing conditions in which they are intended to be cultivated. These conditions include the local soil type, natural weather patterns, water availability, and pest pressures. Unlike research station trials, where conditions are often optimized and controlled, on-farm trials present a more diverse and realistic set of challenges. This ensures that the varieties selected through PPB will be resilient and productive in real farming scenarios.

Diverse Agro-Ecological Zones: Farmers across different regions often face varying environmental conditions. Decentralized trials in PPB allow the same crop variety to be tested across different agro-ecological zones, helping breeders develop varieties that are not only high-yielding but also versatile across different environments. For instance, a variety that performs well in high-altitude areas may not thrive in lowland regions with higher temperatures. By decentralizing trials, PPB allows for varieties tailored to specific geographical regions and farming systems.

Farmer Feedback: Farmers play a crucial role in evaluating the performance of crops during decentralized trials. Their feedback on traits like yield, drought tolerance, pest resistance, and taste helps breeders understand the strengths and weaknesses of different varieties. This real-time feedback allows for quicker adjustments and refinements in the breeding process, ensuring that the final product meets the expectations of the local farming community.

3. Selection of Locally Adapted Varieties

The core objective of PPB is to develop crop varieties that are well-adapted to the local environmental conditions in which they will be grown. The conventional breeding process often produces varieties that are meant to be universally grown across broad geographic areas, but these varieties may not perform well in specific, localized environments. PPB addresses this gap by focusing on the development of varieties specifically suited to local agro-ecological conditions.

Adaptation to Local Conditions: Different regions have unique combinations of climatic conditions, soil types, water availability, and pest and disease pressures. PPB leverages local knowledge and real-world trials to develop varieties that are better suited to handle these local challenges. For example, in regions prone to drought, farmers might select for varieties that have deeper root systems or are more tolerant to water stress. In areas with specific pest pressures, varieties with natural pest resistance might be prioritized.

Micro-Environment Focus: PPB's strength lies in its ability to create varieties that thrive in micro-environments, which are often overlooked by conventional breeding. Smallholder farmers who operate in marginal lands, high-altitude areas, or regions prone to erratic weather can benefit significantly from varieties that have been tailored to the unique challenges of their specific environment.

Resilience to Local Stressors: Local adaptation not only increases yields but also ensures long-term resilience in the face of biotic and abiotic stressors. By prioritizing traits like pest resistance, drought tolerance, and heat tolerance, farmers are equipped with varieties that are more robust in the face of changing climatic conditions and fluctuating environmental stressors.

4. Farmer-Led Selection

One of the distinguishing features of PPB is the emphasis on farmer-led selection. Farmers take an active role in selecting superior crop lines at different stages of the breeding cycle. Their selection criteria may differ significantly from the criteria typically prioritized by breeders in formal breeding programs, which makes farmer-led selection an invaluable aspect of the PPB process.

Farmer Priorities: Traditional breeding programs often emphasize traits like yield, disease resistance, or seed uniformity, but these may not always align with the priorities of farmers. For example, farmers may prioritize traits such as taste, cooking quality, shelf life, or the ability to withstand particular environmental challenges, such as drought or salinity. In PPB, farmers directly select varieties based on these traits, ensuring that the final product meets their specific needs.

Hands-On Selection: Farmer-led selection allows farmers to physically inspect and choose the most promising plants in real time. They evaluate the plants based on their practical experience, assessing traits like growth habit, fruit quality, maturity time, and resilience to pests. This hands-on approach empowers farmers to shape the breeding process according to their preferences and ensures that the selected varieties are those that will thrive under their management practices.

Collaboration with Breeders: Farmer-led selection is often conducted in collaboration with plant breeders. While farmers bring practical knowledge and preferences, breeders bring scientific expertise in genetics and plant physiology. This collaboration enhances the overall quality of the breeding program, combining both traditional and scientific knowledge to develop superior crop varieties.

Challenges and Limitations of Participatory Plant Breeding (PPB)

While Participatory Plant Breeding (PPB) offers a collaborative and inclusive approach to crop improvement, it is not without its challenges. These limitations can sometimes hinder the efficiency and scalability of PPB efforts. Below are some of the major challenges associated with PPB:

1. Time-Consuming Process

PPB tends to be more time-consuming than conventional breeding methods due to its decentralized and participatory nature. While traditional breeding often occurs in controlled environments or research stations, PPB involves on-farm trials, which require continuous feedback from farmers. This interaction between multiple stakeholders at various stages of the breeding process can slow down the overall timeline of variety development.

- » **Extended Time Frame for Evaluation:** PPB relies heavily on real-world conditions to test and evaluate crops. Unlike traditional methods where breeding can be accelerated through controlled environments, on-farm trials expose crops to varying agro-climatic conditions, which can lengthen the time needed to gather consistent data.
- » **Collaborative Decision-Making:** Since PPB emphasizes collaboration, decisions about which varieties to prioritize or traits to focus on must be made collectively. The need to align the interests of multiple parties (e.g., breeders, farmers, and local communities) can add time to the decision-making process.

Impact: The longer development time can be a challenge in regions where new varieties are urgently needed to address immediate concerns like food insecurity, disease outbreaks, or rapidly changing environmental conditions. The slower pace of PPB can make it difficult to respond swiftly to emerging agricultural threats.

2. Coordination Challenges

Another limitation of PPB is the difficulty in coordinating among multiple stakeholders spread across different regions. Effective communication and collaboration between breeders, farmers, agricultural institutions, and possibly even policymakers are critical for the success of PPB. However, this is easier said than done, especially when working across diverse regions with different farming systems, languages, and cultural practices.

- » **Geographical Disparities:** Farmers and breeders participating in PPB may come from different geographical regions, each with distinct environmental conditions. Coordinating decentralized on-farm trials and ensuring that feedback from all regions is incorporated into the breeding program can be logistically complex.
- » **Cultural and Linguistic Barriers:** In diverse regions, language differences and varying

cultural practices can make communication more difficult. Farmers may have distinct priorities based on local customs and needs, while breeders may have scientific goals that don't always align with those of the farmers. Bridging these cultural and linguistic divides can pose a significant challenge.

» **Different Farming Systems:** Smallholder farmers, commercial farmers, and subsistence farmers may all participate in PPB programs. Each farming system may have different needs, making it difficult to develop varieties that work across these various scales.

Impact: The coordination challenges can lead to delays and inefficiencies in decision-making, making it harder to maintain a coherent breeding program. Miscommunications or disagreements between stakeholders can disrupt the progress of PPB, particularly when it comes to setting breeding priorities or evaluating trial results.

3. Limited Resources

PPB often requires more resources compared to traditional breeding methods due to the need for decentralized on-farm trials, transportation of experts and materials, farmer training programs, and regular community meetings. These additional costs can be a barrier to implementing or scaling PPB programs, particularly in resource-limited settings.

» **Infrastructure and Logistics Costs:** Conducting on-farm trials across multiple locations can be expensive, requiring transportation of seeds, equipment, and personnel. In regions with poor infrastructure, these costs can increase further, as logistical challenges make it harder to move materials and people between locations.

» **Farmer Training and Education:** Many smallholder farmers may not be familiar with breeding techniques or scientific evaluations of crops. PPB requires that farmers be trained and equipped to monitor trials, provide data, and offer meaningful feedback. This training can require time and financial resources.

» **Scaling Issues:** In resource-constrained environments, it can be difficult to expand PPB programs beyond pilot projects due to the cost of running multiple trials across diverse regions. This challenge is exacerbated in large or ecologically diverse regions, where different microclimates may require separate trials and breeding efforts.

Impact: Limited resources can restrict the scope and scale of PPB projects, reducing their potential impact. Resource constraints may also result in uneven implementation, where certain regions or groups of farmers are more actively involved than others, leading to unequal benefits from the breeding program.

4. Balancing Scientific and Practical Goals

One of the unique challenges in PPB is aligning the scientific objectives of plant breeders with the practical needs of farmers. Breeders may focus on agronomic traits that have broad applicability, such as disease resistance, yield potential, or abiotic stress tolerance. Meanwhile, farmers may prioritize different traits that are critical to their day-to-day operations, such as ease of harvest, taste, or suitability for local markets. Balancing these sometimes divergent goals can be challenging.

» **Differing Priorities:** Breeders may prioritize traits based on scientific criteria, such as improving genetic resistance to pests or diseases. Farmers, on the other hand, may place more emphasis on traits like taste, cooking quality, or even market preferences (e.g., color or size of the crop). This divergence in priorities can create tension between breeders and farmers when selecting lines for further development.

» **Complexity of Traits:** Some of the traits farmers value, such as flavor or post-harvest shelf life, can be more complex to breed for than straightforward agronomic traits like height or yield. This adds an extra layer of complexity to PPB, as breeders must work to integrate these complex traits while still addressing broader agronomic goals.

» **Perception of Value:** While breeders often rely on quantitative data such as yield per hectare or resistance ratings, farmers may assess crops based on more subjective criteria, such as taste or ease of management. Integrating these qualitative assessments into a formal breeding process can be challenging but is critical to the success of PPB.

Impact: When scientific and practical goals are misaligned, the success of PPB can be compromised. Breeding programs may produce varieties that excel in agronomic tests but fail to meet farmers' expectations or market demands. Conversely, farmers' selections might not always align with broader scientific goals like long-term disease resistance or climate adaptability, leading to short-term gains but potential long-term drawbacks.

Examples of Successful Participatory Plant Breeding (PPB) Projects

Participatory Plant Breeding (PPB) has proven to be a successful approach in addressing local agricultural challenges by directly involving farmers in the crop breeding process. Below are some notable examples of successful PPB projects around the world:

1. Rice in Nepal

In the mountainous regions of Nepal, a PPB initiative was launched to develop rice varieties better suited to the high-altitude conditions where temperatures are cooler, and growing seasons are shorter. Traditionally, these regions faced challenges with rice production due to environmental constraints, and conventional varieties were often not suited for such difficult conditions. The introduction of PPB allowed breeders and farmers to collaboratively develop locally adapted varieties.

» **Farmer Involvement:** Farmers in these regions have invaluable knowledge of local agro-ecological conditions. They worked with breeders to select rice varieties based on traits such as cold tolerance, early maturity, and adaptability to shorter growing seasons. Their direct involvement helped identify rice varieties that were highly suited to their specific climatic and soil conditions.

» **Outcomes:** This program resulted in the development of several new rice varieties, which are now widely cultivated by smallholder farmers. The varieties produced through PPB showed improved cold tolerance and productivity under the challenging high-altitude conditions, enhancing food security for the region's rural households.

» **Impact:** These varieties have improved yields for farmers in mountainous areas, where traditional varieties struggled. The success of this PPB project also led to the broader adoption of PPB practices in other crops, and more farmers are now participating in the selection process for new varieties.

2. Sorghum in Mali

In Mali, sorghum is a staple food crop for millions of rural households, particularly in arid and semi-arid regions. Sorghum cultivation faces challenges due to drought, pests, and variable soil fertility. A PPB program was initiated in Mali to develop more resilient sorghum varieties that could withstand drought and resist common pests. Farmers in the region, who had deep knowledge of the challenges faced in sorghum cultivation, played a key role in this process.

» **Farmer Participation:** Farmers were actively involved in selecting sorghum lines that demonstrated resilience to drought and pests. Their experience with local environmental conditions and farming systems allowed them to identify varieties that performed well in real-world conditions.

» **Results:** The PPB process led to the development of several sorghum varieties that exhibited strong resistance to drought and local pests, helping to secure the food supply in rural areas prone to environmental stresses. Additionally, farmers reported improved yields and crop reliability, even in marginal growing seasons.

» **Impact:** These new sorghum varieties have been widely adopted by local farmers, contributing to improved food security in rural communities. The success of the program also demonstrated the value of PPB in developing crops suited to the harsh environmental conditions in the Sahel region.

3. Maize in Mexico

Mexico is the center of origin for maize and boasts a rich diversity of maize landraces. However, many regions, especially in the south, experience marginal growing conditions characterized by drought, poor soil fertility, and erratic rainfall. A PPB initiative was introduced to help smallholder farmers in these areas develop maize varieties that could perform better in challenging environments. Farmers were crucial in identifying the best-performing varieties based on traits relevant to their needs.

» **Involvement of Farmers:** Farmers collaborated with breeders to select maize varieties that demonstrated drought tolerance, improved yield, and high kernel quality. They also evaluated the maize for other important traits such as cooking quality and taste, ensuring that the varieties were not only agronomically superior but also met local preferences for food use.

» **Outcomes:** Several maize varieties were successfully developed and are now grown in southern Mexico. These varieties exhibit improved drought tolerance, which has been crucial in helping farmers cope with water scarcity and irregular rainfall. The varieties also show better adaptability to marginal soils and have been embraced by smallholder farmers across the region.

» **Impact:** The introduction of these maize varieties has contributed to increased food security in drought-prone areas of southern Mexico. The PPB process has also strengthened the relationship between farmers and researchers, promoting a more inclusive approach to agricultural development.

D. Plant Breeders' Rights and Regulations for Plant Variety Protection and Farmers' Rights

The protection of plant varieties through legal frameworks is essential for promoting innovation in agriculture, encouraging plant breeders to develop new varieties, and ensuring that farmers' rights are respected. These frameworks are designed to balance the interests of breeders, farmers, and the public while supporting sustainable agricultural practices and biodiversity conservation.

1. Plant Breeders' Rights (PBR)

Plant Breeders' Rights (PBR) serve as a specialized form of intellectual property protection designed to promote innovation in plant breeding. These rights incentivize breeders to invest in the development of new plant varieties by granting them control over the commercial use of their varieties. With PBR, breeders have the exclusive authority to produce, reproduce, commercialize, and distribute their plant varieties. This legal framework encourages the advancement of agricultural productivity by supporting the creation of disease-resistant and environmentally adaptable crops, which is crucial for sustaining global food security and meeting the challenges posed by climate change.

Scope of PBR

The scope of PBR encompasses the breeder's exclusive rights over both the **propagating material** and **harvested material** of the protected variety. Propagating material includes seeds, cuttings, or bulbs, which are essential for reproducing the plant, while harvested material refers to products like fruits, flowers, and even grains produced from the variety. This protection ensures that no one can produce, reproduce, market, or distribute the variety without the breeder's consent, thereby providing a mechanism for breeders to financially benefit from their innovation.

For example, a breeder who develops a disease-resistant rice variety is entitled to exclusive rights over the sale of its seeds or the rice harvested from it. Unauthorized use of these materials would violate PBR, allowing the breeder to take legal action to protect their rights.

Requirements for PBR Protection

To qualify for PBR, a variety must meet specific criteria that ensure its distinctiveness and utility in agricultural systems. These include:

- **Novelty**: The variety must be new, meaning it has not been exploited commercially prior to the application. This ensures that breeders are rewarded for true innovations rather than existing varieties that have already been widely distributed and used.
- **Distinctness**: The variety must exhibit clearly distinguishable traits from any other known variety. This ensures that the new variety is not simply a replication of an existing variety but has unique characteristics, such as a new flower color, improved drought tolerance, or pest resistance.
- **Uniformity**: The variety must show uniformity in its characteristics across all individual plants. This consistency is critical for commercial viability, as it allows for predictable performance in the field or marketplace.
- **Stability**: The variety must remain stable across multiple generations of propagation,

maintaining its key characteristics. Breeders must ensure that the traits for which the variety is being protected do not change significantly with each cycle of reproduction.

The ability to meet these criteria ensures that only innovative and useful varieties receive PBR protection, fostering continuous advancements in agricultural productivity.

Duration of Protection

PBR protection is not indefinite; it typically lasts for 20 years for most crops and 25 years for trees and vines. Once this protection period expires, the variety enters the public domain, allowing anyone to freely use it. During the protection period, the breeder can benefit financially by licensing the variety or selling its seeds, creating an incentive for breeders to continue investing in research and development.

For example, a high-yielding wheat variety may enjoy PBR protection for 20 years, after which farmers, seed companies, and other breeders can freely use it without the original breeder's permission, contributing to the dissemination of improved varieties in the agricultural landscape.

Exceptions to PBR

PBR frameworks often include specific exceptions to balance breeders' rights with the broader agricultural needs of farmers and researchers. Two key exceptions are:

» **Farmers' Privilege:** In some countries, small-scale farmers are allowed to save and use seeds from PBR-protected varieties on their own farms without needing the breeder's permission. This exception is particularly important in developing regions where farmers depend on saving seeds from their harvests for the next planting season. However, the sale of these seeds to others may be restricted under the PBR laws, ensuring that breeders still retain control over large-scale commercial seed distribution.

» **Research and Breeding Exception:** PBR does not prevent other breeders or researchers from using protected varieties to conduct research or to develop new varieties. This exception is vital for fostering continued innovation in plant breeding. For instance, a breeder may use a protected wheat variety as a parent in a crossbreeding program aimed at improving disease resistance without infringing on the original breeder's rights. This ensures that agricultural progress is not hindered and that genetic diversity can continue to evolve through new breeding programs.

Global Importance and Harmonization

The global nature of agriculture means that PBR protection has been harmonized across many countries through frameworks like the **International Union for the Protection of New Varieties of Plants (UPOV)**. UPOV provides international standards for plant variety protection, enabling breeders to seek protection in multiple countries. This global framework encourages cross-border collaboration in plant breeding and facilitates the exchange of germplasm, which is critical for addressing global challenges such as food security, climate change, and sustainability.

2. Regulations for Plant Variety Protection (PVP)

Plant Variety Protection (PVP) is a comprehensive legal framework designed to grant plant breeders exclusive rights over new varieties they develop. This framework ensures breeders are rewarded for their innovations by controlling the use, commercialization, and distribution of their new varieties. PVP is a key driver of agricultural innovation, allowing breeders to invest in research and development while safeguarding their intellectual property. These regulations are often aligned with international treaties to create a harmonized approach to plant breeding and variety protection across countries.

UPOV Convention

The **International Union for the Protection of New Varieties of Plants (UPOV)** is the central body governing plant variety protection worldwide. The UPOV Convention was established in 1961 to create a standardized international framework for protecting plant breeders' rights. It has been revised several times, notably in 1978 and 1991, to address changing agricultural and legal landscapes.

The **UPOV Convention** provides the foundational legal guidelines that countries must follow when developing their national PVP systems. By signing the UPOV Convention, member countries agree to harmonize their laws with UPOV standards, which ensure that breeders' rights are protected across borders. The UPOV Convention is instrumental in encouraging global collaboration and innovation in plant breeding, as it creates a predictable legal environment for breeders seeking to protect their intellectual property in multiple countries.

Key Features of PVP Under UPOV

The UPOV framework ensures a robust legal foundation for plant variety protection while promoting continued access to genetic resources. The key features of PVP under the UPOV Convention include:

- **Protection for All Plant Genera and Species**: Under the 1991 Act of the UPOV Convention, PVP protection extends to all plant genera and species. This is a significant expansion from earlier versions of the treaty, ensuring that breeders working on minor crops or ornamental species are afforded the same protection as those developing major food crops like wheat, rice, or maize.
- **Exclusive Breeders' Rights**: Breeders are granted exclusive rights over the propagation and commercialization of their protected varieties. This includes the production, reproduction, sale, and marketing of both propagating and harvested material derived from the protected variety. For example, a breeder who develops a new apple variety can control its use, from the sale of its seeds to the marketing of the harvested apples.
- **Farmers' Privilege**: In some cases, **farmers' privilege** allows small-scale farmers to save seeds of protected varieties for use on their own farms without infringing on breeders' rights. However, this privilege is limited and varies by country, depending on national laws. For example, under the UPOV 1991 Act, farmers' privilege is restricted, and seeds from protected varieties cannot be sold or exchanged commercially without the breeder's consent.
- **Breeders' Rights Over Harvested Material**: The UPOV 1991 Act extends breeders' rights beyond propagating material to include harvested material, such as fruits, grains, or vegetables, derived from the protected variety. This means that even after harvesting, the breeder retains control over the use of the produce, providing further incentives for investment in plant breeding.

- **Research and Breeding Exemptions**: The UPOV Convention allows for the use of protected varieties in further breeding or research without the breeder's permission. This **breeding exemption** ensures that innovation and crop improvement can continue, as breeders and researchers can use protected varieties to develop new cultivars. For instance, a protected wheat variety can be used in a breeding program to develop new wheat lines with enhanced traits such as drought tolerance or disease resistance.

Two Versions of the UPOV Convention

The UPOV Convention has two key versions, the **1978 Act** and the **1991 Act**, each offering different levels of protection for breeders:

- **The UPOV 1978 Act**: This version provides more lenient protection, allowing farmers greater flexibility in saving and reusing seeds without violating breeders' rights. It is considered more favorable to small-scale farmers, as it grants them broader rights to use the seeds they harvest.
- **The UPOV 1991 Act**: The 1991 Act offers broader and more comprehensive protection for breeders. It extends breeders' control to harvested material and imposes tighter restrictions on farmers' ability to save and reuse seeds without the breeder's consent. Most new UPOV member countries are required to adopt the 1991 Act, which prioritizes breeders' rights more heavily.

Countries that signed the 1991 Act are required to implement national laws that offer stricter PVP protections, reducing the scope for farmers to save seeds of protected varieties and increasing breeders' control over their plant varieties.

National PVP Systems

Many countries have developed their own **national PVP systems**, aligned with UPOV guidelines, to protect breeders' rights while balancing the needs of local farmers. These national systems often incorporate additional provisions to safeguard the interests of small-scale and subsistence farmers, particularly in developing countries where agriculture plays a critical role in food security and livelihoods.

Key components of national PVP systems include:

- **Farmers' Rights**: National laws often include provisions to protect farmers' rights to save, use, and exchange seeds for non-commercial purposes. This is particularly important for subsistence farmers who rely on saved seeds for their next planting season. For example, in India, the **Protection of Plant Varieties and Farmers' Rights Act (PPVFR)** balances breeders' rights with farmers' rights, allowing farmers to save seeds of protected varieties for their own use.
- **Benefit-Sharing**: Some national PVP systems include **benefit-sharing** provisions, ensuring that local communities or indigenous farmers benefit from the commercial exploitation of genetic resources they have developed or conserved over time. This is particularly important for ensuring equitable access to the benefits derived from plant breeding, especially in biodiversity-rich regions.
- **Registration and Enforcement**: Breeders must register their new varieties with national PVP offices to receive protection. These offices are responsible for enforcing PVP laws and adjudicating any disputes between breeders and users of protected varieties. In cases of

infringement, breeders can take legal action to prevent unauthorized use or commercialization of their varieties.

The Role of PVP in Global Agriculture

PVP regulations play a crucial role in driving global agricultural innovation by providing breeders with the legal framework to protect their intellectual property. This incentivizes continuous investment in research and development, leading to the creation of new varieties that improve crop yield, pest and disease resistance, and adaptability to changing environmental conditions. In a world facing growing food security challenges due to climate change, the PVP framework is essential for ensuring that breeders continue to develop the varieties needed to meet future agricultural demands.

3. Farmers' Rights

Farmers' rights are crucial legal and ethical principles that aim to recognize and safeguard the invaluable contributions of farmers, particularly small-scale and indigenous farmers, to the conservation and improvement of plant genetic resources. These rights ensure that farmers maintain control over their traditional agricultural practices and biodiversity, even as global intellectual property regimes increasingly focus on plant breeders' rights (PBR) and plant variety protection (PVP). The recognition of farmers' rights serves as a counterbalance to the privatization of plant genetic resources, allowing farmers to continue their role as stewards of agricultural biodiversity.

Recognition of Farmers' Contributions

Farmers have, for centuries, played a pivotal role in the selection, conservation, and improvement of plant varieties. Their traditional agricultural practices have shaped much of the biodiversity found in crops today, contributing to the development of numerous landraces and varieties that are vital for food security. Farmers' rights recognize these historical and ongoing contributions, ensuring that farmers continue to have access to plant genetic resources and are not marginalized by modern breeding practices that prioritize intellectual property over shared heritage. By recognizing their role in biodiversity conservation, farmers' rights foster equitable access to genetic resources for sustainable agriculture.

Key Aspects of Farmers' Rights

The core components of farmers' rights focus on preserving traditional agricultural practices while ensuring farmers benefit from their contributions to plant breeding and biodiversity conservation. These rights are vital to maintaining agricultural diversity and ensuring farmers' ability to continue sustainable agricultural practices.

1. Right to Save, Use, Exchange, and Sell Farm-Saved Seeds

One of the most important elements of farmers' rights is the ability to save seeds from harvested crops for future planting. This practice has been the foundation of traditional farming systems, allowing farmers to adapt crops to local environmental conditions and cultural preferences over generations. Farmers' rights ensure that they can continue saving, using, and exchanging seeds without being overly restricted by breeders' rights under PVP laws.

While many national PVP laws, particularly under the UPOV 1991 Act, restrict the commercial sale of farm-saved seeds, smallholder farmers are often allowed to use seeds for their own farms. However, there may be limitations on selling these seeds in the commercial market. By allowing farmers to retain the ability to save and exchange seeds, farmers' rights protect traditional farming systems from becoming entirely dependent on purchased seeds from commercial breeders.

2. Protection of Traditional Knowledge

Farmers' rights also extend to the protection of traditional knowledge associated with plant genetic resources. Indigenous and smallholder farmers often possess critical knowledge about how to cultivate, improve, and conserve diverse plant varieties under different environmental conditions. This knowledge is essential for adapting crops to new challenges such as climate change, pests, and diseases. Legal frameworks ensure that this knowledge is respected and that farmers are compensated when it is used in commercial breeding programs.

Mechanisms such as **Access and Benefit-Sharing (ABS)** agreements, governed by the Nagoya Protocol under the Convention on Biological Diversity (CBD), ensure that when traditional knowledge or plant genetic resources are used for commercial purposes, farmers and indigenous communities receive equitable benefits. This helps prevent the exploitation of farmers' knowledge and genetic resources by corporations or breeders who might otherwise patent those resources without providing compensation to the communities from which they originated.

3. Participation in Decision-Making

Farmers have the right to participate in decision-making processes that affect agricultural policies, plant variety protection laws, and biodiversity conservation. This participation is essential for ensuring that laws and policies reflect the needs and realities of smallholder and indigenous farmers, who are often most affected by changes in agricultural systems.

Participation in national and international policy discussions empowers farmers to advocate for policies that support their traditional farming practices and protect their access to plant genetic resources. International frameworks such as the **International Treaty on Plant Genetic Resources for Food and Agriculture (ITPGRFA)** promote farmer participation in discussions related to plant breeding, seed systems, and biodiversity conservation. Farmer-led initiatives also ensure that agricultural systems remain diverse and resilient to new challenges.

International Treaty on Plant Genetic Resources for Food and Agriculture (ITPGRFA)

The **International Treaty on Plant Genetic Resources for Food and Agriculture (ITPGRFA),** commonly referred to as the Seed Treaty, is one of the most significant international agreements recognizing farmers' rights. Adopted by the Food and Agriculture Organization (FAO) of the United Nations in 2001, the ITPGRFA provides a global framework for the conservation and sustainable use of plant genetic resources while ensuring that farmers share in the benefits derived from their use. The treaty emphasizes the critical role of plant genetic resources in ensuring global food security and recognizes the essential role of farmers in maintaining agricultural biodiversity.

Farmers' Rights Under the ITPGRFA

The ITPGRFA establishes several key principles to promote farmers' rights and ensure that they benefit from the use of plant genetic resources:

1. Protect Traditional Knowledge Relevant to Plant Genetic Resources

The treaty recognizes that traditional knowledge held by farmers, especially in indigenous communities, is critical for the conservation and sustainable use of plant genetic resources. Farmers' traditional knowledge helps preserve genetic diversity in crops, which is crucial for adapting to changing environmental conditions and for crop improvement efforts. The ITPGRFA promotes the protection of this knowledge and ensures that it is respected in agricultural systems.

2. Equitable Sharing of Benefits Arising from the Use of Plant Genetic Resources

One of the treaty's key provisions is the equitable sharing of benefits derived from the use of plant genetic resources. This includes financial rewards, capacity-building opportunities, technology transfer, and access to improved seeds. Farmers' rights under the ITPGRFA ensure that farmers and local communities benefit when their plant varieties or traditional knowledge are used in commercial breeding programs.

3. Promote Farmers' Participation in Relevant Decision-Making Processes

The ITPGRFA promotes the active participation of farmers in national and international decision-making processes concerning plant genetic resources. By including farmers in these discussions, policies can be shaped in ways that protect their interests and ensure their access to the resources needed for sustainable agricultural practices. This participation helps bridge the gap between local farming communities and policymakers.

4. Provide Access to Plant Genetic Resources for Research and Breeding

The treaty ensures that farmers have access to plant genetic resources for research, breeding, and other purposes. This access is critical for smallholder farmers who depend on diverse plant varieties to manage risks associated with changing environmental conditions, pests, and diseases. Ensuring access to plant genetic resources allows farmers to continue their role as innovators and custodians of agricultural biodiversity.

4. Challenges and Future Directions

Despite the progress in creating legal frameworks that balance **Plant Breeders' Rights (PBR)** with **Farmers' Rights**, numerous challenges persist. These challenges highlight the complexities in managing intellectual property over plant genetic resources, particularly in a way that ensures both innovation and equity in agricultural systems. To ensure a fair balance between the commercial interests of breeders and the historical contributions of farmers, especially smallholders, and indigenous communities, several areas need to be addressed. Below are the key challenges and potential future directions to strengthen these frameworks:

1. Access to Genetic Resources

One of the primary challenges faced by farmers, particularly in developing countries, is access to improved seeds and plant varieties. Despite the advancements in plant breeding, many smallholder farmers are often unable to access high-quality, improved varieties due to financial barriers, lack of infrastructure, and sometimes restrictive PVP laws.

» **Barriers to Access**: In regions where poverty is high, the cost of improved seeds may be prohibitive for small-scale farmers. Commercial seed companies may charge prices that smallholders cannot afford, leading to an overreliance on saved seeds, which may not perform as well as newer varieties in terms of yield, disease resistance, or environmental adaptability.

» **Inadequate Infrastructure**: Many rural farming communities lack the necessary infrastructure, such as seed distribution networks or access to information on improved varieties. This makes it difficult for farmers to access the latest seed technologies, even when these technologies are available.

Future Directions:

» **Strengthening PVP Laws for Farmers' Access**: There is a need for PVP laws that balance the protection of breeders' rights with provisions that allow farmers, especially smallholders, to access these improved varieties affordably. This could involve subsidies for small-scale farmers, government programs to distribute seeds, or policies that encourage the sharing of benefits from seed innovation.

» **Investment in Infrastructure**: Governments, international organizations, and private companies should invest in building the infrastructure needed for seed distribution and education. Providing rural communities with better access to seeds, coupled with extension services to educate farmers on their use, would ensure the benefits of plant breeding reach all agricultural producers.

2. Balancing Breeders' and Farmers' Interests

Another significant challenge lies in balancing the interests of plant breeders with those of farmers. While PBRs incentivize innovation by granting breeders exclusive rights over their new varieties, these rights can sometimes conflict with the traditional farming practices of saving, using, and exchanging seeds. This is particularly problematic in developing regions where informal seed systems are the norm and farmers are accustomed to sharing seeds as part of their cultural heritage.

» **Potential Conflicts**: In some countries, stricter PVP regulations under the UPOV 1991 Act limit farmers' ability to save and exchange seeds from protected varieties. These restrictions, while benefiting breeders, can severely impact smallholder farmers who cannot afford to purchase new seeds every season.

Future Directions:

» **Creating Flexible Legal Frameworks**: One approach to address this challenge is to create more flexible PVP frameworks that allow farmers to save and exchange seeds under certain conditions, while still protecting the rights of breeders. The UPOV 1978 Act, which permits more farmer-friendly provisions, could serve as a model for balancing these interests in regions where smallholder farming dominates.

» **Farmers' Privilege**: Expanding and legally recognizing the concept of "farmers' privilege," which allows farmers to save seeds from protected varieties for use on their own farms, could help preserve traditional practices while still promoting seed innovation. Farmers' privilege provisions should be explicitly included in national PVP laws to protect smallholder farmers from legal actions when they engage in seed-saving activities.

3. Ensuring Benefit Sharing

The **International Treaty on Plant Genetic Resources for Food and Agriculture (ITPGRFA)**, along with other frameworks, promotes the **equitable sharing of benefits** arising from the use of plant genetic resources. However, putting these principles into practice remains challenging. There are numerous cases where farmers, especially indigenous communities, have not been adequately compensated when their genetic resources or traditional knowledge have been used in commercial plant breeding programs.

» **Inadequate Implementation**: The implementation of benefit-sharing mechanisms is often inconsistent. For example, despite commitments to share benefits arising from the commercialization of new plant varieties developed from traditional knowledge or landraces, many farmers and local communities do not receive fair compensation.

» **Lack of Awareness**: Many farming communities are unaware of their rights under international treaties like the ITPGRFA, which means they are less likely to assert these rights when their genetic resources are commercialized. This lack of awareness prevents farmers from engaging in benefit-sharing agreements that could support local agricultural development.

Future Directions:

» **Strengthening Implementation of Benefit-Sharing Mechanisms**: Governments and international bodies need to ensure the full implementation of access and benefit-sharing (ABS) agreements, particularly under the **Nagoya Protocol**. These mechanisms must be transparent, enforceable, and include clear provisions for compensating farming communities that contribute genetic resources or knowledge to plant breeding programs.

» **Building Capacity and Awareness**: Raising awareness among farming communities about their rights under frameworks like the ITPGRFA is essential for empowering them to negotiate fair benefit-sharing agreements. Training programs and legal support should be provided to farmers and indigenous groups so they can better protect their resources and knowledge in dealings with commercial breeders.

» **Innovative Benefit Sharing Models**: Developing innovative models of benefit sharing that involve both monetary and non-monetary rewards could be another step forward. For example, communities could receive not only financial compensation but also access to improved seeds, farming technology, and educational opportunities in exchange for their contributions to plant breeding.

4. Adapting to Climate Change and Future Agricultural Challenges

The changing global climate presents significant challenges to agricultural systems, and plant breeding will need to adapt accordingly. Participatory plant breeding, which involves farmers in the selection of locally adapted varieties, may play a crucial role in developing crops that can withstand new environmental stresses like drought, flooding, and heat.

» **Breeding for Climate Resilience**: One of the biggest future challenges for plant breeding will be to develop varieties that can thrive in increasingly unpredictable and harsh climates. To ensure that these varieties benefit smallholder farmers, breeding programs will need to consider the local needs and knowledge of farming communities.

Future Directions:

- » **Integrating Climate Adaptation into PVP and PBR**: PVP laws and plant breeding programs should increasingly focus on climate-resilient varieties. Governments and international bodies could incentivize breeding for traits like drought tolerance, flood resistance, and salinity tolerance, ensuring that these varieties are accessible to smallholder farmers.
- » **Incorporating Farmers' Knowledge**: Participatory plant breeding (PPB) can be an important tool in addressing climate change, as it allows farmers to contribute their local knowledge to the breeding process. This collaboration can help develop varieties that are better suited to specific agro-ecological environments and that can cope with the new challenges posed by climate change.

E. DUS Testing

(Distinctness, Uniformity, and Stability Testing)

DUS testing is a fundamental process used in the evaluation of new plant varieties to determine if they meet the essential criteria of **distinctness, uniformity**, and **stability** (DUS). It is a key component in the protection of plant breeders' rights (PBR) and for the registration of new plant varieties in many countries under Plant Variety Protection (PVP) laws. The results from DUS testing are essential for ensuring that new varieties possess novel characteristics and are true to type across different growing conditions.

Key Components of DUS Testing:

1. Distinctness

Distinctness refers to the requirement that a new variety must clearly differ from all other existing or commonly known varieties in at least one or more important traits. These traits may include morphological, physiological, or biochemical characteristics.

- » **Traits Evaluated for Distinctness**: A new variety is compared to existing varieties for attributes such as plant height, leaf shape, flower color, seed size, and growth habits. Distinctness ensures that the new variety is unique and recognizable.
- » **Importance**: The distinctness criterion guarantees that new plant varieties are not simple replications of existing varieties, thereby promoting true innovation in plant breeding.

2. Uniformity

Uniformity means that the plants within the variety must be sufficiently consistent in their traits, both visually and genetically. This consistency must be observed among plants grown in different locations and over successive growing seasons.

- » **Traits Evaluated for Uniformity**: The uniformity of a variety is typically assessed by examining the similarity of traits such as plant form, flowering time, and fruit size across a sample population of plants. Minor variations may be allowed, but the variation should not affect the essential characteristics of the variety.

- **Importance:** Uniformity is critical for commercial production, ensuring that a variety produces consistent results, making it easier for farmers to predict performance and for consumers to receive a reliable product.

3. Stability

Stability requires that the new variety remains unchanged with respect to its essential characteristics after repeated propagation over generations or cycles of reproduction.

- **Traits Evaluated for Stability:** The stability of a variety is assessed by observing whether its key traits remain consistent when propagated through seeds or vegetative means over successive generations or seasons.
- **Importance:** Stability ensures that the variety remains true to type and performs consistently, regardless of how many times it is propagated. This is important for maintaining the integrity of the variety over time.

The DUS Testing Procedure

The DUS testing procedure is a systematic and scientific process designed to evaluate new plant varieties based on their **Distinctness, Uniformity, and Stability**. This testing is crucial for obtaining Plant Variety Protection (PVP) under various national and international regulations, such as those administered by the **International Union for the Protection of New Varieties of Plants (UPOV)**. The testing procedure ensures that the variety is genuinely new, consistent, and stable enough to be eligible for intellectual property rights. The process follows a structured series of steps to guarantee that the plant variety meets the required criteria.

1. Application for Testing

The first step in the DUS testing process is the submission of an application to the relevant authority responsible for plant variety protection in the breeder's country or an international body like UPOV. This formal request initiates the evaluation of the new variety.

- **Application Requirements:** The breeder must submit detailed information about the new variety, including its name, origin, and a description of its key traits. In addition, the breeder may need to provide propagation material, such as seeds or cuttings, to be used in the trials. Some jurisdictions may also require a description of how the variety was developed.
- **Application Review:** The application is reviewed by the testing authority to ensure completeness and to check whether the variety meets the basic criteria for DUS testing. Once the review is successful, the authority accepts the application and moves to the next stage.

2. Trial Setup

Once the application is approved, the variety enters the trial phase. These trials are typically conducted over two or more growing seasons, depending on the crop species. The trials are performed under controlled and natural conditions to assess the variety's performance in various environments.

- **Field and Greenhouse Trials:** The trials are set up in fields, greenhouses, or both, depending on the crop type and growing conditions. For example, crops that are typically grown in open fields, such as wheat or maize, are tested in field conditions, while plants like tomatoes or peppers might also undergo greenhouse testing.

- **Multi-location Trials**: To ensure that the variety is evaluated for adaptability, trials are conducted across multiple locations with varying environmental conditions. This helps to assess the variety's uniformity and stability under diverse climates, soil types, and growing conditions.

3. Comparison with Reference Varieties

An essential part of the DUS testing process is the comparison of the new variety with already existing varieties, known as reference varieties. These are varieties with similar characteristics and performance to the new variety.

- **Selection of Reference Varieties**: Reference varieties are carefully selected based on their similarity to the new variety in terms of morphology, agronomic traits, and growth habits. This comparison ensures that the distinctiveness of the new variety is accurately assessed.
- **Standardized Guidelines**: The testing authority uses internationally or nationally recognized guidelines, such as those developed by UPOV, to ensure the traits of the new variety are consistently evaluated. These guidelines provide detailed instructions on how to measure and record various traits, such as leaf size, flower color, plant height, fruit weight, and other agronomic characteristics.

4. Data Collection

During the trial, comprehensive data is collected on the performance of the new variety. These observations are taken throughout the growing season to assess how the variety develops from seedling to maturity.

- **Measurement of Traits**: Detailed measurements of key characteristics are taken regularly during the growth cycle. For instance, plant height, leaf area, flowering time, fruit size, and yield are common metrics that are measured to assess the variety's distinctness and uniformity.
- **Morphological, Physiological, and Agronomic Data**: The collected data covers a wide range of traits, including morphological (e.g., flower shape, leaf length), physiological (e.g., drought tolerance, nutrient use efficiency), and agronomic (e.g., yield, harvest index) parameters. This ensures a comprehensive understanding of how the variety compares with reference varieties.
- **Data Analysis**: Once the data is collected, it is analyzed using statistical methods to determine whether the new variety meets the DUS criteria. This analysis is used to confirm whether the variety is clearly distinct from existing varieties and whether it shows the required uniformity and stability across different environments.

5. Final Evaluation and Report

At the end of the trial period, the data is compiled into a final evaluation report, which provides a detailed assessment of the variety's performance based on the DUS criteria. The report serves as the official document to support the variety's eligibility for protection under plant variety laws.

- **Evaluation for Distinctness:** The variety is evaluated to ensure that it shows clear differences from reference varieties in one or more critical traits. Distinctness is a vital criterion for granting breeders' rights, as it ensures that the new variety is truly innovative.
- **Evaluation for Uniformity:** The trial data is examined to verify that the new variety is sufficiently uniform. Uniformity ensures that plants of the same variety perform consistently across different locations and under varying conditions, which is crucial for commercial production.
- **Evaluation for Stability:** Stability is assessed to confirm that the variety's key traits remain unchanged over time and across different cycles of propagation. Stability is critical for ensuring that the variety will maintain its unique characteristics for many generations.
- **Final Report:** The final evaluation report details the performance of the variety in all stages of testing. If the variety passes the DUS test, the breeder is granted a certificate of protection. This report is essential for breeders to secure intellectual property rights and register their variety under Plant Variety Protection (PVP) laws.

Importance of DUS Testing in Plant Breeding

DUS (Distinctness, Uniformity, and Stability) testing plays a critical role in modern plant breeding by providing a rigorous framework for the evaluation, protection, and commercialization of new plant varieties. Its importance spans innovation, intellectual property rights, global standardization, and confidence in the agricultural value chain. Here are key aspects of its importance:

1. Ensuring Innovation

DUS testing encourages innovation in the field of plant breeding by setting a high standard for what constitutes a "new" plant variety. For a variety to pass the DUS test, it must show clear differences from existing varieties in at least one measurable characteristic (distinctness), display uniformity in those traits (uniformity), and maintain consistency across generations (stability).

- **Incentive for Breeders:** By ensuring that only genuinely distinct varieties are granted protection, DUS testing incentivizes breeders to pursue groundbreaking innovations rather than incremental changes. This encourages the continuous development of new varieties with enhanced characteristics, such as better yields, disease resistance, improved nutritional content, or adaptability to climate change.
- **Encouraging Diversity:** Through DUS testing, breeders are motivated to explore genetic diversity to find novel traits, increasing the genetic base available for future breeding efforts. This expands the genetic pool and enhances the resilience of agricultural systems, which is critical for addressing global challenges like food security and climate change.

2. Intellectual Property Protection

One of the primary goals of DUS testing is to **provide intellectual property (IP) protection** to plant breeders under frameworks like Plant Breeders' Rights (PBR) or Plant Variety Protection (PVP) laws. Without such protection, breeders would be reluctant to invest in the costly and time-consuming research needed to develop new varieties.

- **Plant Breeders' Rights (PBR)**: Successfully passing DUS tests is a prerequisite for securing PBR, which gives breeders the exclusive right to control the commercialization and reproduction of their new variety. This legal protection ensures that breeders can monetize their innovations by granting them control over the sale and distribution of seeds or plants.
- **Recouping R&D Investments**: The high cost of breeding new varieties, particularly in terms of research and development (R&D), makes IP protection essential. DUS testing ensures that breeders can recoup their investments by giving them exclusive rights to commercialize their new variety for a fixed period, typically 20 to 25 years, depending on the crop type.
- **Fostering Global Competitiveness**: In an increasingly globalized market, securing IP protection through DUS testing allows breeders to compete internationally, knowing their innovations are safeguarded under consistent international laws, such as the UPOV Convention.

3. Standardization

DUS testing provides a **standardized, scientific, and objective framework** for evaluating plant varieties. This standardization is crucial for maintaining the integrity of the plant breeding industry, ensuring that all varieties meet the same rigorous criteria before they are released to the market.

- **Consistency in Evaluation**: The use of consistent DUS testing protocols ensures that new varieties are evaluated fairly and uniformly across different regions and jurisdictions. This is particularly important for international trade and the sharing of genetic resources, as it allows varieties to be compared and assessed using the same scientific benchmarks, regardless of where they were developed.
- **Global Trade and Collaboration**: Standardization also facilitates international collaboration and trade. Countries that adhere to UPOV standards can freely exchange germplasm and plant varieties, knowing that the varieties have been subject to the same level of scrutiny. This is essential for global food security, as it allows the movement of crop varieties that can help address local food challenges or adapt to regional climates.

4. Consumer and Farmer Confidence

DUS-tested varieties offer both farmers and consumers the assurance that the plant varieties they are using or consuming will perform as advertised. This reliability is essential for maintaining trust in the agricultural supply chain and ensuring that crops meet the demands of the market.

- **Performance Reliability**: For farmers, the reliability of DUS-tested varieties means they can confidently invest in seeds, knowing that the plants will exhibit the uniformity, stability, and distinctiveness promised by the breeder. Whether it's yield performance, disease resistance, or harvest timing, these characteristics are critical to a farmer's success and profitability.
- **Predictable Characteristics**: Farmers benefit from the consistency of DUS-tested varieties, especially in large-scale operations where predictability in growth, maturity, and yield is essential for efficient crop management. DUS testing ensures that the variety will maintain uniform growth patterns, ripen at the same time, and be resistant to specific pests or diseases, reducing uncertainty in crop production.

» **Market Demand and Consumer Trust**: From a consumer perspective, DUS-tested varieties lead to improved quality and uniformity in food products, whether it's the flavor of a fruit, the size of a vegetable, or the nutritional content of grains. This consistency enhances consumer trust in the agricultural industry and drives market demand for high-quality produce. In addition, it helps meet the quality standards of the food industry, where uniformity is key for processing, packaging, and distribution.

5. Enhancing Food Security

By ensuring the development of high-performing, adaptable, and reliable varieties, **DUS testing indirectly enhances food security**. New varieties that pass the DUS test can contribute to higher yields, better resistance to biotic and abiotic stresses, and improved adaptability to diverse growing conditions.

» **Tackling Climate Change**: With the increasing unpredictability of climate change, DUS testing is crucial for the development of varieties that can withstand extreme weather conditions, such as drought, flooding, or temperature extremes. Breeders can focus on developing resilient varieties that meet the needs of farmers facing these challenges.

» **Contribution to Global Agriculture**: DUS-tested varieties are not just beneficial for individual farmers but also play a key role in improving the overall agricultural landscape. They contribute to the development of food systems that are more resilient, productive, and sustainable, ensuring that the growing global population can be fed efficiently.

UPOV Guidelines and DUS Testing

The **International Union for the Protection of New Varieties of Plants (UPOV)** plays a crucial role in standardizing the legal framework and procedures for **Distinctness, Uniformity, and Stability (DUS)** testing across its member countries. UPOV's guidelines ensure that plant breeders have a consistent, transparent, and internationally recognized system for the protection of their new varieties. This harmonized approach to plant variety protection (PVP) facilitates innovation in plant breeding and promotes global agricultural development.

UPOV Test Guidelines

UPOV has developed detailed **crop-specific test guidelines** that outline the traits to be evaluated, the conditions for testing, and the specific procedures that should be followed during DUS testing. These guidelines ensure consistency in the DUS testing process, regardless of the location or the type of crop being tested. The standardization provided by UPOV is essential for protecting plant varieties in multiple countries and for promoting uniformity in plant breeding and variety registration.

1. Traits to be Observed

Each UPOV test guideline specifies a range of **morphological, physiological, and agronomic traits** that should be observed to assess the distinctness, uniformity, and stability of a new variety. These traits vary depending on the species but typically include characteristics such as plant height, leaf shape, flower color, seed size, and fruit quality.

» **Distinctness:** Traits that make the new variety distinct from existing, commonly known varieties are highlighted. This ensures that breeders are introducing truly new and innovative varieties to the market.

» **Uniformity:** The variety must show uniformity in the expression of these traits across multiple plants. This is important to ensure that the variety performs consistently in the field.

» **Stability:** The stability of these traits is evaluated over multiple growing seasons, ensuring that the variety will maintain its characteristics across generations.

2. Testing Conditions

UPOV guidelines outline the **specific conditions under which DUS testing should be conducted** to ensure accurate results. This includes factors such as:

» **Location of trials:** Trials may be conducted in different geographic regions to account for environmental variation.

» **Number of growing seasons:** DUS tests are often conducted over two or more growing seasons to assess the stability of the variety's traits.

» **Planting density and field layout:** Proper spacing and management practices are described to ensure that the variety is grown under standardized conditions, which helps in the comparison of its performance with reference varieties.

3. Standardization

UPOV's guidelines standardize the DUS testing process globally, ensuring that **consistent methods and criteria** are applied in all member countries. This enables plant breeders to secure intellectual property rights in multiple countries by following the same DUS test procedures. Standardization also reduces the risk of discrepancies or variations in the testing results, promoting fairness and transparency in plant variety protection.

» **International Recognition:** Because UPOV guidelines are recognized by its member countries, a variety that passes DUS testing in one country may be eligible for protection in others, simplifying the process for breeders seeking international protection.

» **Harmonization Across Nations:** UPOV's guidelines ensure that testing procedures are harmonized across different countries, facilitating global trade in seeds and plant varieties and contributing to food security by allowing the dissemination of high-quality, improved crop varieties.

4. Implementation by National Authorities

National authorities in UPOV member countries implement the test guidelines provided by UPOV to carry out the DUS testing for plant varieties within their jurisdictions. These authorities are responsible for conducting field trials, collecting data, and ultimately granting plant breeders' rights based on the DUS testing results.

» **National Plant Variety Offices:** These offices are tasked with overseeing the entire DUS testing process, from accepting applications to conducting field trials and issuing plant breeders' rights once the variety passes the test.

- **Collaboration and Exchange**: In some cases, national authorities may collaborate with other countries or research institutions to share information or conduct joint trials, especially for crops grown across borders. This collaboration strengthens the consistency and reliability of the DUS testing system globally.

5. Crop-Specific Guidelines

UPOV's test guidelines are tailored to each crop or plant species, ensuring that the specific characteristics relevant to that species are tested. For example, the traits evaluated in wheat (e.g., grain quality, ear shape) will differ significantly from those in ornamental plants like roses (e.g., flower color, petal number).

- **Example: Wheat**: In the case of wheat, UPOV guidelines may require the evaluation of traits such as spike length, grain shape, growth habit, and disease resistance, ensuring that new wheat varieties are distinct and stable.
- **Example: Ornamentals**: For ornamental plants such as roses, traits like flower size, petal shape, and color are important for determining distinctness and uniformity. These traits ensure that the variety stands out in the commercial flower market.

Importance of UPOV Guidelines in Global Agriculture

UPOV's guidelines for DUS testing have a profound impact on plant breeding, agriculture, and food security. Some of the key contributions include:

- **Encouraging Breeding Innovation**: By providing a clear and standardized pathway for breeders to protect their new varieties, UPOV incentivizes the development of innovative crops that are better suited to modern agricultural challenges, such as climate change and food insecurity.
- **Facilitating Global Trade**: UPOV's standardized DUS testing guidelines enable breeders to protect and market their varieties across multiple countries, promoting international trade in seeds and plant genetic resources. This fosters the exchange of improved plant varieties, benefiting farmers, consumers, and the agricultural industry worldwide.
- **Promoting Biodiversity**: Through the development of new and distinct varieties, UPOV guidelines contribute to the diversification of the global agricultural gene pool. This diversity is essential for developing resilient cropping systems that can withstand biotic and abiotic stresses.
- **Supporting Sustainable Agriculture**: By encouraging the development of varieties that are disease-resistant, drought-tolerant, or more efficient in nutrient use, UPOV guidelines help promote sustainable agricultural practices. The testing and protection of these varieties ensure that farmers have access to high-quality, productive plants that contribute to environmental sustainability.

Challenges in DUS Testing

While **Distinctness, Uniformity, and Stability (DUS) testing** is a critical component in plant breeding and variety protection, several challenges complicate its execution. These challenges stem from the inherent complexities of plant genetics, environmental interactions, and the technical aspects of the testing process. Addressing these challenges is essential to ensure the efficiency and accuracy of DUS testing.

1. Complexity for Polyploid Species

Polyploid species, such as **wheat** (Triticum aestivum), **sugarcane** (Saccharum spp.), and certain **fruit crops**, present a significant challenge in DUS testing due to their **multiple sets of chromosomes**. Unlike diploid species (with two sets of chromosomes), polyploid plants often exhibit **greater genetic variability** and more complex inheritance patterns.

- **Uniformity and Stability in Polyploids**: Achieving the required levels of **uniformity and stability** in polyploid crops can be more difficult because these plants may display **more variation in traits** such as plant height, leaf morphology, or fruit size. This variability makes it harder to ensure that the crop meets the standard DUS criteria.
- **Gene Expression in Polyploids**: Polyploid species may exhibit different expressions of the same genes due to the interaction between multiple chromosome sets. This complexity can affect the expression of key traits like disease resistance, yield, or drought tolerance, leading to additional challenges in evaluating distinctness and stability.
- **More Rigorous Observation**: For crops with complex genetic structures, DUS testing often requires **more stringent observations** across multiple traits and environments to ensure that the variety is consistent. This can lead to **longer testing periods** and **increased resource requirements** for breeders working with polyploid species.

2. Environmental Variability

Plant performance is heavily influenced by environmental factors, such as climate, soil type, and water availability. This variability poses a significant challenge for DUS testing, especially for traits like **yield**, **disease resistance**, and **abiotic stress tolerance** (e.g., drought or heat resistance). The interaction between the **genotype** of a plant and its **environment** can make it difficult to assess these traits consistently.

- **Environmental Interactions**: Plants often express different traits in response to varying environmental conditions, making it challenging to distinguish between genetic traits and environmental effects. For instance, a variety may perform well in one location but poorly in another due to differences in temperature, humidity, or soil conditions.
- **Multi-Location Trials**: To address environmental variability, DUS testing is usually conducted across multiple locations and seasons. However, this approach increases the complexity and cost of testing. Even with multi-location trials, significant environmental differences between regions can still lead to **variability in results**.
- **Abiotic and Biotic Stress Factors**: Varieties often face different levels of **biotic stress** (pests and diseases) or **abiotic stress** (drought, salinity) depending on the testing location. This makes it challenging to evaluate a variety's resistance to these stresses, as its performance could fluctuate based on the local environmental pressures.

3. Cost and Time

DUS testing is typically a **long-term process** that requires rigorous multi-season and multi-location trials to accurately assess the distinctness, uniformity, and stability of a new variety. This extensive testing is necessary to ensure that the variety performs consistently across different environments and seasons, but it comes with **significant financial and time costs**.

- **Resource-Intensive Trials**: Conducting DUS trials involves setting up experimental plots, managing crops through multiple growing seasons, and gathering extensive data. The costs associated with land, labor, data analysis, and trial replication can be prohibitive, especially for **small-scale breeders** and researchers in **developing countries**.
- **Time Delay**: For certain crops, especially those that have long growing cycles (e.g., fruit trees), DUS testing can take several years to complete. This delay can slow down the process of variety development and commercialization, particularly for breeders who are under pressure to release new varieties quickly in response to changing market demands or environmental conditions.
- **Barrier for Small Breeders**: The high cost and time requirements of DUS testing can act as a **barrier for smaller breeders** or institutions with limited resources. While larger commercial breeders may have the infrastructure to handle the complexity and cost, smaller breeders may struggle to compete, potentially stifling innovation and the development of locally adapted varieties.

4. Subjectivity in Trait Evaluation

Despite the existence of **standardized guidelines** for DUS testing, the evaluation of certain traits can still involve a degree of **subjective judgment**, particularly for characteristics like color, shape, and texture. Traits that rely on visual assessment may vary depending on the **observer's interpretation**, leading to **inconsistencies** in the testing process.

- **Visual Traits**: Characteristics such as flower color, fruit shape, or leaf texture may be subject to human interpretation, which introduces an element of **subjectivity** into the DUS testing process. Even with established guidelines, two different observers may not assess the same variety in exactly the same way.
- **Reducing Subjectivity**: Advances in technology, such as **image analysis, digital phenotyping**, and **sensor-based measurements**, are helping to reduce subjectivity by providing **quantifiable data** on traits that were traditionally evaluated visually. However, these technologies are not yet universally accessible, especially in **resource-constrained regions**.
- **Training and Calibration**: To minimize subjectivity, DUS testers require **extensive training** and calibration to ensure consistency in trait evaluation. National authorities conducting DUS testing must ensure that their staff are well-versed in the guidelines and that their assessments are in line with **international standards**, such as those provided by UPOV.

While DUS testing is a critical tool for ensuring the distinctness, uniformity, and stability of new plant varieties, it is not without its challenges. The **complexity of testing polyploid species, environmental variability, cost, and time constraints**, and the **subjectivity in evaluating certain traits** all present hurdles that plant breeders and testing authorities must navigate.

As the demand for **new and improved plant varieties** continues to grow, especially in response to **climate change** and **global food security needs**, it is crucial to refine the DUS testing process. This could involve **leveraging new technologies**, improving **cost-efficiency**, and enhancing **international cooperation** to ensure that breeders worldwide can contribute to the development of resilient, high-yielding, and sustainable crops.

Chapter - 6

Molecular Breeding of Crops

A. Molecular Breeding: Molecular Markers

Molecular breeding represents a significant advancement in the field of plant breeding, integrating molecular biology tools to enhance the precision and efficiency of selecting desirable traits in crops. The core concept behind molecular breeding is the use of molecular markers, which serve as vital tools in detecting and tracking the presence of specific genes or genomic regions associated with important agronomic traits. Unlike traditional breeding methods, which rely heavily on phenotypic selection, molecular breeding allows breeders to focus on the genetic makeup of plants, speeding up the development of improved varieties.

Molecular markers have revolutionized the breeding process by providing a way to monitor genetic diversity and guide selection decisions based on genetic information rather than just observable traits. This allows breeders to identify favorable genetic variations that may not yet be expressed at the phenotypic level but hold great potential for future generations of plants. By utilizing molecular markers, breeders can target traits such as drought tolerance, disease resistance, and improved nutritional content, leading to crop varieties that are better suited to meet the growing demands of global agriculture.

Introduction to Molecular Markers

Molecular markers are specific sequences of DNA that are tightly linked to particular genes or regions within the genome, which can be used to identify genetic differences between individuals. These markers act as flags or indicators that help breeders determine which plants carry desirable genes, even when those traits are not yet visible. This technology has become an indispensable tool in modern plant breeding programs, particularly for traits that are difficult to select for through traditional methods, such as those influenced by multiple genes or those affected by environmental factors. The advent of molecular markers began with the discovery of Restriction Fragment Length Polymorphisms (RFLPs) in the 1980s. RFLPs marked the beginning of a new era in plant breeding, providing a method to detect genetic variation at the molecular level. This breakthrough was followed by the development of other marker systems, including Simple Sequence Repeats (SSRs), Amplified Fragment Length Polymorphisms (AFLPs), and Single Nucleotide Polymorphisms (SNPs), each offering unique advantages in different breeding contexts.

DOI: 10.1201/9781003757412-6

Molecular markers have the unique advantage of being able to track traits that are difficult or time-consuming to evaluate through phenotypic selection alone. For example, disease resistance traits, which may only become apparent after several growing seasons, can be identified at the seedling stage using molecular markers. This enables breeders to make selection decisions much earlier in the breeding cycle, saving time and resources while also increasing the accuracy of trait selection. The precision offered by molecular markers also allows for the pyramiding of multiple desirable traits into a single plant variety, further enhancing the potential of modern crop breeding. The use of molecular markers in plant breeding also helps in preserving genetic diversity. By identifying and cataloging different genetic variations, breeders can maintain a wider pool of genetic material, which is essential for the long-term sustainability of breeding programs. This is particularly important in the context of climate change and the need to develop crops that are resilient to new and evolving environmental stresses.

Types of Molecular Markers

There are several types of molecular markers used in plant breeding, each with its own unique characteristics and applications. The choice of marker depends on the breeding objectives, the genetic complexity of the trait being studied, and the resources available. The most commonly used molecular markers in plant breeding include Restriction Fragment Length Polymorphisms (RFLPs), Simple Sequence Repeats (SSRs), Single Nucleotide Polymorphisms (SNPs), and Amplified Fragment Length Polymorphisms (AFLPs).

1. **Restriction Fragment Length Polymorphisms (RFLPs):** RFLPs were the first type of molecular markers used in plant breeding. These markers are based on the variation in DNA fragment sizes generated by restriction enzymes, which cut the DNA at specific recognition sites. The resulting fragments are then separated by gel electrophoresis, revealing differences in fragment lengths between individuals. RFLPs are co-dominant markers, meaning they can distinguish between heterozygous and homozygous individuals, providing valuable information for breeding programs. However, the use of RFLPs is limited by the labor-intensive nature of the technique and the need for high-quality DNA samples. Despite these limitations, RFLPs are still used in certain contexts, particularly for constructing genetic linkage maps.
2. **Simple Sequence Repeats (SSRs):** SSRs, also known as microsatellites, are short, repetitive sequences of DNA found throughout the genome. These markers are highly polymorphic, meaning they show significant variation between individuals, making them particularly useful for detecting genetic diversity. SSRs are co-dominant markers and are widely used in plant breeding for a variety of purposes, including gene mapping, population genetics, and marker-assisted selection (MAS). One of the key advantages of SSRs is their high reproducibility and their ability to detect both qualitative and quantitative traits.
3. **Single Nucleotide Polymorphisms (SNPs):** SNPs are the most abundant type of molecular markers in the genome. These markers represent single base-pair changes in the DNA sequence and are highly desirable for molecular breeding due to their widespread distribution across the genome. SNPs are ideal for high-throughput genotyping, and their high density makes them particularly useful for fine-scale mapping and genome-wide association studies (GWAS). The ease of detecting SNPs using advanced sequencing technologies has made them one of the most popular markers in modern breeding programs. SNPs are co-dominant

markers, making them highly informative for breeding applications.

4. **Amplified Fragment Length Polymorphisms (AFLPs):** AFLPs are a type of molecular marker that combines the principles of RFLP and polymerase chain reaction (PCR) technologies. AFLPs generate a large number of markers per assay and are highly sensitive, making them useful for detecting genetic diversity and constructing linkage maps. One of the key advantages of AFLPs is their ability to generate a large number of polymorphisms in a relatively short amount of time, making them suitable for high-throughput applications. However, AFLPs are dominant markers, meaning they do not differentiate between homozygous and heterozygous individuals, which can limit their usefulness in certain contexts.

Applications of Molecular Markers in Plant Breeding

Molecular markers have a wide range of applications in plant breeding, allowing breeders to make more informed and efficient selection decisions. Some of the most significant applications of molecular markers include marker-assisted selection (MAS), genetic mapping, gene pyramiding, and the assessment of genetic diversity.

1. **Marker-Assisted Selection (MAS):** Marker-assisted selection (MAS) is one of the most powerful applications of molecular markers in plant breeding. MAS allows breeders to select individuals with desirable traits based on the presence of molecular markers linked to those traits, rather than relying on phenotypic selection alone. This is particularly useful for traits that are difficult to evaluate phenotypically, such as disease resistance, drought tolerance, or nutritional content. MAS accelerates the breeding process by enabling early selection of favorable alleles, reducing the need for extensive field testing. This technology has been widely adopted in breeding programs for crops such as rice, maize, wheat, and soybeans, where it has been used to improve traits such as yield, pest resistance, and stress tolerance.
2. **Genetic Mapping:** Genetic mapping is another critical application of molecular markers. Genetic maps provide a framework for understanding the relationship between different genes and traits, allowing breeders to identify the genomic regions associated with desirable traits. Molecular markers are used to construct linkage maps, which show the relative positions of genes or markers on a chromosome. This information is essential for identifying quantitative trait loci (QTLs), which are regions of the genome that control complex traits such as yield, disease resistance, or abiotic stress tolerance. QTL mapping, combined with MAS, has been instrumental in developing improved crop varieties.
3. **Gene Pyramiding:** Gene pyramiding is the process of combining multiple desirable genes into a single plant. Molecular markers play a crucial role in this process by enabling breeders to track the inheritance of multiple genes simultaneously. This is particularly important for traits such as disease resistance, where pyramiding several resistance genes can provide more durable protection against pathogens. Molecular markers allow breeders to ensure that all desired genes are present in the same individual, making the process of gene pyramiding more efficient and accurate.
4. **Genetic Diversity Analysis:** Assessing genetic diversity is essential for maintaining healthy and resilient breeding populations. Molecular markers provide a way to measure the genetic

variation within and between populations, which helps breeders identify valuable genetic resources for future breeding programs. Understanding the genetic diversity of a breeding population is also important for preventing genetic erosion, which can occur when breeding programs focus too narrowly on a small number of elite varieties. Molecular markers enable breeders to maintain a broader genetic base, ensuring the long-term sustainability of breeding efforts.

Impact of Molecular Markers on Crop Improvement

The use of molecular markers in plant breeding has had a profound impact on crop improvement, enabling breeders to develop new varieties with greater precision and speed. By allowing breeders to make selection decisions based on genetic information, molecular markers have transformed the way breeding programs are conducted, making them more efficient and cost-effective.

One of the most significant benefits of molecular markers is their ability to accelerate the breeding process. Traditional breeding methods can take many years to develop new varieties, particularly for complex traits that are influenced by multiple genes or environmental factors. Molecular markers allow breeders to select these traits much earlier in the breeding cycle, reducing the time and resources required to bring new varieties to market. In addition to accelerating the breeding process, molecular markers have also improved the accuracy of selection. By providing a direct link between genotype and phenotype, molecular markers allow breeders to select individuals with the most favorable genetic combinations, even if those traits are not yet visible. This has led to the development of crop varieties with improved yields, better nutritional content, and greater resilience to environmental stresses. The use of molecular markers has also facilitated the development of crops that are better suited to meet the challenges of climate change. By identifying genes associated with traits such as drought tolerance, heat resistance, and disease resistance, molecular markers have enabled breeders to develop crops that can thrive in a changing environment. This has important implications for global food security, as it allows breeders to develop crops that are more resilient to the stresses imposed by climate change.

B. Fundamental Concepts in the Development of Molecular Markers

The development of molecular markers is a key advancement in the field of plant breeding and genetics, providing tools that allow scientists and breeders to study and manipulate the genetic makeup of organisms with high precision. Molecular markers serve as identifiable DNA sequences that are associated with particular traits, allowing researchers to map genomes, identify gene locations, and make informed decisions in breeding programs. The foundation of molecular marker development involves a deep understanding of the genome structure, variability among populations, and the underlying genetic principles that govern inheritance.

1. Concept of DNA Polymorphism

The concept of DNA polymorphism serves as the cornerstone of molecular marker development. In the context of genetics and plant breeding, polymorphism refers to the natural variation that exists in DNA sequences between individuals or populations. These variations occur as differences in nucleotide sequences and can manifest in different forms, such as single nucleotide changes, insertions or deletions, or variations in repeated sequences. Polymorphisms are critical because they contribute to the genetic diversity found within species, providing the raw material for natural selection, evolution, and the development of traits that breeders seek to enhance or manage.

Role of DNA Polymorphisms in Genetic Diversity

DNA polymorphisms play a crucial role in generating the genetic diversity that underpins biological variation. Without polymorphism, populations would be genetically uniform, and the potential for adaptation to environmental pressures would be severely limited. In evolutionary biology, polymorphisms are essential for natural selection because they create variations in traits that may confer a selective advantage in a given environment. For plant breeders, these variations are equally important, as they allow breeders to select for specific traits, such as drought tolerance, disease resistance, or yield potential. The diversity created by DNA polymorphisms also plays a pivotal role in maintaining the long-term viability of breeding populations. By ensuring a wide pool of genetic variation, plant breeders can prevent issues like inbreeding depression, which can occur when breeding programs rely too heavily on a limited number of elite varieties. Inbreeding depression often leads to reduced fitness, lower yields, and increased susceptibility to diseases. Thus, the study of DNA polymorphism is essential for developing molecular markers that guide breeding programs and help manage genetic resources effectively.

Exploiting Polymorphisms for Molecular Marker Development

Molecular markers are designed to exploit these naturally occurring polymorphisms to distinguish between different individuals or genotypes. The markers serve as flags that indicate the presence of specific genetic variations in the genome, allowing breeders to track and select plants that carry desirable traits. By tagging polymorphisms, researchers can trace the inheritance of genes associated with these traits, even before they are expressed phenotypically. This ability to link DNA polymorphisms with phenotypic traits forms the foundation of marker-assisted selection, one of the most powerful tools in modern plant breeding.

Three major types of DNA polymorphisms are commonly exploited for the development of molecular markers: Single Nucleotide Polymorphisms (SNPs), Simple Sequence Repeats (SSRs), and Insertions/Deletions (Indels). Each type of polymorphism offers unique advantages depending on the breeding context and the specific goals of the genetic study.

i. Single Nucleotide Polymorphisms (SNPs)

Single Nucleotide Polymorphisms (SNPs) represent the most common form of genetic variation in both plant and animal genomes. A SNP occurs when a single nucleotide in the DNA sequence is altered between different individuals or populations. These changes can take place in any part of the genome, from coding (exonic) regions that directly influence protein structure and function to non-coding regions (intronic or intergenic) that may affect gene regulation. SNPs are a vital resource in molecular breeding, providing precise tools to identify and map the genetic basis of complex traits.

SNPs in Coding and Non-Coding Regions

SNPs are versatile and can occur across both coding and non-coding regions of the genome. In coding regions, the implications of SNPs can vary depending on whether they are synonymous or non-synonymous. **Synonymous SNPs** do not result in a change in the amino acid sequence of the encoded protein. Although they don't affect protein structure, they can still influence gene expression levels, mRNA stability, or other regulatory processes, impacting important phenotypic traits.

In contrast, **non-synonymous SNPs** cause an amino acid substitution in the protein, potentially altering its function. This can have direct implications for traits controlled by protein-coding genes, such as disease resistance, drought tolerance, or pest resistance in crops. For instance, a single non-synonymous SNP in a gene responsible for encoding a disease resistance protein could lead to an enhanced ability to recognize and defend against pathogens, providing a significant advantage in agricultural systems.

In non-coding regions, SNPs may not directly influence protein sequences but can have an important regulatory role. These SNPs might affect how genes are expressed, influencing the activity of promoters, enhancers, or silencers, and therefore indirectly impacting the development of traits. SNPs located in introns might influence splicing events, while those in intergenic regions could play a role in regulating gene networks critical for complex agronomic traits such as yield stability or environmental adaptability.

SNPs in Molecular Breeding

SNPs are immensely valuable in molecular breeding because of their abundance and uniform distribution across the genome. Their high frequency allows breeders and researchers to conduct fine-scale genetic mapping, providing deeper insights into the genetic control of complex traits. Moreover, the relatively simple nature of SNPs makes them highly amenable to high-throughput technologies, allowing for the genotyping of thousands, or even millions, of SNPs simultaneously. This capability facilitates large-scale genetic studies, which are essential for the rapid advancement of crop improvement programs.

One of the most powerful applications of SNPs in breeding is in **Genome-Wide Association Studies (GWAS)**. GWAS is a method that associates specific genetic variants (like SNPs) with phenotypic traits across a population. By scanning the genome for SNPs linked to traits such as yield, resistance to pests, or tolerance to environmental stresses, breeders can identify the regions of the genome that control these traits and focus their breeding efforts on individuals that carry the favorable alleles.

The high density of SNPs in the genome also makes them ideal markers for **genetic linkage mapping**. Through linkage mapping, breeders can identify regions of the genome that are closely linked to desirable traits and use this information to make targeted crosses, selecting plants with the best combination of genetic traits for further breeding. The use of SNPs in this context accelerates the breeding process, reducing the number of generations required to achieve the desired traits in a crop.

Co-Dominant Nature of SNPs

A major advantage of SNPs is their **co-dominant** nature, which means they can distinguish between homozygous and heterozygous individuals. In plant breeding, this information is crucial because it allows breeders to track both alleles of a gene. Knowing whether an individual is homozygous or heterozygous for a specific trait helps breeders make informed decisions about which individuals to select for crossing, as it reveals the true genetic makeup of the plant.

For example, if a SNP is linked to a gene that controls disease resistance, breeders can use SNP markers to determine whether an individual is homozygous (both alleles confer resistance) or heterozygous (one allele confers resistance, the other does not). In cases where a homozygous resistant genotype is preferred, breeders can prioritize individuals with two copies of the resistance allele for further breeding.

This level of genetic precision is invaluable in marker-assisted selection (MAS), where breeders can use SNP markers to identify individuals carrying desirable genetic combinations early in the breeding process, even before those traits are visibly expressed. This allows breeders to make faster and more accurate selection decisions, significantly reducing the time and resources required to develop improved crop varieties.

Application of SNPs in Major Crops

SNP markers have become indispensable tools in the breeding of major crops like maize, rice, and wheat. These crops form the foundation of global food security, and the ability to enhance their productivity, resilience, and nutritional value through molecular breeding is essential.

» **Maize:** In maize, SNP markers have been used to improve traits such as drought tolerance, pest resistance, and nitrogen use efficiency. Maize breeding programs have also utilized SNPs to enhance nutritional qualities, such as increasing provitamin A content to combat vitamin A deficiency in developing countries. The high-throughput genotyping capabilities of SNP arrays have enabled breeders to rapidly identify and select for favorable traits in large breeding populations.

» **Rice:** In rice, SNP markers have played a crucial role in improving both yield and resilience to environmental stresses such as salinity, submergence, and drought. SNPs linked to major genes controlling these traits have been incorporated into breeding programs through marker-assisted selection. For example, the identification of SNPs linked to the **Sub1** gene, which confers submergence tolerance, has enabled the development of rice varieties that can survive flooding for extended periods.

» **Wheat:** In wheat, SNPs are commonly used to improve traits related to disease resistance, such as resistance to rusts and other fungal pathogens. SNP markers have also been employed to enhance the nutritional quality of wheat, including the development of varieties with higher protein content or improved gluten quality for baking.

ii. Simple Sequence Repeats (SSRs)

Simple Sequence Repeats (SSRs), commonly referred to as microsatellites, are short stretches of DNA composed of repetitive nucleotide sequences, typically 2 to 6 base pairs long, that are found throughout the genome. These repetitive sequences occur in tandem, and the number of repeat units at a given SSR locus can vary significantly between individuals. This variation in repeat number makes SSRs highly polymorphic, which is a key feature that renders them useful as molecular markers in plant breeding and genetic research. SSRs have become a cornerstone in molecular breeding due to their versatility, high level of polymorphism, and ease of use. They are co-dominant markers, meaning they can distinguish between homozygous and heterozygous genotypes, which provides valuable information about the genetic composition of individuals. Their widespread use in plant breeding programs is driven by their capacity to track genes related to important agronomic traits and to assess genetic diversity, population structure, and parentage.

Polymorphism and Utility in Genetic Studies

The high degree of polymorphism exhibited by SSRs stems from the variability in the number of repeat units at specific loci within the genome. For example, one individual might have six repeats of a particular sequence (e.g., CA), while another individual might have eight. This variability can be easily detected through polymerase chain reaction (PCR) amplification, followed by gel electrophoresis or capillary electrophoresis, which separates the DNA fragments based on their size. The differences in fragment lengths between individuals indicate genetic variation, which can be used to study relationships between individuals, populations, or even species.

SSRs are widely used for assessing genetic diversity, as they provide a robust measure of variation both within and between populations. This is particularly important for plant breeding programs that seek to maintain genetic diversity while selecting for desirable traits. Genetic diversity is crucial for the long-term sustainability of breeding programs because it ensures that breeding populations are resilient to environmental changes, diseases, and pests. Without sufficient diversity, breeding populations may suffer from inbreeding depression, which can lead to reduced vigor, fertility, and adaptability. In studies of population structure, SSRs can reveal patterns of gene flow and genetic differentiation between populations. By analyzing the distribution of SSR alleles, researchers can determine how genetically distinct different populations are from one another, which can be useful for understanding the evolutionary history of a species or for conserving genetic resources. Additionally, SSRs are highly informative in parentage analysis, allowing breeders to track the inheritance of traits across generations and to confirm the pedigree of breeding lines.

SSRs in Marker-Assisted Selection (MAS)

One of the most powerful applications of SSRs is their use in **marker-assisted selection (MAS)**, where breeders can track the inheritance of genes associated with important agronomic traits. SSR markers can be linked to genes that control traits such as disease resistance, drought tolerance, pest resistance, and yield, allowing breeders to make informed selection decisions based on genetic information rather than relying solely on phenotypic evaluations. This is especially valuable for traits that are controlled by multiple genes (quantitative traits) or traits that are influenced by environmental factors, which can make phenotypic selection challenging. For example, SSR markers linked to resistance genes for pests and diseases have been used to develop crop varieties that are more resilient to biotic stresses. In crops like tomatoes and soybeans, SSR markers have been used to track resistance to pathogens such as **Phytophthora**, **Fusarium**, and **Verticillium**. By identifying plants that carry the desired resistance alleles, breeders can efficiently select individuals that are better equipped to withstand these pathogens, reducing the need for chemical interventions and increasing the sustainability of crop production. In addition to pest and disease resistance, SSRs have been used to track genes associated with abiotic stress tolerance, such as drought, salinity, and temperature extremes. These markers are particularly valuable in regions where crops are regularly exposed to environmental stresses, as they allow breeders to develop varieties that are more resilient to challenging growing conditions. By selecting for genotypes that carry favorable alleles for stress tolerance, breeders can ensure that crops maintain high productivity even in adverse environments.

SSRs are also useful for tracking traits related to yield and quality, such as fruit size, flavor, and nutritional content. In grapes, for example, SSR markers have been used to improve fruit quality traits, including sugar content, acidity, and aroma. Similarly, in soybeans, SSRs have been employed to enhance traits such as protein content, oil quality, and seed size. The ability to select for these traits at the molecular level accelerates the breeding process and ensures that new varieties meet the market demands for high-quality products.

Advantages of SSRs in Plant Breeding

One of the primary advantages of SSRs is their **co-dominant** nature, which allows for the differentiation between homozygous and heterozygous genotypes. This capability provides breeders with detailed information about the genetic makeup of individuals, which is critical for making precise selection decisions. For instance, if a breeder is selecting for a dominant trait, it is important to know whether an individual is homozygous for the desired allele (which would ensure the trait is passed on to all offspring) or heterozygous (which may result in segregation in the next generation). SSRs provide this level of detail, enabling breeders to make more accurate predictions about the inheritance of traits.

Another key advantage of SSRs is their **high reproducibility**. Once developed, SSR markers can be consistently used across different laboratories and breeding programs, ensuring that the results are comparable and reliable. This reproducibility is particularly important for large-scale breeding programs that involve collaboration between multiple institutions or countries. The consistency of SSR data allows breeders to track genetic progress over time and to make informed decisions about the direction of their breeding programs. The ease of detection and amplification of SSR markers using **polymerase chain reaction (PCR)** techniques further enhances their utility in plant breeding. PCR is a relatively simple and cost-effective method for amplifying DNA, and it can be performed in most molecular biology laboratories without the need for specialized equipment. The ability to amplify SSR loci using PCR means that breeders can quickly and efficiently screen large numbers of individuals for the presence of desirable alleles, accelerating the breeding process.

Moreover, the high level of polymorphism in SSRs makes them particularly valuable for **genetic diversity studies**. Because SSR loci can exhibit a large number of alleles, they provide a fine-scale measure of genetic variation within populations. This level of detail is crucial for identifying elite varieties that possess unique combinations of alleles for desirable traits. SSR markers can also be used to monitor the genetic diversity of breeding populations over time, ensuring that genetic erosion does not occur as a result of intense selection pressure.

Applications of SSRs in Major Crops

SSRs have been extensively used in a wide range of crops to improve traits related to yield, quality, and resilience. Some of the key crops where SSRs have had a significant impact include grapes, soybeans, and tomatoes.

» **Grapes:** In grape breeding, SSR markers have been used to improve fruit quality traits such as sugar content, acidity, and flavor. By tracking the inheritance of genes associated with these traits, breeders have been able to develop grape varieties that produce high-quality wine and table grapes. SSR markers have also been employed in studies of grapevine diversity and parentage analysis, helping to conserve important genetic resources.

» **Soybeans:** In soybean breeding, SSRs have been used to enhance disease resistance, yield stability, and seed quality. SSR markers linked to genes controlling resistance to fungal pathogens, such as **Phytophthora sojae**, have enabled the development of resistant soybean varieties, reducing the reliance on chemical treatments. Additionally, SSR markers have been used to select for traits related to seed protein and oil content, improving the nutritional value of soybeans for both human consumption and livestock feed.

» **Tomatoes:** In tomatoes, SSR markers have been used to track genes associated with disease resistance, fruit size, and flavor. By selecting for SSR alleles linked to resistance genes, breeders have developed tomato varieties that are more resilient to pests and pathogens, reducing crop losses and increasing productivity. SSRs have also been employed to improve fruit quality traits, such as sweetness and acidity, enhancing the appeal of tomatoes to consumers.

iii. Insertions and Deletions (Indels)

Insertions and Deletions (Indels)

Insertions and deletions (Indels) are a type of DNA polymorphism characterized by the addition (insertion) or loss (deletion) of small segments of DNA within a genome. These variations can be as small as a single nucleotide or span several base pairs. Indels contribute to genetic diversity and can have significant effects on the phenotype of an organism. They are considered an important source of genetic variation in both coding and non-coding regions of the genome. Depending on their location and size, Indels can impact gene function, protein structure, or gene regulation, making them valuable markers in molecular breeding programs.

Indels are used in plant breeding as molecular markers to track desirable traits such as yield, disease resistance, and abiotic stress tolerance. Their ease of detection using polymerase chain reaction (PCR) and gel electrophoresis techniques has made them a popular choice for genetic studies and marker-assisted selection (MAS). In some cases, Indels may cause a frameshift mutation in a coding sequence, potentially leading to loss-of-function mutations. This can be beneficial in breeding programs where breeders aim to select against undesirable alleles. Indel markers are frequently used in crops like rice, where they have contributed to improving key agronomic traits.

Indels in Coding and Non-Coding Regions

Indels can occur in both coding (exonic) and non-coding (intronic, regulatory, or intergenic) regions of the genome. The effects of Indels depend largely on where they occur. In **coding regions**, Indels can have a profound impact on protein function. For example, a single nucleotide insertion or deletion in a protein-coding gene may cause a **frameshift mutation**, altering the reading frame of the gene. This change often leads to the production of a truncated or non-functional protein, potentially resulting in a **loss-of-function** mutation. Such mutations can influence critical agronomic traits, such as disease resistance or tolerance to environmental stresses. For instance, loss-of-function mutations in resistance genes could make a plant more susceptible to pathogens, while mutations in regulatory genes may affect growth or development.

In **non-coding regions**, Indels may have more subtle but equally important effects. These polymorphisms can occur in **promoter** or **enhancer** regions, impacting the regulation of gene expression. For instance, an Indel in a promoter region might alter the binding of transcription factors, leading to changes in the timing, location, or level of gene expression. Such changes could affect traits that are regulated at the transcriptional level, such as flowering time, fruit ripening, or response to environmental stress. Additionally, Indels in **intronic regions** can affect the splicing of pre-mRNA, potentially creating alternative splice forms of a gene that may confer novel functions or regulatory control.

Indels in **regulatory elements** or **epigenetic regions** of the genome can also have an impact on gene expression and phenotypic traits. For example, an Indel in a non-coding RNA gene might alter the production or function of small RNA molecules, which can regulate the expression of other genes. Understanding the role of Indels in these non-coding regions is crucial for unraveling the complex genetic networks that control important traits in crops.

Detection and Utility of Indels in Molecular Breeding

Indels are valuable molecular markers due to their high **polymorphic nature** and ease of detection. They can be identified using simple and cost-effective techniques such as **PCR** and **gel electrophoresis**. This simplicity makes Indel markers accessible to many breeding programs, especially those operating on limited budgets. The size differences between alleles, caused by insertions or deletions, are easily visualized as bands of different lengths on an electrophoresis gel. Breeders can then determine the presence or absence of specific alleles associated with desired traits in their breeding populations. Indels can be **dominant** or **co-dominant** markers, depending on the nature of the polymorphism. Co-dominant markers, like Indels, allow breeders to distinguish between homozygous and heterozygous individuals. This is important for tracking the inheritance of specific traits and for ensuring that selected individuals possess the desired genetic makeup for breeding purposes. Indels are frequently used in **genetic linkage mapping** and **quantitative trait loci (QTL) analysis**, where they help identify regions of the genome that control complex traits such as yield, stress tolerance, or disease resistance.

In some cases, Indels may cause **frameshifts** in coding sequences. A frameshift changes the reading frame of the gene, potentially resulting in an altered protein product. Frame shifts caused by Indels can be beneficial or detrimental, depending on the specific gene and trait involved. For example, in a disease-resistance gene, a frameshift that causes a loss of function might eliminate the plant's ability to resist a pathogen. In such cases, breeders may use Indel markers to select against individuals carrying the undesirable mutation. Conversely, frameshifts in regulatory genes might create novel regulatory mechanisms that could be exploited for trait improvement.

Application of Indels in Crop Breeding

Indels have been widely used in molecular breeding programs for a range of crops, with significant applications in crops like **rice, maize**, and **wheat**. One of the key advantages of Indel markers is their ability to be used in **marker-assisted selection (MAS)** to track important agronomic traits. In **rice breeding**, Indel markers have been instrumental in improving traits such as grain size, disease resistance, and **tolerance to flooding**. For example, the introduction of **Indel markers** linked to the **Sub1** gene has allowed breeders to develop rice varieties that can survive prolonged submergence during flooding, which is a major challenge in rain-fed lowland rice systems. Similarly, Indel markers linked to genes controlling grain size and quality have been used to enhance yield potential and market value in rice.

Indels have also been employed to improve **disease resistance** in crops like maize and wheat. In maize, for instance, Indel markers associated with **resistance to corn leaf blight** have been used to develop varieties that are less susceptible to this devastating fungal disease. In wheat, Indels have been linked to genes conferring resistance to **rust diseases**, which are major threats to wheat production worldwide. By using Indel markers to track the inheritance of these resistance genes, breeders can more efficiently develop wheat varieties that are resilient to rust pathogens. Another important application of Indels is in improving **tolerance to abiotic stresses** such as drought, salinity, and temperature extremes. For instance, in rice, Indel markers linked to genes controlling **salt tolerance** have been employed to develop varieties that can thrive in saline environments. This is particularly important for areas affected by soil salinity, where traditional rice varieties struggle to survive. The use of Indel markers in this context has allowed breeders to introduce salt-tolerant genes into elite rice varieties without compromising yield or grain quality.

Indels in Gene Expression and Epigenetics

The ability of Indels to affect **gene expression** and **epigenetic modifications** further enhances their utility in molecular breeding. Indels in regulatory regions of the genome, such as promoters or enhancers, can influence the expression of genes involved in important traits. For example, an Indel in the promoter region of a **drought-responsive gene** could alter the plant's ability to withstand water stress by modulating the level of gene expression. This could lead to the development of varieties with enhanced drought tolerance, an increasingly important trait in the face of climate change. Indels can also influence **epigenetic modifications**, which involve changes to the genome that affect gene expression without altering the underlying DNA sequence. For instance, an Indel in a region associated with DNA methylation could change the pattern of methylation and, consequently, the expression of nearby genes. These epigenetic modifications can have a profound impact on complex traits, such as flowering time, stress tolerance, and yield stability. By studying Indels in regulatory and epigenetic regions, breeders can gain a better understanding of the genetic mechanisms that control these traits and develop strategies to manipulate them for crop improvement.

The Significance of DNA Polymorphisms in Marker Development

The discovery and exploitation of DNA polymorphisms form the foundation of molecular marker development. By identifying and cataloging these polymorphisms, researchers can develop markers that are closely linked to genes of interest. These markers provide breeders with a powerful tool for tracking the inheritance of traits across generations and for making more informed breeding decisions. The ability to associate specific polymorphisms with desirable traits, such as improved yield, pest resistance, or tolerance to abiotic stress, is a critical aspect of modern plant breeding. The continued advancement of sequencing technologies and bioinformatics tools has greatly enhanced our ability to identify and utilize DNA polymorphisms. Whole-genome sequencing and genome-wide association studies are uncovering new polymorphisms at an unprecedented rate, expanding the pool of molecular markers available for breeding programs. As a result, breeders can now select for complex traits more efficiently, paving the way for the development of crop varieties that are better suited to the challenges of modern agriculture, including climate change and food security.

2. Genome Mapping and Marker Development

Genome mapping is a crucial process in the development of molecular markers, as it helps in determining the precise location and relative distances of genes or molecular markers on chromosomes. The primary objective of genome mapping is to create a roadmap of the genome that shows the positions of genes or markers, which is essential for understanding the genetic control of important traits. This mapping process is the backbone of many modern breeding programs, especially those utilizing **marker-assisted selection (MAS)**, where molecular markers are used to select for desirable traits in crops. Genome mapping is broadly divided into two categories: **genetic mapping** and **physical mapping**. Both approaches are vital for developing an accurate understanding of genome structure and for linking genetic markers to traits of interest. By combining these two types of mapping, researchers and breeders can create high-resolution maps that help improve the efficiency and precision of plant breeding programs.

Genetic Mapping

Genetic mapping is a method that determines the relative positions of genes or markers on chromosomes based on the frequency of **recombination events** during meiosis. When genes or markers are located close to each other on a chromosome, they tend to be inherited together because the likelihood of a recombination event occurring between them is low. Conversely, genes or markers that are farther apart are more likely to be separated by recombination. This principle allows geneticists to estimate the distance between markers, which is measured in **centiMorgans (cM)**. One centiMorgan represents a 1% chance of recombination occurring between two genes or markers in a single generation.

Molecular markers such as **Restriction Fragment Length Polymorphisms (RFLPs)**, **Simple Sequence Repeats (SSRs)**, and **Single Nucleotide Polymorphisms (SNPs)** are commonly used to construct genetic maps. These markers provide a way to track specific regions of the genome during the breeding process, helping researchers identify the loci associated with important traits. For instance, a genetic map can help locate **quantitative trait loci (QTLs)**, which are genomic regions that control complex traits such as drought tolerance, disease resistance, or yield potential.

The process of constructing a genetic map typically involves crossbreeding individuals that differ in the trait of interest and then analyzing the recombination frequencies between molecular markers. By studying the offspring and determining which markers are inherited together, geneticists can infer the order and relative distances between markers on the chromosome. The resulting genetic map provides a valuable tool for breeders who want to select individuals with desirable combinations of alleles. In the context of plant breeding, genetic maps are especially useful for identifying and selecting traits that are controlled by multiple genes. For example, traits like yield or disease resistance often involve complex interactions between many genes, and genetic mapping helps breeders pinpoint the specific regions of the genome that contribute to these traits. Once the QTLs are identified, breeders can use molecular markers linked to these QTLs to track the inheritance of the traits in subsequent generations, making the selection process more efficient and precise.

Physical Mapping

While genetic maps provide an estimate of the relative distances between genes or markers, **physical mapping** gives the actual distance between them in terms of base pairs. Physical maps are essential for genome sequencing projects, where the goal is to determine the complete nucleotide sequence of an organism's DNA. Physical mapping relies on molecular markers to anchor sequenced fragments of the genome to their correct positions on the chromosome.

The process of physical mapping involves breaking the genome into smaller, overlapping fragments, which are then sequenced and assembled to create a continuous representation of the genome. Molecular markers such as RFLPs, SSRs, and SNPs are used as reference points to ensure that the fragments are assembled in the correct order. The result is a high-resolution physical map that provides a detailed picture of the genome, including the locations of genes, regulatory elements, and other functional regions.

Advances in **sequencing technologies**, such as **next-generation sequencing (NGS)** and **whole-genome sequencing (WGS)**, have made it possible to develop high-resolution physical maps for many crop species. These technologies allow researchers to sequence entire genomes quickly and at relatively low cost, providing a complete and accurate representation of the genome. Physical maps are particularly important for studying complex traits that are influenced by multiple genes located on different chromosomes, as they provide the resolution needed to identify the specific genes responsible for these traits. Physical mapping also plays a crucial role in identifying **structural variations** in the genome, such as insertions, deletions, duplications, and translocations, which can have a significant impact on gene function and phenotype. By providing a detailed view of the genome, physical maps help researchers understand the genetic basis of these variations and their effects on traits like disease resistance, stress tolerance, and yield.

Integration of Genetic and Physical Mapping

The integration of genetic and physical mapping is essential for fully understanding the relationship between molecular markers and traits of interest. While genetic maps provide a relative framework based on recombination frequencies, physical maps offer a more precise and detailed view of the genome. By combining these two approaches, researchers can link specific molecular markers to traits of interest with greater accuracy. This integration is particularly important in **marker-assisted selection (MAS)**, where molecular markers are used to select individuals with desirable traits based on their genetic makeup. MAS relies on the identification of markers that are tightly linked to genes controlling important traits, and the combination of genetic and physical maps provides the resolution needed to make these associations.

For example, in a breeding program aimed at improving drought tolerance, researchers might use genetic mapping to identify QTLs associated with the trait. Once the QTLs are identified, physical mapping can be used to pinpoint the specific genes within the QTL region that are responsible for drought tolerance. Molecular markers linked to these genes can then be used in MAS to select individuals that carry the favorable alleles for drought tolerance, even before the trait is expressed phenotypically. The combination of genetic and physical mapping also enables **fine mapping** of QTLs, which involves narrowing down the QTL region to identify the specific genes that control the trait. Fine mapping is essential for understanding the genetic architecture of complex traits and for developing molecular markers that are highly predictive of the trait. By combining the strengths of both genetic and physical mapping, breeders can make more informed and accurate decisions, leading to the development of crop varieties with improved traits.

3. Future Perspectives in Molecular Marker Development

As sequencing technologies continue to advance, the development and application of molecular markers in plant breeding will become even more efficient and widespread. Whole-genome sequencing (WGS) and next-generation sequencing (NGS) technologies are driving the discovery of new markers, particularly SNPs, at an unprecedented scale. These technologies enable the rapid identification of markers across entire genomes, facilitating genome-wide association studies (GWAS) that can uncover novel trait-marker relationships.

In the future, molecular markers will likely play an even more significant role in precision breeding. The integration of genomic selection (GS), where markers are used to predict the breeding value of individuals based on their entire genome, represents the next frontier in molecular breeding. This approach allows breeders to make more informed decisions at an earlier stage, reducing the breeding cycle time and enhancing the accuracy of selection. Additionally, advances in bioinformatics and computational biology are expected to improve the analysis of large datasets generated by molecular marker studies. Machine learning algorithms and artificial intelligence (AI) are already being applied to predict complex trait inheritance, analyze genetic interactions, and optimize breeding strategies based on molecular marker data.

C. Marker Types and Development

The choice of molecular marker type depends on several factors, including the level of polymorphism, ease of detection, and the nature of the breeding program or genetic study. The following are fundamental types of markers that have been developed over the years:

1. Restriction Fragment Length Polymorphisms (RFLPs)

Restriction Fragment Length Polymorphisms (RFLPs) were among the first molecular markers developed and are based on variations in DNA fragment lengths caused by the presence or absence of specific **restriction enzyme cutting sites**. These enzymes recognize and cut DNA at particular sequences, and the resulting DNA fragments can differ in length between individuals due to genetic variation. This variation forms the basis of RFLP analysis, allowing researchers to detect polymorphisms between individuals or populations.

RFLPs became a cornerstone of molecular genetics and breeding in the 1980s and were used extensively for **genetic mapping** in many plant and animal species. Despite being labor-intensive and requiring significant quantities of high-quality DNA, RFLPs provided a solid foundation for the development of other molecular markers and played a critical role in early genome mapping projects.

Mechanism of RFLP

Restriction Fragment Length Polymorphisms (RFLPs) function by utilizing the action of **restriction enzymes**, which are specialized proteins that cleave DNA at specific recognition sequences. These enzymes are naturally found in bacteria, where they act as part of a defense system against foreign DNA, such as that of invading viruses. In RFLP analysis, restriction enzymes scan the DNA molecule and cut it at designated sequences, which are often **short palindromic motifs** (such as **GAATTC**). These cutting sites are typically consistent across genomes unless genetic variations like **single nucleotide**

polymorphisms (SNPs) or **insertions and deletions (Indels)** alter the recognition site, preventing the enzyme from making a cut.

If a mutation occurs in the recognition sequence, the enzyme is unable to cleave the DNA at that particular location. This results in different fragment lengths between individuals, as their DNA will be cut at different points. These variations in fragment lengths are the basis for RFLP analysis and can be used to distinguish between individuals based on their genetic makeup.

Step-by-Step Process of RFLP:

1. **DNA Extraction and Restriction Enzyme Digestion**: The first step in the RFLP process is the extraction of genomic DNA from the individuals being studied. Once the DNA is isolated, it is subjected to treatment with a specific restriction enzyme. This enzyme recognizes a particular sequence in the DNA and cuts it at that site. If a mutation, such as an SNP or an Indel, has altered the sequence of the recognition site in certain individuals, the enzyme will not cut the DNA at that location, leading to differences in the fragment lengths generated.

   The DNA from different individuals is thus digested into multiple fragments of varying lengths, depending on the presence or absence of restriction enzyme recognition sites. These fragment length differences represent the **polymorphisms** that are central to the RFLP analysis.

2. **Gel Electrophoresis**: After digestion by the restriction enzyme, the DNA fragments are separated by size using **gel electrophoresis**. In this process, the digested DNA is loaded into wells in an agarose gel, which is placed in an electric field. DNA molecules are negatively charged due to their phosphate backbone, so they migrate toward the positive electrode when an electric current is applied. Smaller DNA fragments migrate faster through the gel matrix than larger fragments, allowing the DNA to be separated by size.

   The result of gel electrophoresis is a series of bands that correspond to DNA fragments of different lengths. Each individual will have a unique banding pattern, depending on whether their DNA was cleaved at all of the expected restriction sites. Individuals with mutations at the restriction sites will show altered banding patterns, reflecting the differences in fragment lengths.

3. **Southern Blotting**: Once the DNA fragments have been separated by gel electrophoresis, they are transferred from the gel onto a more durable support, such as a **nylon** or **nitrocellulose membrane**, through a process known as **Southern blotting**. This step is necessary because the gel is fragile and does not allow for long-term analysis. The transfer is done by capillary action, which moves the DNA from the gel to the membrane while preserving the relative positions of the fragments.

   During Southern blotting, the DNA is denatured (converted from a double-stranded form to a single-stranded form), making it accessible for hybridization with a labeled DNA probe in the next step.

4. **Hybridization with a Labeled Probe**: To identify specific fragments of interest, a **labeled DNA probe** is used. A probe is a short, single-stranded segment of DNA that is complementary to the sequence of interest in the DNA fragments. The probe is labeled with either a radioactive isotope or a fluorescent dye, allowing it to be detected later.

   When the probe is applied to the membrane, it hybridizes (binds) to the complementary

DNA fragments that contain the target sequence. This step ensures that only the fragments containing the sequence of interest are detected, reducing background noise from other, unrelated DNA fragments.

5. **Detection of Polymorphisms**: After hybridization, the membrane is washed to remove any unbound probe. The labeled probe that remains bound to the DNA fragments of interest is then visualized using X-ray film (if radioactive) or through a fluorescence detector (if fluorescent). The resulting image shows distinct bands that represent the DNA fragments containing the probe's target sequence.

    The differences in fragment lengths between individuals, which result from mutations that prevent the restriction enzyme from cutting, are seen as **different banding patterns**. Individuals with different genetic variants at the restriction site will show different band positions, allowing researchers to detect the **polymorphisms** that distinguish them. For example, one individual may have a longer fragment (because the enzyme failed to cut at a mutation-altered site), while another individual may have shorter fragments due to normal cleavage at the restriction site.

Applications in Genetic Studies

The differences in DNA fragment lengths visualized through RFLP can be used for a variety of genetic analyses, including:

1. **Genetic Mapping**: RFLPs were one of the first types of molecular markers used for **genetic linkage mapping**, which involves determining the relative positions of genes or markers on a chromosome based on their recombination frequency. In the early days of genome mapping, RFLPs were instrumental in creating the first **linkage maps** for many crops, including rice, maize, and wheat. These maps provided critical information for understanding the genetic control of important traits, such as disease resistance, yield, and abiotic stress tolerance.

    The advantage of RFLPs lies in their **co-dominant nature**, meaning they can distinguish between homozygous and heterozygous individuals. This characteristic made RFLPs particularly valuable for studying inheritance patterns and tracking the transmission of alleles through generations. For example, in a breeding program focused on disease resistance, RFLP markers linked to resistance genes could be used to identify plants that carry one or two copies of the resistance allele, allowing breeders to make informed decisions about which individuals to select for further breeding. RFLPs also proved useful for **quantitative trait loci (QTL) mapping**, which aims to identify regions of the genome that control complex traits influenced by multiple genes. By analyzing the co-segregation of RFLP markers and phenotypic traits in mapping populations, researchers were able to locate QTLs associated with important agronomic traits. This information has been crucial for improving crop varieties, as it allowed breeders to focus on specific genomic regions associated with enhanced yield, stress tolerance, or disease resistance.

2. **Disease Diagnostics**: RFLP analysis can be used to detect specific genetic mutations linked to inherited diseases. By analyzing RFLP patterns in affected and unaffected individuals, researchers can identify genetic markers associated with diseases, aiding in diagnosis and carrier detection.
3. **Forensic and Paternity Testing**: RFLP can distinguish between individuals based on their unique DNA profiles, making it useful for forensic investigations and paternity testing. The

differences in RFLP banding patterns between individuals serve as a genetic fingerprint, allowing for identification in legal cases.

Limitations of RFLPs

Despite their initial success, RFLPs have several limitations that have led to their gradual replacement by newer, more efficient molecular markers. One of the main drawbacks of RFLP analysis is that it requires **large quantities of high-quality DNA**. Since the technique relies on the cutting of genomic DNA with restriction enzymes, any degradation of the DNA can significantly impact the results, making it essential to obtain pure, intact samples. This requirement can be challenging in some species or environments, where obtaining sufficient amounts of high-quality DNA is difficult.

Another limitation of RFLPs is that the technique is **labor-intensive** and time-consuming. The process of restriction enzyme digestion, gel electrophoresis, Southern blotting, and hybridization with probes involves multiple steps, each of which requires careful handling and optimization. This complexity limits the scalability of RFLP analysis, particularly for large-scale breeding programs that require the genotyping of thousands of individuals. In contrast, newer marker systems like **SNPs** and **SSR markers** are much faster and can be automated, making them more suitable for high-throughput applications. Moreover, while RFLPs offer high reliability, they tend to have lower **polymorphism information content (PIC)** compared to other marker systems, such as SSRs or SNPs. This means that fewer polymorphisms can be detected with RFLPs, reducing their ability to provide detailed information about genetic diversity within and between populations.

Impact and Legacy of RFLPs

Although RFLPs are no longer the primary molecular marker system used in plant breeding and genetic research, their historical significance cannot be understated. RFLPs provided the foundation for the development of modern genetic mapping techniques and molecular marker-assisted breeding. The principles underlying RFLP analysis, such as the detection of DNA polymorphisms and thc construction of linkage maps, are still relevant today and continue to inform the use of newer marker systems.

RFLPs also played a critical role in some of the earliest **genome mapping** projects, contributing to the development of genetic maps for many important crop species. These maps were instrumental in identifying key genes and QTLs associated with agronomic traits, helping to improve crop productivity and resilience. Additionally, RFLP markers were used in **comparative mapping studies**, where the genomes of different species were compared to identify conserved regions of the genome. This work laid the groundwork for **genome-wide association studies (GWAS)** and other modern genetic approaches that are widely used today.

2. Random Amplified Polymorphic DNA (RAPD)

Random Amplified Polymorphic DNA (RAPD) is a molecular marker technique used to detect genetic polymorphism by amplifying random segments of an organism's genome through the **polymerase chain reaction (PCR)**. This technique employs short, arbitrary primers to bind randomly across the genome, which results in the amplification of multiple, random DNA fragments. The amplified fragments differ in length between individuals due to the presence of genetic variations, which serve as the basis for identifying polymorphisms.

RAPD is known for being relatively simple and cost-effective compared to other molecular marker techniques, as it does not require prior knowledge of the DNA sequence to be analyzed. However, it has some significant drawbacks, the most prominent of which is its **low reproducibility**. Because the primers bind to random locations in the genome, the results can vary from one experiment to the next, making it difficult to obtain consistent data. Despite this limitation, RAPD remains a valuable tool in certain contexts, particularly when sequence information is unavailable or when cost constraints make other techniques less viable.

Mechanism of RAPD

The **Random Amplified Polymorphic DNA (RAPD)** technique is based on the random amplification of DNA segments using PCR (Polymerase Chain Reaction). This method detects genetic polymorphisms by amplifying random sections of the genome without prior knowledge of the DNA sequence. The random nature of primer binding and subsequent amplification makes RAPD a useful tool for studying genetic diversity and assessing relationships between individuals or populations. Below is a detailed, step-by-step explanation of how RAPD works:

1. DNA Isolation and Primer Selection

The first critical step in the RAPD process is **DNA isolation** from the organism under study. The quality and purity of DNA play an essential role in the success of the RAPD technique. The genomic DNA is extracted using standard isolation protocols, which involve breaking down the cell walls or membranes (in the case of plants and animals) to release the genetic material. The DNA must be sufficiently pure, and free from contaminants like proteins, RNA, or chemical impurities, which can interfere with the PCR process.

Once high-quality DNA is isolated, **random primers** are introduced. These primers are short, arbitrary sequences, typically 10 base pairs in length, and they are designed to bind to multiple, random sites across the genome. Unlike techniques such as SSR (Simple Sequence Repeats) or SNP (Single Nucleotide Polymorphisms), RAPD primers are not based on known DNA sequences, meaning they do not target specific loci. This randomness allows RAPD to detect polymorphisms even in species where genomic sequence information is not available. Each primer anneals to different locations across the genome, depending on where its sequence matches a region of the genome. These primer sequences are short enough that they will find multiple complementary sites in the genome, thus generating multiple amplification products. Because the primers are arbitrary, the DNA segments that get amplified differ from one individual to another. The variation in primer binding between individuals is one of the foundations of the polymorphisms detected by RAPD. These amplified fragments form the basis for analyzing genetic differences between individuals or populations.

2. PCR Amplification

The core of RAPD is the **PCR amplification** process, which enables the replication of DNA segments starting from where the primers have bound. PCR consists of three major steps, and these steps are repeated in cycles to ensure multiple copies of the DNA are generated. Here is a detailed breakdown of how PCR operates in RAPD:

» **Denaturation**: The PCR process begins by heating the DNA sample to a high temperature, typically around 94-96°C. This heat causes the two strands of the double-stranded DNA to separate or "denature," creating single strands that can serve as templates for the next stage of amplification.

» **Annealing**: The temperature is then lowered to about 36-40°C, allowing the random primers to anneal (bind) to complementary sequences in the DNA template. The primers are arbitrary, meaning they bind wherever their sequence happens to match a segment of the DNA. Because RAPD primers are short, they anneal at multiple random sites throughout the genome. The annealing temperature in RAPD is generally lower than in other PCR techniques, to allow primers to bind even if the match is not perfect, further contributing to the random nature of the binding.

» **Extension**: After the primers have annealed, the temperature is raised to about 72°C, which is the optimal temperature for the DNA polymerase enzyme. This enzyme synthesizes new DNA strands by adding nucleotides to the primer, starting from the point where the primer has annealed. This extension process continues until the enzyme reaches the end of the DNA strand or another binding site. The result is the amplification of various fragments of DNA, with each fragment having a different length depending on where the primers have bound.

This cycle of **denaturation, annealing, and extension** is repeated 30-40 times, with each cycle doubling the amount of DNA produced. As a result, millions of copies of specific DNA fragments are generated. Since the binding of primers is random and depends on the genome of the individual, the amplified fragments will differ between individuals, creating polymorphisms. These differences are critical for detecting genetic diversity. The key advantage of RAPD PCR is that it can amplify multiple regions of the genome simultaneously, even without prior knowledge of the DNA sequence. This makes RAPD particularly useful in species with little or no genomic data, as it does not require the design of specific primers based on known sequences.

3. Gel Electrophoresis

After the DNA has been amplified, the next step is to **separate the amplified DNA fragments** by size using **gel electrophoresis**. This technique is essential for visualizing the different DNA fragments produced during the RAPD process.

» **Gel preparation**: A gel is made from agarose, a substance derived from seaweed. The gel acts like a molecular sieve, allowing smaller DNA fragments to move through it more easily than larger fragments. The gel is placed in a buffer solution that maintains the pH and conductivity needed for electrophoresis.

» **DNA loading and electrophoresis**: The amplified DNA samples are loaded into wells in the agarose gel. An electric field is applied to the gel, with the DNA moving from the negative electrode to the positive electrode, since DNA molecules are negatively charged due to their phosphate backbone. Smaller DNA fragments move faster through the gel matrix, while larger fragments move more slowly, resulting in a separation based on size.

» **Visualization of bands**: Once electrophoresis is complete, the gel is stained with a DNA-binding dye such as **ethidium bromide** or **SYBR Green**. These dyes fluoresce under ultraviolet (UV) light, allowing the DNA fragments to be seen as bands in the gel. The pattern of bands produced by each sample represents the various DNA fragments that were amplified during PCR.

Each individual or population produces a unique banding pattern based on the size of the amplified DNA fragments, and these patterns can be compared to identify polymorphisms. The bands correspond to the random DNA segments amplified by the arbitrary primers, and differences between individuals in the number or position of the bands indicate genetic variation.

4. Polymorphism Detection

The differences in the banding patterns between individuals or populations are analyzed to detect **genetic polymorphisms**. Polymorphisms arise because genetic variations—such as **mutations, insertions, deletions**, or **rearrangements**—can alter the binding sites for the primers. These variations lead to the amplification of different fragments of DNA, resulting in distinct banding patterns for each individual.

» **Genetic diversity**: RAPD is widely used to assess genetic diversity within and between populations. By comparing the presence or absence of specific bands across individuals, researchers can estimate the level of genetic variation in a population. High genetic diversity is important for the survival of populations, especially in conservation biology, where it is critical to maintain variability to ensure long-term species adaptability.

» **Marker discovery**: RAPD is also used in **marker discovery** for identifying molecular markers associated with specific traits. By identifying polymorphisms linked to desirable traits, such as disease resistance or stress tolerance, RAPD markers can be used in plant and animal breeding programs to track and select for these traits.

Applications of RAPD

Despite its low reproducibility, RAPD has been widely used in various research fields, including **genetic diversity studies, evolutionary biology, population structure analysis**, and **taxonomy**. RAPD is particularly valuable in contexts where genomic information is limited or where low-cost methods are required.

1. **Genetic Diversity and Population Studies**: RAPD has been commonly used to measure genetic diversity within and between populations. By comparing the banding patterns of individuals, researchers can assess the level of genetic variation in a population. This information is crucial for understanding the genetic health of populations, especially in conservation efforts where maintaining genetic diversity is important for species survival. In crop breeding, genetic diversity studies help breeders identify genetically distinct parents to combine desirable traits and maintain variability.
2. **Phylogenetic and Evolutionary Studies**: RAPD has been utilized in phylogenetic studies to infer the evolutionary relationships among different species or populations. By analyzing the similarities and differences in banding patterns, researchers can construct phylogenetic trees that illustrate the genetic relatedness of the samples. This is particularly useful in species where detailed genomic information is unavailable.
3. **Taxonomy and Species Identification**: RAPD can be used to identify genetic markers

that differentiate species or even subspecies. In taxonomy, RAPD helps distinguish closely related organisms by providing genetic fingerprints based on their unique banding patterns. This application is especially useful in plant and microbial taxonomy, where morphological differences are often subtle.

4. **Marker-Assisted Selection in Breeding**: Although RAPD is less commonly used for **marker-assisted selection (MAS)** due to its low reproducibility, it has been applied in some breeding programs to identify molecular markers linked to desirable traits, such as disease resistance, yield, or stress tolerance. RAPD markers can help breeders select individuals with favorable traits for further breeding.

Advantages and Limitations of RAPD

Advantages:

» **No prior sequence knowledge required**: RAPD does not require knowledge of the target organism's DNA sequence, making it suitable for use in species where genomic information is scarce or unavailable.

» **Simplicity and low cost**: RAPD is a straightforward and cost-effective technique compared to other molecular marker systems, making it accessible for laboratories with limited resources.

» **Useful in a wide range of species**: RAPD can be applied to almost any species, including plants, animals, and microorganisms, without the need for species-specific primers.

Limitations:

» **Low reproducibility**: One of the major drawbacks of RAPD is the low reproducibility of the results. The random binding of primers can lead to variation in the amplified fragments between different experiments, making it difficult to replicate results.

» **Dominant markers**: RAPD markers are dominant, meaning they cannot distinguish between heterozygous and homozygous individuals. This limits the amount of genetic information that can be derived from RAPD compared to co-dominant markers, such as SSRs or SNPs.

» **Randomness of amplification**: Because the primers bind randomly, the amplified regions of the genome may not be directly linked to the traits of interest, reducing the precision of the technique for marker-assisted selection.

Comparison with Other Molecular Markers

Compared to other molecular marker systems like **SNPs, SSRs**, or **AFLPs**, RAPD is less precise and less reproducible, but it remains advantageous in cases where resources are limited or where sequence information is unavailable. **SSR markers** offer greater reproducibility and co-dominance, allowing for more detailed genetic analysis. **SNP markers**, on the other hand, provide high precision and are suited for high-throughput genotyping, but they require significant genomic information for primer design.

While RAPD is not as widely used in modern breeding programs as more advanced molecular markers, it still has value in specific research contexts, particularly for initial genetic diversity screening or when studying species with limited genomic data.

3. Simple Sequence Repeats (SSRs)

Simple Sequence Repeats (SSRs), also referred to as **microsatellites**, are short, tandemly repeated DNA sequences typically ranging from 2 to 6 base pairs in length. These repeats occur throughout the genome and are highly polymorphic, meaning the number of repeat units can vary greatly between individuals within a species. Due to their high level of polymorphism, co-dominant nature, and ease of detection, SSRs have become one of the most widely used molecular markers in genetic research, plant and animal breeding, population genetics, and conservation studies.

The development and application of SSR markers involve identifying and analyzing these repeat regions within the genome, making them highly valuable tools for studying genetic diversity, population structure, and the inheritance of traits. Below is a more detailed breakdown of SSRs, including their development, mechanisms, and applications.

Characteristics of SSRs

1. **Polymorphism**: SSR markers are highly polymorphic because the number of repeat units in a given microsatellite region can vary between individuals. For example, in one individual, a microsatellite region might contain six repeats of the sequence **CA**, while another individual might have ten repeats. This variability in the number of repeats makes SSRs excellent markers for distinguishing between closely related individuals or populations. The high level of polymorphism provides a rich source of genetic variation, which is essential for breeding programs and genetic diversity studies.
2. **Widespread Distribution**: SSRs are widely distributed across the genome, occurring in both coding and non-coding regions. This widespread occurrence allows researchers to use SSR markers for a broad range of applications, from gene mapping to evolutionary studies. Their presence in coding regions also makes SSRs useful for studying traits directly related to gene function, while their presence in non-coding regions helps in understanding regulatory mechanisms and genetic diversity.
3. **Co-dominance**: One of the most important characteristics of SSRs is their **co-dominant** nature. This means that SSR markers can distinguish between **homozygous** and **heterozygous** individuals. In a homozygous individual, both alleles at a locus have the same number of repeats, while in a heterozygous individual, the two alleles have different numbers of repeats. This ability to differentiate between genotypes provides more detailed genetic information than dominant markers like RAPD, which only indicate the presence or absence of a specific allele.

The co-dominant nature of SSRs is particularly useful in **marker-assisted selection (MAS)** and **breeding programs**, where it is essential to track the inheritance of specific traits and alleles over generations.

Development and Mechanism of SSRs

The development of **Simple Sequence Repeat (SSR)** markers, also known as microsatellites, involves identifying highly variable, short tandem repeat regions within the genome. These SSR markers are widely used in genetic studies for various purposes, including population genetics, breeding programs, and genetic diversity assessments. SSRs are co-dominant markers, meaning they can distinguish between homozygous and heterozygous individuals. Below is a detailed explanation of the steps involved in the development and mechanism of SSRs.

1. Microsatellite Discovery

The first step in the development of SSR markers is identifying the microsatellite regions within the genome, which are composed of short tandem repeats, typically 2 to 6 base pairs in length. Microsatellites are distributed across the genome, making them ideal for use as molecular markers. The identification of microsatellite regions can be performed using the following methods:

- **In Silico Analysis:** For species with reference genomes, **in silico analysis** is the most efficient method for identifying microsatellites. This involves using bioinformatics tools to search for regions containing tandem repeats within the genome. Specialized software scans the genomic sequence for repetitive motifs and flags potential SSR loci. The flanking regions around these repeats are then identified, allowing for the design of primers that can be used to amplify the SSR loci.
- **Screening Genomic Libraries:** For species that do not have a reference genome, microsatellite discovery can be more challenging. In such cases, **genomic libraries** are constructed by randomly fragmenting the genome and cloning the fragments into vectors. The fragments are then screened for sequences that contain repeated motifs using hybridization techniques or sequencing. Once the microsatellite-containing fragments are identified, the sequences flanking the microsatellite repeats are determined, and primers can be designed for subsequent amplification.

In both methods, the goal is to identify repeat motifs that vary in length between individuals, as this variation forms the basis for detecting polymorphisms using SSR markers.

2. Primer Design

Once the microsatellite regions have been identified, the next step is to design **specific primers** to amplify the SSR loci. These primers bind to the flanking regions of the microsatellite, ensuring that the repeated region is efficiently amplified through **polymerase chain reaction (PCR)**.

- **Flanking Region Selection:** The primers must bind to **conserved flanking regions** surrounding the microsatellite repeats. These regions are typically stable and less prone to mutations compared to the microsatellite region itself, which ensures that the primers will reliably bind to the same location across different individuals.
- **Specificity and Accuracy:** Primers are designed to be highly specific to the microsatellite locus of interest. This specificity ensures accurate amplification of the region, preventing non-specific binding and amplification of unintended DNA regions. The length of the amplified DNA fragment is determined by the number of repeat units in the microsatellite region.

The design of high-quality primers is essential for the success of SSR marker development, as it ensures consistent and reproducible amplification of the target region in different individuals.

3. PCR Amplification and Gel Electrophoresis

The core of SSR marker analysis involves the **PCR amplification** of the microsatellite region using the primers designed in the previous step. PCR is a technique used to replicate specific DNA segments, making it possible to analyze even small quantities of DNA.

» **PCR Amplification**: During PCR, the DNA sample is subjected to cycles of heating and cooling, which cause the DNA to denature (separate into single strands), anneal (primers bind to the flanking regions), and extend (DNA polymerase synthesizes a new strand starting from the primer). This process results in the amplification of the microsatellite region. The size of the amplified DNA fragment depends on the number of repeats within the SSR locus, and this size will vary between individuals based on their genetic variation.

» **Gel Electrophoresis**: Once the microsatellite region is amplified, the next step is to separate the DNA fragments by size using **gel electrophoresis**. In this technique, the amplified DNA is loaded into wells of an agarose or polyacrylamide gel and subjected to an electric field. Since DNA molecules are negatively charged, they migrate through the gel towards the positive electrode. Smaller DNA fragments migrate faster through the gel than larger ones, resulting in a separation based on size.

» **Visualization of Bands**: After electrophoresis, the DNA fragments are visualized by staining the gel with a DNA-binding dye such as **ethidium bromide** or **SYBR Green**. The stained DNA fragments appear as distinct bands, with each band corresponding to an amplified DNA fragment of a specific size. These bands represent the number of repeat units in the SSR region for each individual.

4. Polymorphism Detection

The primary goal of SSR marker analysis is to detect **polymorphisms**—variations in the length of the microsatellite region between individuals. These polymorphisms are identified by comparing the size of the amplified DNA fragments.

» **Homozygosity vs. Heterozygosity**: SSR markers can distinguish between **homozygous** and **heterozygous** individuals. Homozygous individuals will have two identical copies of the microsatellite region, resulting in a single band of a specific size in gel electrophoresis. Heterozygous individuals, on the other hand, will have two different alleles with different numbers of repeats, resulting in two distinct bands of different sizes.

» **Polymorphism Analysis**: The differences in fragment sizes among individuals reflect the **polymorphisms** present in the microsatellite regions. These polymorphisms are analyzed to study **genetic diversity**, **population structure**, and **trait inheritance**. The high level of variability in SSR loci makes them particularly useful for distinguishing between individuals, even within closely related populations.

» **Applications**: Since SSRs can be highly variable, they are ideal for a wide range of applications in genetics and breeding, such as:

- **Genetic diversity studies**: SSRs are widely used to measure genetic diversity within populations, which is essential for maintaining the genetic health of species, especially in breeding programs and conservation efforts.

- **Marker-assisted selection (MAS):** In breeding programs, SSR markers are used to track the inheritance of desirable traits, such as disease resistance or drought tolerance, by selecting individuals with specific alleles associated with these traits.
- **Population structure and phylogenetic studies:** SSRs are used to assess the genetic structure of populations and to understand the evolutionary relationships between different species or populations.

Applications of SSRs

SSRs have a wide range of applications in genetics and breeding programs, owing to their high polymorphism, co-dominance, and reproducibility. Below are some of the key applications of SSRs:

1. **Genetic Diversity Studies:** SSR markers are commonly used to assess genetic diversity within and between populations. By analyzing the polymorphisms present in microsatellite loci, researchers can measure the genetic variation present in a population. High genetic diversity is essential for the long-term survival of populations, as it allows for greater adaptability to changing environmental conditions. SSRs are frequently used in the conservation of endangered species, where maintaining genetic diversity is critical for species recovery.

    For example, in plant breeding programs, SSR markers are used to measure the genetic diversity of breeding lines, ensuring that breeders maintain a wide gene pool while selecting for desirable traits. SSRs have been widely used in crop species such as maize, rice, wheat, and barley to monitor genetic diversity and track the inheritance of valuable agronomic traits.

2. **Marker-Assisted Selection (MAS):** SSRs are extremely valuable in **marker-assisted selection (MAS)**, where they are used to track the inheritance of specific traits in breeding programs. Because SSRs can differentiate between homozygous and heterozygous individuals, they provide precise information about the alleles present at a given locus. This allows breeders to select individuals that carry desirable alleles for traits such as disease resistance, drought tolerance, or increased yield.

    SSR markers linked to important agronomic traits have been used to accelerate breeding programs in crops such as rice and wheat. For instance, SSR markers linked to genes controlling disease resistance in wheat have allowed breeders to develop new varieties with enhanced resistance to pathogens such as rusts and mildews.

3. **Population Structure and Phylogenetic Studies:** SSR markers are also used to study **population structure** and **phylogenetic relationships**. By comparing the variation in SSR loci across individuals or populations, researchers can infer the genetic relationships and evolutionary history of different groups. This is particularly useful in understanding the geographic distribution of genetic variation and in identifying distinct populations for conservation or breeding purposes.

    In evolutionary biology, SSRs are used to construct **phylogenetic trees** that illustrate the genetic relationships between species or populations. The high mutation rate of SSR loci provides sufficient variation to resolve the relationships between closely related species or populations, making them a valuable tool for studying speciation and population divergence.

4. **Parentage Analysis and Forensic Genetics**: Due to their high level of polymorphism, SSR markers are frequently used in **parentage analysis** and **forensic genetics**. In parentage testing, SSRs can confirm or exclude potential parents based on the comparison of allele sizes between offspring and potential parents. Similarly, SSRs are used in forensic science to establish genetic fingerprints, which can be used for individual identification in criminal investigations or wildlife poaching cases.

Advantages and Limitations of SSRs

Simple Sequence Repeats (SSRs), also known as microsatellites, are widely used molecular markers that have both significant advantages and some limitations. SSRs are highly polymorphic, co-dominant markers, making them valuable in genetic diversity studies, population structure analysis, and marker-assisted selection (MAS). However, there are also some practical challenges associated with their use, particularly in the time and resources required for their development.

Advantages of SSRs

1. **High Polymorphism**: SSRs exhibit high levels of polymorphism due to the variability in the number of repeat units at each microsatellite locus. This variability allows researchers to distinguish between individuals, populations, or species with a high degree of precision. The polymorphic nature of SSRs makes them excellent for studying **genetic diversity**, **population structure**, and **evolutionary relationships**. Because the repeat number can vary significantly among individuals, SSRs provide a rich source of genetic variation, which is critical in **breeding programs** and **conservation genetics**.
   » **Example**: In plant breeding, SSR markers can differentiate between closely related crop varieties, allowing breeders to select for desirable traits while maintaining genetic diversity.
2. **Co-dominant Nature**: SSRs are **co-dominant markers**, meaning they can distinguish between homozygous and heterozygous individuals at a particular locus. In homozygous individuals, both alleles at a locus have the same number of repeats, resulting in a single band during gel electrophoresis. In heterozygous individuals, the two alleles have different numbers of repeats, resulting in two distinct bands. This ability to differentiate between genotypes provides detailed genetic information that is particularly valuable in **breeding** and **genetic mapping**.
   » **Benefit for Breeding**: The co-dominant nature of SSRs enables breeders to track the inheritance of specific alleles and allows for more accurate selection of individuals carrying desirable traits, such as disease resistance or drought tolerance.
3. **Reproducibility**: SSR markers are known for their high **reproducibility**. Once developed, SSR markers provide consistent and reliable results across different experiments, laboratories, and populations. This reproducibility makes SSRs particularly useful for **large-scale genetic studies** and **comparative studies** across different research groups or environments.
   » **Example**: In conservation genetics, SSR markers are often used to compare genetic diversity across different populations or species, and their reproducibility ensures that results are comparable regardless of the lab conducting the analysis.

Limitations of SSRs

1. **Development Time**: One of the main limitations of SSR markers is the time-consuming nature

of their development. The process of identifying microsatellite regions within the genome and designing specific primers to amplify these regions can take considerable time, particularly in species that lack well-characterized genomic resources.

» **Microsatellite Discovery**: For species without reference genomes, the identification of microsatellites often requires the construction of **genomic libraries** and **screening for repeat motifs**, which adds time and complexity to the process. Even with bioinformatics tools, designing primers that specifically target the flanking regions of microsatellites involves trial and error to ensure specificity and reliability.

» **Challenge for Non-Model Species**: For organisms that have not been extensively studied or sequenced (i.e., non-model species), this process is even more challenging, as it may require sequencing and assembling large portions of the genome before SSR development can begin.

2. **Moderate Cost**: Although SSR analysis is less expensive than **SNP genotyping**, it is still more costly than simpler marker systems such as **RAPD**. The costs associated with SSR analysis stem from the need for **species-specific primer design**, the requirement for precise PCR conditions, and the use of **gel electrophoresis** or **capillary electrophoresis** for fragment size analysis. These factors make SSR analysis more labor-intensive and technically demanding compared to dominant markers like RAPD.

» **Cost of Primer Design**: Each SSR locus requires the design of specific primers to amplify the microsatellite region, which increases both the time and expense involved in marker development. In some species, hundreds of SSR loci may need to be developed for comprehensive genetic studies or breeding programs.

» **PCR and Electrophoresis Costs**: SSR analysis often requires precise PCR conditions to ensure specific amplification of the microsatellite region, and the resulting DNA fragments must be accurately sized using gel electrophoresis or capillary electrophoresis. This step requires equipment and reagents that add to the overall cost of SSR genotyping.

Summary

Advantages:

» **High Polymorphism**: SSRs are highly polymorphic, making them ideal for distinguishing between individuals and populations.

» **Co-dominant Nature**: SSRs can differentiate between homozygous and heterozygous individuals, providing detailed genetic information for breeding and diversity studies.

» **Reproducibility**: SSR markers are highly reproducible, making them reliable for large-scale genetic studies.

Limitations:

» **Development Time**: Developing SSR markers is time-consuming, as it involves identifying microsatellite regions and designing specific primers.

» **Moderate Cost**: SSR analysis is more expensive than RAPD due to the need for species-specific primer design, precise PCR conditions, and gel electrophoresis.

4. Single Nucleotide Polorphisms (SNPs)

Single-nucleotide polymorphisms (SNPs) are the most common and abundant type of genetic variation found within a species' genome. They occur at a single base-pair position where the nucleotide (A, T, C, or G) varies between individuals of the same species. SNPs have revolutionized molecular marker development due to their high abundance, stability, and distribution across the genome. They are a fundamental tool in genetic research, especially in **high-resolution genetic mapping** and **genome-wide association studies (GWAS)**, and have become a cornerstone in modern plant and animal breeding programs.

SNPs are highly valued because of their **co-dominant nature**, which means they can differentiate between **homozygous** and **heterozygous** genotypes, making them highly informative markers. Additionally, SNP genotyping technologies have advanced to the point where high-throughput platforms can analyze **thousands of SNPs** simultaneously, enabling large-scale genetic studies with unprecedented precision.

Characteristics of SNPs

1. **Abundance and Distribution**: SNPs are the most abundant form of genetic variation, occurring approximately once every 100-300 base pairs in the genome, depending on the species. This high frequency of SNPs allows for dense and comprehensive coverage of the genome, making them ideal for **genetic mapping** and **association studies**. The widespread distribution of SNPs across both coding (exonic) and non-coding regions of the genome provides researchers with an extensive array of markers for investigating different traits and genetic architectures.
2. **Co-Dominance**: SNPs are co-dominant markers, which means they can distinguish between homozygous (both alleles are the same) and heterozygous (the alleles differ) individuals at a particular locus. This is a significant advantage over dominant markers like RAPD, which cannot distinguish between heterozygous and homozygous states. The co-dominance of SNPs allows for a more detailed analysis of genetic variation, aiding in the precise tracking of alleles in breeding populations and enabling more accurate selection.
3. **High Stability**: SNPs are relatively stable compared to other types of markers, such as microsatellites (SSRs), because mutations at a single nucleotide position are less likely to occur than changes in the number of repeat units at an SSR locus. This stability makes SNPs particularly useful for long-term studies on genetic inheritance, population structure, and evolutionary history.

Mechanism of SNP Genotyping

Single Nucleotide Polymorphisms (SNPs) are single base-pair variations in the genome and are the most abundant form of genetic variation. Detecting and analyzing SNPs requires specialized **genotyping platforms** that can accurately identify these variations across the genome. Modern SNP genotyping technologies use different methods, such as **microarrays, next-generation sequencing (NGS)**, and **Kompetitive Allele-Specific PCR (KASP)**. These platforms are designed to handle large-scale SNP detection, making SNP genotyping an indispensable tool in genetic studies, breeding programs, and **genome-wide association studies (GWAS)**. Below is a detailed breakdown of the primary SNP genotyping methods:

1. Microarray-Based Genotyping

Microarray technology is one of the most commonly used methods for SNP genotyping. It relies on the hybridization of **DNA samples** to a **SNP microarray chip**, which contains thousands to millions of pre-designed probes that are complementary to known SNPs across the genome. Each probe is allele-specific and designed to bind only to a specific variant of the SNP (either one allele or the other). Once hybridized, the binding of the sample DNA to these probes generates a **fluorescent signal**, which is detected and analyzed to determine the SNP genotype.

» **How it works**:

1. **DNA Extraction**: Genomic DNA is extracted from the samples being analyzed.
2. **Fragmentation and Labeling**: The extracted DNA is fragmented and labeled with a fluorescent dye.
3. **Hybridization**: The labeled DNA is then hybridized to the microarray chip, which contains thousands of allele-specific probes. Each probe on the chip is designed to target a specific SNP, and complementary binding occurs if the sample DNA contains the matching allele.
4. **Fluorescent Detection**: After hybridization, the chip is washed to remove unbound DNA, and the remaining hybridized DNA emits a fluorescent signal. The intensity of the fluorescence is proportional to the amount of DNA that has hybridized to each probe, and the signal pattern is read by a scanner.
5. **Genotype Calling**: The software analyzes the fluorescent signals to determine the genotype at each SNP locus. By comparing the intensity of the signals for different probes, the software can determine whether the individual is homozygous for one allele, homozygous for the other allele, or heterozygous.

» **Applications**:

- Microarrays are ideal for **large-scale SNP genotyping**, where the goal is to detect **hundreds of thousands to millions of SNPs** in a single experiment. This method is especially useful in **GWAS**, where researchers seek to identify associations between specific SNPs and traits such as disease susceptibility or yield improvement in crops.
- Microarrays are highly effective for **pre-defined SNP panels**, where the researcher is analyzing known SNPs that have been previously identified in the species' genome.

2. Next-Generation Sequencing (NGS)

Next-generation sequencing (NGS) platforms are highly flexible and provide base-by-base sequencing of the genome, allowing for the detection of both known and novel SNPs. NGS offers unparalleled precision and can be used for either **whole-genome sequencing (WGS)** or **targeted sequencing**, depending on the research objectives and budget.

» **How it works:**

1. **Library Preparation**: Genomic DNA is fragmented into small pieces, and adapters are added to both ends of the fragments to create a library.
2. **Sequencing**: The DNA fragments are sequenced using NGS platforms such as **Illumina** or **PacBio**, which provide high-throughput, massively parallel sequencing. Millions of short reads are generated, covering the genome comprehensively.
3. **Alignment**: The sequencing reads are aligned to a reference genome using bioinformatics tools. The alignment helps to identify positions in the genome where the sample DNA differs from the reference genome, which are flagged as potential SNPs.
4. **Variant Calling**: Specialized software tools identify the single nucleotide variants (SNPs) by comparing the sample sequence to the reference genome. These tools also assess the quality of the SNP calls and remove any low-confidence variants.

» **Advantages**:

- NGS allows for **de novo SNP discovery**, which is particularly useful for species with poorly characterized genomes or when researchers aim to discover novel SNPs not previously reported.
- NGS provides **high-resolution data** across the entire genome, allowing for the analysis of both coding and non-coding regions. This is crucial in studies focusing on **complex traits** influenced by regulatory or structural variations in the genome.

» **Applications**:

- NGS is ideal for **population genomics**, **comparative genomics**, and **evolutionary studies**.
- It is also used in **genomic selection** and **GWAS**, where researchers are interested in understanding the genetic basis of important traits in crops or livestock.

3. Competitive Allele-Specific PCR (KASP)

KASP (Kompetitive Allele-Specific PCR) is a **low-cost, high-throughput** method for SNP genotyping. It is particularly useful for **targeted genotyping**, where a limited number of known SNPs need to be analyzed across a large population. KASP technology uses **allele-specific PCR primers** that are designed to compete for binding to the SNP alleles.

» **How it works**:

1. **Primer Design**: For each SNP, two allele-specific primers and one common reverse primer are designed. The allele-specific primers have a tail sequence that corresponds to a specific fluorescent reporter dye (typically two different colors).
2. **PCR Amplification**: During PCR, the allele-specific primers bind competitively to the SNP locus, with each primer binding to one of the two possible alleles. If the sample DNA contains a specific allele, the corresponding primer will bind and initiate PCR amplification.

3. **Fluorescent Detection**: As PCR proceeds, the primers generate a signal based on which allele has been amplified. The fluorescent signal is read, and the genotype is determined based on which dye was incorporated into the PCR product.

» **Advantages**:

- KASP is a **cost-effective** method that provides high-throughput SNP genotyping for **known SNPs**. It is particularly useful in **plant breeding programs** where specific SNPs are linked to desirable traits such as disease resistance or drought tolerance.
- KASP is highly scalable and can be used to genotype **thousands of samples** in a short period, making it ideal for large population studies.

» **Applications**:

- KASP is widely used in **marker-assisted selection (MAS)** in crop and livestock breeding, where breeders need to track specific SNPs linked to important traits.
- It is also used in **targeted genotyping** for validating SNP-trait associations discovered through GWAS.

4. High-Throughput Genotyping

One of the key strengths of SNP markers is their compatibility with **high-throughput genotyping platforms**. These platforms can analyze **hundreds of thousands to millions of SNPs simultaneously**, making them highly efficient for large-scale genetic studies and breeding programs. Some of the most widely used high-throughput platforms include **TaqMan assays, Illumina Infinium arrays**, and **Affymetrix GeneChips**.

» **How it works**:

- These platforms combine **high-density microarrays** with advanced bioinformatics tools to genotype large numbers of SNPs across many individuals. The SNP data generated by these platforms can be used for **genome-wide association studies (GWAS), genomic selection**, and **genetic diversity analysis**.

» **Applications**:

- High-throughput SNP genotyping is essential in **GWAS** for identifying genetic markers associated with diseases or important agronomic traits.
- It is also crucial for **genomic selection**, where breeders use genomic data to predict the breeding value of individuals based on their SNP profiles.

Applications of SNPs

SNPs have become indispensable in various genetic applications, especially in plant and animal breeding programs, genetic diversity studies, and genome-wide association studies. Below are the key applications of SNPs:

1. **Genome-Wide Association Studies (GWAS)**: One of the most significant applications of SNP markers is in **genome-wide association studies (GWAS)**, where thousands to millions of SNPs are analyzed across the genome to identify genetic variations associated with particular traits. GWAS

are particularly powerful for identifying the genetic basis of complex traits, such as yield, disease resistance, or drought tolerance in crops, or milk production and disease resistance in livestock.

In a GWAS, SNP markers are used to compare the genomes of individuals with different phenotypes to identify regions of the genome that are statistically associated with the trait of interest. This allows researchers to pinpoint specific loci or genes that control important traits, enabling more targeted breeding and selection strategies.

2. **High-Resolution Genetic Mapping**: SNPs provide unparalleled precision in **genetic mapping** due to their abundance and wide distribution throughout the genome. SNPs are often used to construct **high-density genetic maps**, which are critical for identifying **quantitative trait loci (QTLs)** associated with important traits. In plant breeding, these maps allow breeders to trace the inheritance of beneficial alleles and make informed decisions about which individuals to cross in order to combine desirable traits.

   SNP-based genetic maps are also useful for studying complex traits controlled by multiple genes, as the high resolution provided by SNP markers enables more precise localization of QTLs compared to other marker systems like SSRs or RAPD.

3. **Marker-Assisted Selection (MAS)**: In **marker-assisted selection (MAS)**, SNPs are used to identify and select for specific alleles associated with desirable traits in plant and animal breeding programs. SNP markers are especially valuable for MAS because they are co-dominant, highly reproducible, and easily integrated into high-throughput genotyping platforms. This allows breeders to screen large populations for specific genetic variants linked to traits such as disease resistance, drought tolerance, or improved yield.

   MAS using SNPs accelerates the breeding process by enabling early selection of individuals carrying favorable alleles, even before the traits are phenotypically expressed. This reduces the time and cost of breeding programs while increasing the accuracy of trait selection.

4. **Genetic Diversity and Population Structure Analysis**: SNPs are commonly used to assess **genetic diversity** within and between populations. Their abundance and distribution across the genome make them ideal for measuring levels of genetic variation and studying **population structure**. In plant conservation, for example, SNP markers are used to evaluate genetic diversity in wild populations to guide conservation efforts and maintain genetic variability.

   SNP markers are also useful for studying the **evolutionary relationships** between species or populations, providing insights into genetic drift, migration patterns, and adaptation to environmental changes.

5. **Genomic Selection (GS)**: SNP markers are central to the practice of **genomic selection (GS)**, a breeding method that uses genome-wide SNP data to predict the genetic potential of individuals. In GS, SNP markers spread across the genome are used to estimate the **genomic estimated breeding value (GEBV)** of individuals, allowing breeders to select the best candidates for breeding without waiting for phenotypic data. This method is revolutionizing breeding programs by increasing the accuracy of selection and speeding up the development of improved varieties.

Advantages and Limitations of SNPs

Single Nucleotide Polymorphisms (SNPs) are among the most widely used molecular markers in genetic research, owing to their abundance and versatility. However, as with any marker system, SNPs have both advantages and limitations, which influence their application in different fields such as genetic mapping, breeding, and genome-wide association studies (GWAS).

Advantages of SNPs

1. **High Abundance**:
   - SNPs are the most abundant type of genetic variation, occurring approximately once every 100 to 300 base pairs in the genome. This high frequency provides **dense genomic coverage**, making them ideal for **genetic mapping**, **GWAS**, and **linkage studies**. Because of their widespread distribution across both coding and non-coding regions, SNPs offer researchers the ability to study a wide range of genetic variation linked to important traits or diseases.
   - **Example**: In plants like maize and wheat, SNPs are used extensively for trait mapping due to their abundance and ability to cover the entire genome comprehensively.
2. **Co-Dominant:**
   - SNPs are **co-dominant markers**, meaning they can distinguish between **homozygous** and **heterozygous** individuals. This ability to differentiate between different genotypes at a single locus provides detailed genetic information that is critical for applications like **marker-assisted selection (MAS)** and **genomic selection (GS)** in breeding programs.
   - **Benefit for Breeding**: The co-dominant nature of SNPs enables breeders to track the inheritance of specific alleles, improving the precision of selecting individuals carrying desirable traits, such as disease resistance or drought tolerance.
3. **High-Throughput Capabilities**:
   - SNP genotyping platforms, such as **microarrays** and **next-generation sequencing (NGS)**, are designed for **high-throughput analysis**. These platforms can simultaneously analyze **thousands to millions of SNPs** across large populations, making SNPs the marker of choice for **large-scale studies** such as GWAS, **genomic selection**, and **genetic diversity analysis**.
   - **Scalability**: The scalability of SNP genotyping allows researchers to screen many individuals in a short period, which is crucial in breeding programs where time and resource efficiency are essential.
4. **Highly Reproducible:**
   - SNP markers are highly reproducible across different experiments, labs, and populations. This **reproducibility** ensures that results are consistent and reliable, making SNPs the preferred marker for **long-term studies**, **comparative genomics**, and **population genetics**.
   - **Consistency**: SNP assays are well-standardized, reducing the likelihood of variability

in results due to environmental factors or technical inconsistencies. This makes SNP markers ideal for genetic studies that require consistent data across multiple laboratories.

Limitations of SNPs

1. **Requires Prior Genomic Information:**
    - One of the main limitations of SNP analysis is that it typically requires **prior genomic information**, such as a **reference genome** or knowledge of the **SNP locations**. This can be a significant barrier for **non-model species** or organisms that have not been well studied, where the genome may not have been fully sequenced or characterized.
    - **Challenge for Non-Model Species**: In species with limited genomic resources, SNP discovery and genotyping can be time-consuming and costly, as the initial steps involve sequencing and identifying SNPs before they can be used in studies or breeding programs.
2. **Cost:**
    - While the cost of SNP genotyping has decreased significantly over the years due to advances in technology, it is still higher compared to simpler marker systems like **RAPD** or **AFLP**. The cost can be a barrier, especially for **small-scale studies** or research groups with limited funding.
    - **Initial Investment**: Although high-throughput platforms like NGS and microarrays offer efficient genotyping for large populations, the **initial investment in equipment** and **bioinformatics resources** can be prohibitive for smaller projects. Additionally, while SNP analysis is cost-effective on a large scale, for smaller studies or when only a few markers are needed, SNP genotyping can be more expensive compared to other marker systems.

Summary

Advantages:

» **High Abundance**: SNPs are abundant in the genome, providing comprehensive coverage for mapping and GWAS.

» **Co-Dominant**: SNPs distinguish between homozygous and heterozygous individuals, offering detailed genetic data for breeding and diversity studies.

» **High-Throughput**: Platforms can analyze thousands to millions of SNPs simultaneously, making SNPs ideal for large-scale studies and breeding programs.

» **Highly Reproducible**: SNP markers produce consistent and reliable results across different experiments and labs.

Limitations:

» **Requires Prior Genomic Information**: SNP analysis typically requires a reference genome or knowledge of SNP locations, which can be a barrier for non-model species.

» **Cost**: While SNP genotyping costs have decreased, it remains higher than simpler marker systems like RAPD or AFLP, especially for small-scale studies.

5. Amplified Fragment Length Polymorphisms (AFLP)

Amplified Fragment Length Polymorphisms (AFLP) is a molecular marker technique used to detect genetic polymorphisms by combining **restriction enzyme digestion** and **polymerase chain reaction (PCR)** amplification. AFLP is a compassionate method that can detect multiple DNA fragments in a single assay, making it valuable for various applications, such as **genetic diversity studies, linkage mapping, phylogenetic analysis**, and **population structure studies**. AFLP is particularly useful in species with limited prior genomic information, as it does not require sequence knowledge to identify genetic variation.

Mechanism of AFLP

Amplified Fragment Length Polymorphism (AFLP) is a widely used molecular marker technique that combines **restriction enzyme digestion** and **polymerase chain reaction (PCR)** amplification to generate a large number of polymorphic markers without requiring prior sequence knowledge of the genome. AFLP is commonly applied in various fields, such as **genetic diversity studies, phylogenetic analysis, population structure analysis**, and **linkage mapping**. The technique is particularly valuable for species with limited genomic information, as it can detect a large number of polymorphisms in a single experiment. Below is a detailed breakdown of each step in the AFLP process.

1. Restriction Enzyme Digestion

The AFLP process begins with the **digestion of genomic DNA** using **two types of restriction enzymes**. The purpose of this step is to generate DNA fragments of various sizes that can be used to assess genetic variation between individuals or populations.

- » **Restriction Enzymes**: Two different restriction enzymes are used:
  - **Rare Cutter (e.g., EcoRI)**: A restriction enzyme that recognizes a specific 6-base pair sequence, cutting the DNA at fewer sites, thus generating larger fragments.
  - **Frequent Cutter (e.g., MseI)**: A second enzyme that cuts the DNA at a more common 4-base pair sequence, generating a larger number of smaller fragments.
- » **Why Two Enzymes?**: The use of two enzymes with different cutting frequencies creates a broad range of fragment sizes, from long fragments cut infrequently by the rare cutter to shorter fragments produced by the frequent cutter. This combination ensures that both large and small DNA fragments are generated, creating a more diverse pool of DNA fragments for subsequent amplification.
- » **Outcome**: The result of this digestion is a collection of DNA fragments that vary in size depending on the location of the restriction sites. These fragments represent a large pool of potential markers, from which a specific subset will be selectively amplified and analyzed in the following steps.

2. Adaptor Ligation

Following the digestion step, **adaptors** are ligated to the ends of the resulting DNA fragments. Adaptors are short, synthetic DNA sequences that are complementary to the overhangs created by the restriction enzymes.

» **Why Use Adaptors?**:

- **Adaptor Sequences**: The adaptors serve as attachment points for the **PCR primers** used in the amplification step. Since the primers are designed to bind specifically to these adaptors rather than the genomic DNA itself, this ligation ensures that only the DNA fragments with adaptors at both ends will be amplified.
- **Increased Specificity**: The ligation of adaptors also allows for greater specificity in the PCR amplification, as the PCR primers will only bind to DNA fragments with the correct adaptor sequences.

» **Ligation Process**: The DNA fragments generated by restriction enzyme digestion have sticky ends (overhangs) that are complementary to the adaptors. The ligation process attaches the adaptors to the ends of the DNA fragments, ensuring that these fragments are ready for selective amplification.

» **Selective Amplification**: By selectively amplifying only the fragments with adaptors, AFLP dramatically reduces the complexity of the DNA mixture, allowing researchers to focus on a smaller, more manageable subset of fragments in the subsequent steps.

3. Selective PCR Amplification

The next step involves **PCR amplification**, where the DNA fragments with adaptors are selectively amplified. This process is carried out in two stages: **pre-selective amplification** and **selective amplification**.

» **Pre-Selective Amplification**: In the first round of PCR, primers that are complementary to the adaptor sequences (with one or two additional nucleotides) are used to amplify a broad range of fragments. This step reduces the overall number of fragments while still maintaining a relatively large pool of amplified products.

- **Primers**: These primers are designed to bind to the adaptor sequences that were ligated to the ends of the DNA fragments in the previous step. The addition of one or two selective nucleotides at the 3' end of the primers ensures that only a subset of the DNA fragments is amplified, as the primer must perfectly match the sequence at the junction of the adaptor and the DNA fragment.

» **Selective Amplification**: In the second round of PCR, **highly selective primers** are used. These primers have additional selective nucleotides (up to three) at their 3' end, further narrowing the number of amplified fragments. The result is a much more specific subset of DNA fragments, which makes it easier to distinguish individual polymorphisms during analysis.

- **Why Selective Nucleotides?**: The inclusion of selective nucleotides in the PCR primers ensures that only a small fraction of the total DNA fragments is amplified. This reduces the complexity of the final DNA mixture, allowing researchers to focus on a manageable number of fragments that are likely to show polymorphisms between individuals.
- **Amplification Process**: During PCR, the DNA is denatured, and the primers anneal to the adaptor regions at the ends of the DNA fragments. DNA polymerase then

extends the primers, amplifying the DNA fragments. By using highly selective primers, AFLP allows researchers to target specific regions of the genome, thereby increasing the likelihood of detecting polymorphisms between individuals.

4. Gel Electrophoresis and Detection

After selective PCR amplification, the next step is to separate the amplified DNA fragments based on their size using **gel electrophoresis**. This step is crucial for visualizing the polymorphisms generated by AFLP.

» **Separation by Size**: The amplified DNA fragments are loaded into wells on an agarose or polyacrylamide gel, and an electric current is applied. Because DNA is negatively charged, the fragments will migrate through the gel toward the positive electrode. Smaller DNA fragments move faster through the gel matrix, while larger fragments move more slowly.

» **Visualization**: The DNA fragments are labeled with **fluorescent dyes** or **radioactive labels** to make them visible on the gel. The bands corresponding to different DNA fragment sizes can be visualized using **fluorescence detection** or **autoradiography**. This allows researchers to analyze the banding patterns and identify genetic differences between individuals.

- **Banding Patterns**: Each band on the gel represents a DNA fragment of a specific size. Differences in the banding patterns between individuals or populations indicate **genetic polymorphisms**. These polymorphisms are detected by comparing the presence or absence of specific bands across different individuals.

5. Polymorphism Detection

The final step in the AFLP process is the detection and analysis of genetic **polymorphisms**. Polymorphisms are variations in DNA sequence that result in differences in the size and number of DNA fragments between individuals.

» **Polymorphism Identification**: By comparing the banding patterns generated through gel electrophoresis, researchers can identify polymorphisms, which represent differences in the genomic DNA between individuals. These polymorphisms may be the result of **insertions**, **deletions**, or **single nucleotide changes** at the restriction sites.

- **Multiple Loci**: AFLP is particularly useful because it can simultaneously detect polymorphisms at multiple loci across the genome. This makes AFLP an efficient tool for studying **genetic diversity**, **population structure**, and **linkage mapping**.

» **Applications**: AFLP has been widely used in **plant and animal breeding**, **conservation genetics**, and **evolutionary biology**. It is especially valuable for studying non-model organisms, where little prior genomic information is available, and for detecting genetic variation in populations.

Applications of AFLP

AFLP is a versatile tool used in various fields of genetics and molecular biology, particularly when researchers need to assess **genetic diversity** or **relationships** between individuals or populations. Some of the most common applications include:

1. **Genetic Diversity Studies**: AFLP is widely used to assess genetic diversity within and between populations. Because AFLP can detect multiple polymorphisms without prior knowledge of the genome, it is particularly useful for studying species with limited genomic information. The technique is applied in plant breeding, conservation biology, and the study of wild species.
   - **Example**: AFLP has been used in crop breeding programs to assess the genetic diversity of breeding lines and identify the best candidates for hybridization.
2. **Population Structure and Phylogenetics**: AFLP is valuable for studying population structure and understanding the **phylogenetic relationships** between species or populations. By analyzing the banding patterns of AFLP markers, researchers can infer the genetic relationships and evolutionary history of different groups.
   - **Example**: In evolutionary biology, AFLP markers have been used to study the genetic differentiation between species and subspecies, providing insights into speciation and population divergence.
3. **Linkage Mapping**: AFLP markers are used in **linkage mapping** to identify chromosomal regions associated with specific traits. The technique helps in identifying **quantitative trait loci (QTLs)** and is applied in both plants and animals.
   - **Example**: AFLP markers have been used to map disease resistance genes in crops like maize and wheat, providing breeders with molecular tools for selecting resistant varieties.
4. **Marker-Assisted Selection (MAS)**: AFLP markers can be used in **marker-assisted selection (MAS)** to track the inheritance of desirable traits in breeding programs. Although AFLP is a dominant marker and less informative than co-dominant markers like SNPs, it can still be useful in early stages of breeding to identify promising candidates for further development.

Advantages of AFLP

1. **No Prior Sequence Information Needed**: One of the most significant advantages of AFLP is that it does not require prior genomic sequence information. This makes it particularly valuable for studying species where the genome has not been sequenced or where detailed genomic information is lacking.
2. **High Polymorphism Detection**: AFLP generates a large number of polymorphic markers in a single experiment. This high level of polymorphism makes AFLP highly effective for assessing genetic diversity, population structure, and genetic mapping.
3. **Applicability Across Species**: AFLP can be applied to almost any species, including plants, animals, fungi, and bacteria. This versatility makes it an essential tool in evolutionary biology, conservation genetics, and crop improvement programs.
4. **Reproducibility**: AFLP is highly reproducible when the same protocols and conditions are followed, making it a reliable technique for comparing genetic data across different laboratories or experiments.

Limitations of AFLP

1. **Dominant Marker**: AFLP is a **dominant marker**, meaning it cannot distinguish between **homozygous** and **heterozygous** individuals. This limits the level of genetic information that can be derived from AFLP markers compared to co-dominant markers like **SNPs** or **SSRs**, which provide more detailed genotype data.
2. **Labor-Intensive**: AFLP is more labor-intensive and technically demanding than some other molecular marker techniques. It requires several steps, including restriction enzyme digestion, adaptor ligation, and multiple rounds of PCR, making the process more time-consuming.
3. **Complex Data Analysis**: The large number of bands generated by AFLP requires complex data analysis and interpretation. The process of scoring bands and analyzing banding patterns can be challenging, especially when dealing with high numbers of samples.
4. **Cost**: Although less expensive than **SNP genotyping** on a large scale, AFLP is more costly than simpler techniques like **RAPD**. The need for specialized equipment (e.g., for gel electrophoresis and visualization) and the use of fluorescent or radioactive labels add to the overall cost of AFLP analysis.

6. Isozymes (Isoenzymes)

Isozymes or **isoenzymes** are multiple forms of an enzyme that catalyze the same biochemical reaction but differ in their **amino acid sequence**, **kinetic properties**, **regulation**, and **tissue distribution**. Isozymes can exist within the same organism and often function in different tissues, developmental stages, or under different environmental conditions. The study of isozymes is an essential aspect of molecular biology, genetics, and biochemistry because it provides insights into how organisms regulate their metabolism and adapt to various physiological demands.

Isozymes are encoded by different **genes** or arise from **post-translational modifications** of the same gene product. They allow for fine-tuned control over biochemical pathways, giving organisms the flexibility to adapt enzyme activity to different tissue-specific or developmental needs.

Characteristics of Isozymes

1. **Multiple Forms of the Same Enzyme**:
   - Isozymes catalyze the same chemical reaction but differ in their structure. This difference in structure may arise from **gene duplication** events, where two or more genes encode enzymes with similar but not identical amino acid sequences. These different forms of enzymes allow for specialization and diversity in metabolic functions across different cell types, tissues, or conditions.
2. **Tissue Specificity:**
   - Isozymes are often found in different **tissues** or **organs** within an organism. For example, one isozyme may be predominant in the liver, while another isozyme catalyzing the same reaction is found mainly in the heart or muscles. This tissue specificity allows for regulation of metabolism according to the functional needs of specific tissues.

3. **Different Kinetic Properties:**

   - Isozymes typically exhibit different **kinetic properties**, such as **substrate affinity (Km)** and **maximum reaction rate (Vmax).** These variations enable tissues or cells to adjust the enzyme's efficiency based on physiological conditions. For instance, one isozyme may work efficiently at low substrate concentrations in a particular tissue, while another isozyme might be activated only under conditions of high substrate availability in a different tissue.

4. **Developmental Regulation:**

   - Isozymes may be expressed at different stages of an organism's **development**. For example, one isozyme might be present in an embryonic stage, while another form becomes more prominent in adult tissues. This developmental regulation ensures that metabolic pathways are optimized for the needs of the organism at each stage of its life cycle.

5. **Regulation and Adaptation:**

   - Isozymes play a key role in adapting to different **environmental conditions**. For instance, some isozymes may be expressed more under conditions of stress, such as hypoxia (low oxygen levels), while others may function optimally under normal physiological conditions. This ability to express different isozymes in response to environmental changes allows organisms to maintain homeostasis in challenging conditions.

Mechanism of Isozyme Formation

Isozymes, or isoenzymes, are different forms of the same enzyme that perform the same biochemical function but differ in their structure, regulatory properties, or tissue distribution. These variations allow organisms to finely regulate enzyme activity to meet specific physiological needs. Isozyme formation is influenced by several genetic and molecular mechanisms, including **gene duplication and divergence**, **alternative splicing**, and **post-translational modifications**. These processes are critical for generating the functional diversity that isozymes provide across tissues and developmental stages. Below is a detailed explanation of each mechanism contributing to the formation of isozymes.

1. Gene Duplication and Divergence

Gene duplication and divergence is one of the most common mechanisms responsible for the formation of isozymes. This evolutionary process occurs when a gene is duplicated in the genome, giving rise to two or more copies of the same gene. Over time, these duplicated genes, known as **paralogs**, accumulate mutations that lead to slight differences in their amino acid sequences. These mutations may alter the protein's structure, function, or regulatory properties, ultimately resulting in the formation of distinct isozymes.

» **Process:**

- **Duplication**: During **gene duplication**, a segment of DNA, including a particular gene, is duplicated due to errors in DNA replication, recombination, or other genetic processes. As a result, the organism now has two copies of the gene.
- **Divergence**: After duplication, the two gene copies evolve independently. **Mutations** accumulate in one or both copies, leading to changes in their amino acid sequences. These

changes can alter the enzyme's function, substrate specificity, or regulatory properties, resulting in the formation of different isozyme forms.

» **Functional Specialization**:

- One of the duplicated genes may retain its original function, while the other evolves to take on a new role or becomes optimized for a different tissue or condition. This specialization allows for the creation of isozymes that are adapted to specific physiological functions. For example, **lactate dehydrogenase (LDH)** exists in different isoforms due to gene duplication, with distinct isoforms functioning optimally in the heart (LDH-H) and muscles (LDH-M).

» **Example**:

- In plants, **gene duplication** has played a significant role in creating isozymes that specialize in distinct biochemical pathways or stress responses. For example, isozymes of **peroxidase** are involved in various functions such as stress response, lignin formation, and hormone metabolism, with each isoform adapted to a specific tissue or developmental stage.

2. Alternative Splicing

Alternative splicing is another mechanism that can lead to the formation of isozymes. In this process, the same gene can produce multiple forms of mRNA by including or excluding different exons (coding regions) during **mRNA processing**. As a result, multiple protein isoforms, each with distinct structural or functional properties, can be produced from a single gene. These protein isoforms may differ slightly in their amino acid sequences, leading to the formation of different isozymes that perform similar enzymatic functions but are specialized for distinct regulatory roles or cellular locations.

» **Process**:

- During the normal processing of pre-mRNA, the **introns** (non-coding regions) are removed, and **exons** (coding regions) are joined together to form a mature mRNA sequence that is translated into a protein. However, in alternative splicing, different combinations of exons can be joined together, resulting in different mRNA transcripts.
- This process allows a single gene to give rise to multiple **protein isoforms**, each potentially having slightly different regulatory properties, localization signals, or functional domains, resulting in the formation of distinct isozymes.

» **Functional Diversity**:

- The protein isoforms generated through alternative splicing may have different **substrate affinities**, **kinetic properties**, or **tissue-specific expression patterns**. This mechanism allows organisms to expand the functional diversity of their enzymes without needing additional genes.

» **Example**:

- **Pyruvate kinase (PK)** is an example of an enzyme that has multiple isoforms produced through alternative splicing. Different isoforms of PK are expressed in different tissues (e.g., muscle, liver, and brain) and play distinct roles in glycolysis. These isoforms have different kinetic properties and regulatory mechanisms, allowing them to meet the specific energy demands of each tissue.

3. Post-Translational Modifications

Post-translational modifications (PTMs) refer to the chemical modifications made to a protein after it has been synthesized. These modifications can significantly alter the activity, stability, or cellular location of the enzyme, resulting in the formation of isozymes. PTMs can include processes such as **phosphorylation, glycosylation**, and **proteolytic cleavage**, each of which can create isoforms of enzymes with different functional characteristics.

» **Types of Post-Translational Modifications**:

1. **Phosphorylation**: The addition of a phosphate group to specific amino acids (usually serine, threonine, or tyrosine) can activate or inactivate an enzyme, or alter its interactions with other proteins or substrates. Phosphorylated versions of enzymes may exhibit different activity levels or regulatory controls, effectively functioning as isozymes.
2. **Glycosylation**: The attachment of carbohydrate groups (glycans) to certain proteins can influence their stability, localization, or interaction with other biomolecules. Glycosylated forms of enzymes may be better suited for specific environments (e.g., secretory pathways or cell surface) and can function differently from their non-glycosylated counterparts.
3. **Proteolytic Cleavage**: Enzymes may undergo proteolytic cleavage, which can activate or deactivate the protein or generate a fragment with a distinct function. Cleavage can result in different isozyme forms with unique functions, often playing roles in regulating biological processes such as apoptosis or hormone maturation.

» **Functional Impact**:

- Post-translational modifications often allow enzymes to be quickly **activated** or **inactivated** in response to specific signals, providing an additional layer of regulation. The modified enzyme isoforms may have different substrate affinities, cellular locations, or half-lives, enabling the cell to fine-tune enzymatic activity to suit particular physiological needs.

» **Example**:

- **Hexokinase**, the enzyme that catalyzes the phosphorylation of glucose to glucose-6-phosphate, exists in multiple isozyme forms due to both alternative splicing and post-translational modifications such as phosphorylation. In the liver, **glucokinase** (a specific isozyme of hexokinase) is phosphorylated, which allows it to respond to high glucose concentrations and regulate blood sugar levels efficiently.

Integration of Isozyme Formation Mechanisms

The formation of isozymes through these mechanisms demonstrates how organisms can evolve and adapt enzyme functions to meet diverse and specialized physiological needs. Often, more than one mechanism contributes to the generation of isozymes within an organism:

» **Combination of Mechanisms**: For example, a gene may undergo duplication, and the resulting paralogs may be further modified through alternative splicing or post-translational modifications, giving rise to multiple distinct isozymes. This layering of mechanisms increases the functional diversity and adaptability of enzymes in response to different developmental stages, tissue types, and environmental conditions.

» **Evolutionary Advantage**: The ability to form isozymes allows organisms to efficiently **expand the functional repertoire** of their enzymes without the need for entirely new genes. This provides an evolutionary advantage, enabling organisms to specialize enzyme activity for different cellular contexts, developmental stages, or environmental challenges.

Examples of Isozymes

Isozymes, or isoenzymes, represent distinct forms of an enzyme that catalyze the same reaction but differ in their amino acid sequences, kinetic properties, or regulatory mechanisms. These differences allow organisms to adapt enzymatic activity to specific tissues, developmental stages, or environmental conditions. Below are three key examples of isozymes that illustrate their functional diversity and physiological significance.

1. Lactate Dehydrogenase (LDH)

Lactate dehydrogenase (LDH) is a classic example of an isozyme system. LDH catalyzes the **interconversion of pyruvate to lactate** during anaerobic glycolysis, a crucial step for energy production in tissues with low oxygen availability. LDH exists in several isozyme forms, each composed of varying combinations of two subunits: **H (heart)** and **M (muscle).** These subunits combine to form different LDH isozymes with distinct functional characteristics suited to different tissues.

» **LDH1 (H4)**:
  - **Location**: Predominantly found in the **heart** and **red blood cells.**
  - **Function**: LDH1 has a higher affinity for pyruvate and is more active in **aerobic conditions.** The heart, which relies primarily on aerobic metabolism to meet its constant energy demand, benefits from LDH1's ability to efficiently convert pyruvate into energy through oxidative phosphorylation.
  - **Significance**: Elevated levels of LDH1 in the bloodstream are often used as a biomarker for **myocardial infarction** (heart attack), as it is released when heart tissue is damaged.

» **LDH5 (M4)**:
  - **Location**: Predominantly found in **skeletal muscle** and **liver.**
  - **Function**: LDH5 is more efficient at converting pyruvate to **lactate** under **anaerobic conditions**, such as during intense muscle activity when oxygen levels are low. The production of lactate allows glycolysis to continue, providing a rapid source of ATP for muscle contraction.
  - **Significance**: LDH5 supports anaerobic glycolysis in muscles during short bursts of high-intensity activity, such as sprinting or weightlifting.

» **Tissue-Specific Distribution**: The tissue-specific distribution of LDH isozymes reflects the different metabolic needs of the heart and muscles. The heart requires aerobic energy production, while skeletal muscles often rely on anaerobic glycolysis during strenuous physical activity. This specialization allows each tissue to optimize energy production under varying oxygen conditions.

2. Glucokinase and Hexokinase

Glucokinase and **hexokinase** are enzymes that catalyze the phosphorylation of glucose to form **glucose-6-phosphate**, the first step of glycolysis. While these enzymes catalyze the same reaction, they exhibit significant differences in their **kinetic properties** and **tissue distribution**, allowing them to fulfill distinct roles in glucose metabolism.

» **Hexokinase**:

- **Location**: Found in most tissues, including the **brain**, **muscles**, and **red blood cells**.
- **Kinetic Properties**: Hexokinase has a **high affinity for glucose** (low Km), meaning it can effectively phosphorylate glucose even at low concentrations. This makes hexokinase well-suited for tissues that require a constant supply of glucose for energy production, such as the brain.
- **Regulation**: Hexokinase is strongly **inhibited by glucose-6-phosphate**, which prevents excessive accumulation of glucose-6-phosphate when glucose is abundant. This regulatory mechanism ensures that hexokinase activity is tightly controlled.

» **Glucokinase**:

- **Location**: Primarily found in the **liver** and **pancreas**.
- **Kinetic Properties**: Glucokinase has a **lower affinity for glucose** (high Km) compared to hexokinase. It becomes active when blood glucose levels are elevated, such as after a meal. This allows the liver to store excess glucose as **glycogen** or use it for energy production.
- **Regulation**: Unlike hexokinase, glucokinase is not inhibited by glucose-6-phosphate. This ensures that the liver can continue to process glucose even when intracellular glucose-6-phosphate levels are high.

» **Functional Role**:

- **Hexokinase** ensures that essential tissues, such as the brain, can continuously produce energy even when glucose levels are low.

» **Glucokinase**, on the other hand, allows the liver to act as a glucose buffer, storing or releasing glucose depending on the body's energy needs. This difference in substrate affinity allows glucokinase to regulate blood sugar levels effectively, especially after a meal.

3. Cytochrome P450 Isozymes

The **Cytochrome P450 (CYP)** family represents a large group of isozymes involved in the **oxidation of organic substrates**, including the metabolism of **drugs**, **toxins**, and **hormones**. These enzymes are located primarily in the **liver**, where they play a central role in **detoxification** and **biotransformation**. The Cytochrome P450 family includes numerous isoenzymes, each specialized for metabolizing different compounds.

- **CYP3A4**:
  - **Location**: Primarily found in the **liver** and **intestines**.
  - **Function**: CYP3A4 is involved in the metabolism of a wide range of drugs, including antibiotics, statins, and immunosuppressants. It accounts for a significant portion of drug metabolism in the human body, making it one of the most clinically important P450 enzymes.
  - **Clinical Significance**: CYP3A4 is involved in drug-drug interactions, as its activity can be induced or inhibited by certain medications, affecting the metabolism of other drugs. For example, **grapefruit juice** can inhibit CYP3A4, leading to elevated levels of drugs metabolized by this enzyme.
- **CYP2D6**:
  - **Location**: Primarily found in the **liver**.
  - **Function**: CYP2D6 is responsible for metabolizing specific drugs, including **antidepressants**, **beta-blockers**, and **opioids**. Unlike CYP3A4, which metabolizes a broad range of substrates, CYP2D6 has more specialized substrate specificity.
  - **Genetic Polymorphism**: CYP2D6 exhibits **genetic polymorphisms** that affect its activity. Some individuals are **poor metabolizers**, meaning they metabolize drugs very slowly, while others are **ultra-rapid metabolizers**, meaning they metabolize drugs more quickly. This variability can influence drug efficacy and the risk of side effects.
- **Functional Diversity**:
  - The diversity of **Cytochrome P450 isozymes** allows organisms to detoxify a wide variety of foreign substances and adapt to different environmental exposures. Each isozyme has a unique substrate specificity, enabling the metabolism of drugs, environmental chemicals, and endogenous compounds like steroids.
  - The Cytochrome P450 system plays a critical role in drug metabolism and detoxification, influencing how individuals respond to medications and environmental toxins.

Applications of Isozyme Analysis

Isozymes are valuable tools in several scientific fields, and their study has numerous practical applications:

1. **Plant and Animal Breeding**:
   - Isozymes are used as **biochemical markers** in **breeding programs**. Isozyme analysis can reveal genetic diversity among populations, aiding in the selection of desirable traits in both plants and animals. Isozyme markers are commonly used to identify genetic relationships, assess inbreeding, and select for disease resistance.
2. **Population Genetics:**
   - In **population genetics**, isozymes serve as genetic markers to study the genetic structure of populations. By analyzing isozyme patterns, researchers can determine gene flow, genetic drift, and other evolutionary processes.

3. **Medical Diagnostics:**
   - In clinical settings, isozyme patterns can be used to diagnose diseases. For example, elevated levels of specific LDH isozymes in the blood may indicate tissue damage, such as myocardial infarction (heart attack), where heart-specific LDH isoforms are released into the bloodstream.
4. **Enzyme Regulation Studies:**
   - Isozyme studies help scientists understand how different forms of enzymes are regulated in various tissues. This knowledge is important for developing therapeutic strategies to target specific isozyme forms involved in diseases, such as cancers where certain isozymes may be overexpressed.

Advantages and Limitations of Isozymes

Advantages:

- **Functional Specialization**: Isozymes allow organisms to fine-tune metabolic reactions in response to tissue-specific needs or environmental conditions.
- **Diagnostic Utility**: Isozymes serve as biomarkers for various diseases, providing valuable information for clinical diagnoses.
- **Diversity Without Redundancy**: Isozymes provide metabolic flexibility without redundant gene expression, as different isozyme forms fulfill specialized roles in different contexts.

Limitations:

- **Complexity in Analysis**: The presence of multiple isozyme forms can complicate genetic and biochemical studies, as distinguishing between isoforms may require specialized techniques.
- **Regulatory Overlap**: In some cases, the functional differences between isozymes may not be well understood, leading to challenges in interpreting their regulatory roles in metabolic pathways.

Chapter - 7

Mapping Populations - MAS

A. Mapping Populations

Mapping populations serve as a fundamental tool in the field of genetics, playing a crucial role in understanding the genetic architecture underlying various traits in both plant and animal species. These populations are deliberately structured genetic groups used to dissect the complex relationships between genes and phenotypic traits, allowing researchers to pinpoint specific **genes** or **quantitative trait loci (QTLs)** responsible for important characteristics. Traits such as **disease resistance**, **yield**, **drought tolerance**, and other agronomic or economically significant attributes often have complex genetic backgrounds, involving the interaction of multiple genes across different loci. Mapping populations enables the detection and association of these genetic elements with observable traits, providing valuable insights for breeding programs, conservation efforts, and genetic improvement strategies.

The creation of mapping populations is a meticulously planned process that involves crossbreeding, selection, and sometimes genetic manipulation. This is necessary to produce populations with well-defined and stable genetic variation. Researchers generate these populations by crossing two genetically distinct parents, followed by controlled inbreeding, backcrossing, or other breeding techniques to establish a population suitable for **QTL mapping**, **linkage analysis**, or **association studies**. Each individual in a mapping population represents a unique combination of genetic material from the parent lines, offering a diverse pool for studying inheritance patterns and identifying genetic markers associated with target traits.

One of the primary goals of using mapping populations is to improve the efficiency and precision of **marker-assisted selection (MAS)** in breeding programs. By identifying markers closely linked to beneficial traits, breeders can accelerate the development of improved plant and animal varieties without relying solely on traditional selection methods, which are time-consuming and often less efficient. Mapping populations provides a means to correlate **genotype** (the genetic makeup) with **phenotype** (the observable traits), thus making it easier to select individuals that carry the desired genetic traits.

Several types of mapping populations are commonly used in genetic studies, each with unique attributes, advantages, and limitations. The most widely used types include **Recombinant Inbred Lines (RILs)**, **Near-Isogenic Lines (NILs)**, **double Haploids (DH)**, and **Backcross (BC)** populations.

DOI: 10.1201/9781003757412-7

These populations differ in their development processes, the level of genetic diversity they capture, and their suitability for specific types of genetic analyses.

1. Recombinant Inbred Lines (RILs)

Recombinant Inbred Lines (RILs) are a widely used mapping population in both plant and animal genetics. They are created through a process of repeated **selfing** (self-pollination or self-fertilization) starting from an **F2 population** generated by crossing two genetically diverse parents. This selfing process is typically continued for six to eight generations, leading to nearly complete **homozygosity** in each line. Each RIL thus represents a unique combination of **parental alleles**, which makes them invaluable for studying the inheritance of traits and for **quantitative trait loci (QTL)** mapping.

RILs are used extensively in plant breeding and research because of their ability to stabilize genetic diversity over generations. The development of RILs enables researchers to systematically dissect the genetic components of complex traits, such as **disease resistance**, **yield**, or **drought tolerance**, by providing a permanent genetic resource that can be reused across multiple experiments. Below, we discuss the **mechanism** by which RILs are created and the **merits** and **demerits** of using this type of population for genetic studies.

Mechanism of Recombinant Inbred Lines (RILs) Formation

The formation of **Recombinant Inbred Lines (RILs)** is a multi-generational and systematic process designed to produce homozygous lines from a cross between two genetically diverse parental lines. These lines are essential for genetic studies, particularly in the mapping of **quantitative trait loci (QTL)**, gene discovery, and understanding the genetic architecture of complex traits. The mechanism behind RIL formation involves controlled breeding, recombination, and the fixation of genetic alleles through successive generations of self-pollination or sibling mating.

Step 1: Initial Parental Cross (P1 × P2)

The creation of RILs begins with the selection of two **genetically distinct parental lines**, designated as **P1** and **P2**. These parental lines are chosen based on their contrasting genetic traits or phenotypes to maximize genetic diversity. The key goal of this step is to introduce a wide array of genetic alleles into the progeny.

- » **Genetic Diversity of Parents**: The parental lines are typically selected based on their differences in traits of interest. These could include differences in yield, disease resistance, stress tolerance, or other agronomically or physiologically important characteristics. The greater the genetic differences between the two parental lines, the more variation there will be in the F1 generation.
- » **First Cross**: The parents are crossed to produce the **F1 generation**. In plants, this usually involves transferring pollen from the male parent (P1) to the female parent (P2) to fertilize the ovules. In animals, controlled mating techniques are used. This first cross introduces genetic variation but does not yet allow for recombination between parental alleles, as the chromosomes from each parent remain intact within the F1 individuals.
- » **Mendelian Segregation**: The initial cross follows **Mendelian inheritance patterns**, meaning

that each offspring receives one allele from each parent for every gene. This hybridization step forms the foundation for future recombination events.

Step 2: Formation of the F1 Generation

The **F1 generation** is the first hybrid generation derived from the cross between the two genetically diverse parents (P1 and P2). The F1 progeny are **heterozygous** at most loci, carrying one allele from each parental line for each gene.

» **Heterozygosity**: Every F1 individual is heterozygous, possessing a mix of genetic material from both parents. For each locus, one allele is inherited from P1, and the other from P2. This heterozygosity is crucial as it ensures that the subsequent generations will have a wide array of genetic recombination during meiosis.

» **Uniform Genetic Composition**: The F1 generation is relatively uniform in its genetic composition because each individual receives an equal mix of parental alleles. However, no recombination between the parental alleles occurs during the F1 stage. At this stage, the alleles from P1 and P2 are still linked in their respective chromosomes.

Step 3: Generation of the F2 Population

The next critical step in the formation of RILs is to produce an **F2 population** through the **selfing (self-pollination or self-fertilization)** of F1 individuals. The F2 generation is the first opportunity for **recombination** to occur between the parental alleles.

» **Recombination**: During meiosis in F1 individuals, **recombination (crossing-over)** occurs between homologous chromosomes. This results in the exchange of genetic material between the parental chromosomes, breaking the linkage between alleles from P1 and P2. This recombination produces **recombinant chromosomes** that carry new combinations of alleles. Each F2 individual, therefore, has a unique genetic makeup, with various combinations of the original parental alleles.

» **Segregation of Alleles**: The F2 population exhibits **segregation of alleles** due to the recombination and independent assortment of chromosomes. Unlike the uniform F1 generation, the F2 generation shows a wide range of genetic variation. Some individuals will be homozygous for certain loci, while others will remain heterozygous. This diversity is critical for the future selection of homozygous lines in subsequent generations.

» **Phenotypic Variation**: Along with genetic variation, the F2 generation also shows considerable **phenotypic variation**. The traits observed in the F2 individuals represent different combinations of parental alleles, providing an opportunity to study the inheritance of traits and their genetic basis.

Step 4: Repeated Selfing or Sibling Mating to Fix Alleles

After the formation of the F2 generation, the next phase in the formation of RILs is **repeated selfing** (or in some cases, sibling mating) over several generations. This step is crucial for fixing the alleles in a homozygous state.

» **Successive Generations of Selfing**: The F2 individuals are selfed or sib-mated for at least six to eight generations. With each generation, the level of **heterozygosity** decreases, and the proportion of **homozygous loci** increases. By the time six to eight generations of selfing are complete, the lines become nearly **completely homozygous** at most loci.

- **Selfing Process**: In each generation, self-pollination involves fertilizing the same plant's ovules with its own pollen. In animals, sibling mating might be used to achieve similar results. These practices gradually increase homozygosity by reducing heterozygous loci.

» **Segregation and Fixation of Alleles**: With each generation of selfing, the offspring have an increasing chance of inheriting identical alleles from both parents at any given locus. As a result, **alleles become fixed** in a homozygous state. The repeated selfing process ensures that, after several generations, most loci in each individual are homozygous, with different lines representing different combinations of parental alleles.

Step 5: Formation of Nearly Homozygous Recombinant Genomes

After several generations of selfing, the resulting RILs are **genetically stable** and almost entirely **homozygous**. These lines carry recombinant genomes formed by the recombination events that occurred in the F2 generation.

» **Recombinant Chromosomes**: Each RIL carries **recombinant chromosomes** that represent unique combinations of alleles from both parental lines. The earlier recombination events in the F2 population have effectively shuffled the parental alleles, and selfing has fixed these combinations in homozygous form in each RIL.

» **Stability and Homozygosity**: The lines are now genetically stable, meaning that their allelic composition will not change in subsequent generations. This homozygosity is a hallmark of RILs and is critical for ensuring reproducibility in genetic studies. Each RIL is essentially a genetic snapshot of a unique combination of parental alleles that will remain unchanged over time, allowing for consistent and repeatable phenotypic evaluations.

Final Step: Completion of RILs

The end result of this process is the production of **Recombinant Inbred Lines** that are **genetically uniform** within each line but differ between lines based on the alleles inherited from the two parents. Each line represents a different **recombinant combination** of the original parental alleles, and these combinations are fixed in a homozygous state.

» **Genetic Resource**: The RILs serve as a permanent genetic resource for researchers. Once developed, these lines can be maintained indefinitely and used in genetic studies, breeding programs, or functional genomics research. The fixed genetic makeup of each line allows researchers to study how different combinations of alleles contribute to traits of interest.

» **Genomic Mapping**: Because RILs have gone through multiple generations of recombination, they are ideal for **fine-scale genetic mapping**. The recombinant genomes in RILs make it possible to identify **quantitative trait loci (QTLs)** with high precision by correlating specific genotypes with phenotypes.

Merits of RILs

1. **High Mapping Resolution**

    - **Explanation**: One of the primary advantages of using **RILs** in genetic studies is their ability to provide **high-resolution mapping**. After multiple generations of recombination, the genetic material from the two parents is broken down into smaller fragments across the genome. This repeated recombination allows for the identification of **quantitative trait loci (QTLs)** with great precision because the linked markers and the loci of interest are separated by recombination events over time.
    - **Benefit**: This high-resolution mapping is particularly valuable for **fine-scale genetic studies**, enabling researchers to pinpoint genes or genetic regions that control complex traits with greater accuracy than populations with fewer recombination events.
    - **Application**: For example, in crop improvement programs, RILs are used to map **QTLs for disease resistance**, **drought tolerance**, or **yield**, allowing breeders to develop varieties with desirable traits more efficiently. By leveraging high mapping resolution, breeders can identify markers closely linked to important traits, facilitating **marker-assisted selection (MAS)**.

2. **Stability**

    - **Explanation**: RILs are highly **genetically stable** due to their nearly complete homozygosity. Each RIL line is effectively an **inbred line**, meaning that the genetic makeup of the line does not change from one generation to the next. This stability allows researchers to use the same RILs in different experiments over time without worrying about genetic drift or segregation of alleles, which can occur in heterozygous populations.
    - **Benefit**: The stability of RILs means that they can be replicated across different environments, years, and experiments, providing consistent and reliable data on how genes and traits interact with environmental factors. This is crucial for **phenotypic evaluations**, as the same genotype can be evaluated repeatedly under different environmental conditions, helping to distinguish genetic effects from environmental influences.
    - **Application**: In multi-environment trials, RILs allow researchers to study **genotype-environment interactions** to understand how genetic factors contribute to trait variation under different growing conditions, such as drought or high salinity.

3. **Long-Term Resource**

    - **Explanation**: Another significant merit of RILs is their potential to serve as a **long-term genetic resource**. Once a RIL population is developed, it can be **preserved** and distributed to other researchers or breeding programs indefinitely. Because the genetic makeup of each line is fixed, RIL populations can be used for **long-term genetic studies**, providing a permanent resource for mapping, breeding, and functional genomics research.
    - **Benefit**: This long-term utility makes RILs particularly valuable for collaborative research and **cross-institutional studies**, where different research groups can evaluate the same population for a variety of traits, making it possible to build large datasets on trait inheritance across different environments and contexts.

- **Application**: In large-scale genetic research projects, RIL populations can be used to study **multiple traits** simultaneously, allowing researchers to explore the genetic basis of **complex traits** that may involve interactions between different genes and environmental factors.

Demerits of RILs

1. **Time-Consuming**
   - **Explanation**: One of the main disadvantages of using RILs is the **time-intensive nature** of their development. Creating a RIL population requires multiple generations of **selfing**, typically six to eight generations, to achieve the necessary level of homozygosity. This process can take several years, depending on the species and the reproductive cycle of the organism.
   - **Limitation**: This extended timeline can be a significant bottleneck for breeding programs or research projects that require rapid results. In species with long generation times, such as certain tree crops or livestock, developing RILs can be particularly challenging.
   - **Application**: In breeding programs where quick results are needed, other mapping populations, such as **doubled haploids (DHs)**, may be preferred because they can achieve homozygosity in a single generation.
2. **Limited Genetic Diversity**
   - **Explanation**: Another limitation of RIL populations is the **limited genetic diversity** they represent. Because RILs are derived from a cross between only two parental lines, the genetic variation present in the population is confined to the alleles contributed by these two parents. This restricts the utility of RILs for detecting **rare alleles** or studying broader genetic variation present in more diverse populations.
   - **Limitation**: As a result, RILs may not be suitable for studying traits controlled by alleles that are rare or absent in the parental lines. Moreover, RILs may not capture the full range of genetic variation required to study **complex traits** influenced by multiple genes and environmental interactions.
   - **Application**: For studies that require a broader genetic base, other mapping populations, such as **backcross populations** or **multi-parent populations**, may offer a more comprehensive view of the genetic diversity present in the species.

2. Near-Isogenic Lines (NILs)

Near-isogenic lines (NILs) are highly valuable resources in genetics and breeding, offering a precise way to study the effects of specific genes or genomic regions. NILs are created through a process of **repeated backcrossing** between a donor parent and a recurrent parent, where the goal is to introduce a small genetic region of interest from the donor parent into the genome of the recurrent parent. After several generations of backcrossing and selection, the NILs differ from the recurrent parent by only a small genomic segment from the donor parent, while the rest of the genome remains virtually identical to the recurrent parent.

This unique characteristic of NILs makes them an excellent tool for **fine mapping** of quantitative trait loci (QTLs), gene discovery, and functional analysis of specific genomic regions. NILs are particularly useful when researchers are interested in the effect of a specific allele or small chromosomal segment on a phenotypic trait, as the minimal genetic differences between the lines reduce background genetic noise, allowing for precise trait analysis.

Mechanism of NIL Formation

The development of NILs is a time-consuming but highly controlled process. The central goal is to replace a small portion of the genome in a recurrent parent with a corresponding region from a donor parent, while keeping the rest of the recurrent parent's genome intact. The process involves multiple generations of **backcrossing** and **selection**, resulting in lines that are nearly isogenic to the recurrent parent, except for the targeted donor region. Below is a detailed description of the mechanism involved in creating NILs.

Step 1: Initial Cross Between Donor and Recurrent Parents

The creation of NILs starts with the selection of two parents: a **donor parent** that carries the allele or genomic region of interest, and a **recurrent parent** that serves as the genetic background.

» **Donor Parent**: The donor parent is chosen because it possesses a specific trait or allele that researchers aim to study or introgress into the genome of the recurrent parent. This allele could be linked to disease resistance, high yield, drought tolerance, or any other important trait.

» **Recurrent Parent**: The recurrent parent is typically a well-established breeding line or variety that researchers want to maintain, except for the small genomic region being replaced by the donor parent. The recurrent parent provides the genetic background for the NILs.

» **First Cross (F1 Generation)**: The donor and recurrent parents are crossed to produce the **F1 generation**. The F1 individuals will carry **50% of their genetic material from the donor parent** and **50% from the recurrent parent**, meaning that the desired allele from the donor parent will be present, but much of the donor genome is also introduced into the F1 progeny.

Step 2: First Backcross to Recurrent Parent

The F1 generation is then **backcrossed** to the recurrent parent to begin the process of reintroducing the recurrent parent's genome while retaining the desired donor allele.

» **Backcrossing**: The F1 individuals are crossed back to the recurrent parent. This process reduces the proportion of the donor genome, as each backcross introduces more of the recurrent parent's genetic material. After the first backcross (BC1), the resulting progeny will have approximately **75% of their genome from the recurrent parent** and **25% from the donor parent**.

» **Selection for the Desired Trait**: After each backcross, **selection** is performed to ensure that the desired donor trait or allele is retained in the offspring. This is done through **phenotypic selection** or **marker-assisted selection (MAS)**. Only individuals that carry the donor allele of interest are selected for further backcrossing.

Step 3: Multiple Generations of Backcrossing

The process of backcrossing is repeated over multiple generations (typically five to six generations), each time crossing the offspring back to the recurrent parent. With each successive backcross, the proportion of the recurrent parent's genome increases, while the proportion of the donor parent's genome decreases.

» **Reduction of Donor Genome**: By the **BC5 or BC6 generation**, the offspring will carry more than **98% of their genome from the recurrent parent**, and only a small segment of the donor genome, including the region of interest, will remain. This ensures that the genomic background is almost entirely that of the recurrent parent, minimizing any background genetic noise from the donor genome.

» **Selection After Each Generation**: Throughout the backcrossing process, **selection** is applied in each generation to ensure that the offspring retain the donor allele or region of interest. Selection can be based on **morphological traits, phenotypic markers**, or more commonly, **molecular markers** that are closely linked to the desired trait. The use of **marker-assisted selection** greatly accelerates the identification and retention of individuals carrying the desired genomic region.

Step 4: Selfing or Sibling Mating to Create Homozygous Lines

Once backcrossing is complete, the selected individuals are **selfed** or mated with their siblings to produce **homozygous lines** for the donor region. This step ensures that the desired donor region is fixed in a **homozygous state**, while the rest of the genome remains homozygous for the recurrent parent alleles.

» **Selfing Process**: The final step is to self-pollinate or sib-mate the selected individuals from the final backcross generation to produce lines that are homozygous for the donor segment. This step can also involve **marker-assisted selection** to ensure that the correct donor region is present in its homozygous form.

» **Final Product**: The resulting **Near-Isogenic Lines (NILs)** will be genetically identical to the recurrent parent except for the **small genomic region introgressed from the donor parent**. The NILs now serve as a valuable resource for studying the specific effects of this small genomic region on traits of interest, as the background genetic variation from the donor parent has been minimized.

Merits of Near-Isogenic Lines (NILs)

1. **High Precision Mapping**

    - **Explanation**: NILs provide high precision for mapping **quantitative trait loci (QTLs)** and genes. This precision is possible because the genetic background of NILs is nearly identical to the recurrent parent, except for a small region introgressed from the donor parent. The uniform genetic background minimizes the impact of **genetic noise**, allowing researchers to attribute observed phenotypic differences directly to the specific genomic region of interest.

- **Benefit**: By focusing on a specific, small genomic region, NILs allow for **fine mapping** and precise identification of the **causal gene or locus** affecting the trait in question. This is particularly useful in breeding programs where pinpointing the exact genetic factors controlling important traits, such as **disease resistance** or **yield**, is essential for marker-assisted selection (MAS).
- **Application**: NILs are commonly used to dissect the genetic basis of **resistance genes** in crops, where a single gene or small genomic region is responsible for conferring resistance to a particular pathogen or environmental stress. This method allows breeders to develop new resistant varieties by introgressing specific resistance genes into elite cultivars.

2. **Efficient Trait Identification**
   - **Explanation**: NILs allow researchers to efficiently identify the **effects of a specific genomic region** on a phenotypic trait. Since the NILs differ from the recurrent parent by only a small genomic region from the donor parent, any phenotypic differences between the NILs and the recurrent parent are directly attributable to the introgressed segment. This makes it much easier to determine how a particular gene or QTL affects the trait.
   - **Benefit**: This focused approach allows researchers to rapidly identify the **role of specific alleles** in complex traits. NILs are highly useful for **trait dissection**, enabling a clear understanding of how one genomic region impacts various aspects of the phenotype, such as plant height, drought tolerance, or seed quality.
   - **Application**: NILs are particularly useful in **plant breeding programs**, where introgressing traits like **pest resistance, herbicide tolerance**, or **nutritional content** into elite varieties is a priority. By studying the effects of the introgressed region, breeders can quickly identify which alleles are most beneficial and then select them in breeding programs.

3. **Uniform Genetic Background**
   - **Explanation**: One of the key advantages of NILs is that they have a nearly identical genetic background, except for the small donor segment. This homogeneity allows for **consistent and reproducible results** in trait analysis, as the effects of other unrelated genes are minimized.
   - **Benefit**: The uniform genetic background in NILs allows for clearer identification of **gene-trait relationships**, as external genetic variability is eliminated. This makes NILs a powerful tool in **functional genomics**, where understanding the function of specific genes or genomic regions is a central goal.
   - **Application**: NILs are commonly used in **functional studies** to understand the impact of **single genes** on phenotypic traits. For example, NILs can be used to study the effect of a single disease-resistance gene in a plant, without interference from other genetic factors.

Demerits of Near-Isogenic Lines (NILs)

1. **Time and Labor Intensive**
   - **Explanation**: The development of NILs is a **time-consuming** and **labor-intensive**

process. The creation of NILs typically requires at least **five to six generations of backcrossing**, followed by **selfing** to achieve homozygosity. Each generation requires careful **selection** for the desired allele or trait, and this process can take many years to complete, especially in species with long generation times.

- **Drawback:** The time and labor required to produce NILs make them less suitable for **rapid trait analysis** or for breeding programs that require immediate results. Additionally, the process demands significant resources, including labor, space, and expertise.
- **Limitation:** This extended timeline can be particularly problematic in breeding programs where **time-to-market** is crucial. For example, in annual crops with shorter life cycles, such as maize or wheat, developing NILs may delay the introduction of new varieties.

2. **Limited Genome Coverage**

- **Explanation:** While NILs provide high precision for studying specific genomic regions, their focus on a small genomic segment limits their utility for **whole-genome studies**. NILs are designed to introgress and study specific alleles or small regions from a donor parent into a recurrent parent's genome, meaning that they cover only a **narrow portion of the genome**.
- **Drawback:** The limited coverage makes NILs unsuitable for **whole-genome association studies** (GWAS) or **QTL mapping** of traits that are influenced by many genes spread across the genome. In such cases, broader mapping populations, such as recombinant inbred lines (RILs), are more appropriate.
- **Limitation:** For example, traits that are **polygenic**, involving contributions from multiple regions across the genome, cannot be fully explored using NILs. They are best suited for **targeted** genetic studies where a specific locus is already known or suspected to play a major role in a trait.

3. **Narrow Focus on Specific Traits**

- **Explanation:** Because NILs are designed to study the effect of a **single gene** or **small chromosomal region**, their utility is confined to traits that are controlled by specific loci. NILs are not well-suited for exploring **complex traits** controlled by many genes, nor are they ideal for **exploratory mapping** when the genetic basis of the trait is unknown.
- **Drawback:** This narrow focus limits their use in understanding the **genetic architecture of complex traits**, such as yield or abiotic stress tolerance, which often involve **polygenic inheritance**. Additionally, NILs cannot provide insights into **epistatic interactions** between different regions of the genome, as their genetic variation is confined to a small region.
- **Limitation:** NILs are therefore less useful in the early stages of genetic research when the goal is to identify novel QTLs or to conduct **genome-wide studies**. They are more appropriate for **fine mapping** or for situations where the researcher is confident about the location of the gene or QTL being studied.

Summary of Merits and Demerits

Merits	Description
High Precision Mapping	NILs provide high-resolution mapping of specific genomic regions with minimal genetic noise from the rest of the genome.
Efficient Trait Identification	NILs enable rapid identification of the role of a particular genomic region or allele in controlling specific traits.
Uniform Genetic Background	The nearly identical genetic background across NILs ensures consistency and clarity in gene-trait relationships.

Demerits	Description
Time and Labor Intensive	Developing NILs requires multiple generations of backcrossing and selection, making the process lengthy and resource-intensive.
Limited Genome Coverage	NILs are focused on specific genomic regions and are not suitable for whole-genome studies or traits influenced by multiple genes.
Narrow Focus on Specific Traits	NILs are best suited for traits controlled by a single gene or small regions, limiting their utility for studying complex traits.

3. Doubled Haploids (DH)

Doubled Haploids (DHs) are a unique and powerful tool in plant breeding and genetic studies. DH populations are created by inducing **haploid cells** (cells that contain a single set of chromosomes) from reproductive cells, such as gametes, and subsequently doubling their chromosome number to restore **diploidy**. This process results in completely **homozygous** individuals in a single generation, bypassing the need for multiple generations of selfing, which is required in traditional breeding methods to achieve homozygosity. Due to their rapid ability to produce homozygous lines, DH populations are widely used in **QTL mapping, marker-assisted selection (MAS)**, and **genetic studies**.

The creation of DH populations allows breeders and researchers to accelerate genetic improvement by producing genetically stable lines more quickly than with conventional breeding techniques. DH populations eliminate the heterozygosity seen in hybrid or segregating populations, allowing for more precise assessment of the relationships between **genotypes** and **phenotypes**.

Mechanism of DH Formation

The formation of **Doubled Haploids (DHs)** involves several key steps. These processes vary depending on the species, but they generally include the induction of haploid cells, followed by chromosome doubling to restore diploidy. Below is a detailed breakdown of the mechanism involved in DH formation.

Step 1: Inducing Haploid Cells

The first and most critical step in the creation of **Doubled Haploids (DHs)** is the induction of **haploid cells**. Haploid cells contain only one set of chromosomes (**n**), unlike the usual diploid cells (**2n**), which contain two sets of chromosomes, one inherited from each parent. By inducing haploid cells from the gametophyte stage, breeders and researchers can create plants or organisms that possess only one set of chromosomes. The ability to manipulate these haploid cells is essential for generating **completely**

homozygous individuals through chromosome doubling. There are several well-established techniques to induce haploid cells, each of which is suited to different species and experimental contexts.

Anther Culture

Anther culture is one of the most widely used techniques for inducing haploid cells, especially in cereals like **rice**, **wheat**, and **barley**. This method targets the **microspores** within the **anthers** (the part of the flower that produces pollen).

» **Mechanism**: In this technique, the **anthers** are collected from the plant before the pollen grains mature. The collected anthers are then cultured in a carefully controlled laboratory environment, where they are induced to undergo **embryogenesis**. The **microspores** (immature pollen grains) inside the anthers can be manipulated to develop into haploid embryos.

» **Embryogenesis**: Instead of following their typical path of pollen development, the microspores undergo **embryogenesis**, where they form embryos rather than developing into gametes. This switch is triggered by stressing the microspores through changes in temperature, hormone levels, or culture medium composition. Once the microspores are induced to become embryos, they are haploid in nature, carrying only a single set of chromosomes derived from one parent.

» **Applications**: **Anther culture** is particularly efficient in species where the microspores respond well to stress and can be easily induced to undergo embryogenesis. This technique is frequently used in breeding programs for major crops like **rice**, **wheat**, and **barley** to accelerate the development of homozygous lines for breeding.

Microspore Culture

Microspore culture is another method used to generate haploid cells, and it is particularly effective for species like **canola** and **tobacco**. Rather than culturing the entire anther, microspore culture isolates and directly manipulates the **microspores** (immature pollen grains) from the anther.

» **Mechanism**: Microspores are collected directly from the anthers and then placed into a **culture medium** under specific environmental conditions. The key difference between microspore culture and anther culture is that in microspore culture, the microspores are isolated first, which allows for more **targeted manipulation** of these gametophytic cells.

» **Embryogenesis**: Like in anther culture, the microspores are induced to undergo **embryogenesis**, where they form embryos instead of developing into mature pollen. This embryogenesis is typically induced by applying specific stresses such as heat shock, cold treatment, or the addition of plant growth regulators to the culture medium.

» **Applications**: Microspore culture allows for greater control over the development of haploid cells compared to anther culture, as it targets the microspores directly. It is especially useful in crops where the anthers themselves may be less responsive to culture but where microspores can still be induced to form haploid embryos.

Ovule Culture

In some plant species, haploid embryos can be produced from **female gametes** (ovules) using **ovule culture techniques**. This method involves culturing **unfertilized ovules** and inducing them to develop into haploid embryos.

» **Mechanism**: The process begins by collecting **unfertilized ovules** from the plant. These ovules contain **egg cells** that, under normal circumstances, would be fertilized by pollen to form a diploid embryo. However, in **ovule culture**, these egg cells are cultured in a controlled environment and are induced to develop into haploid embryos **without fertilization**.

» **Parthenogenesis**: The development of haploid embryos in ovule culture occurs through a process known as **parthenogenesis**, where the egg cell develops into an embryo without being fertilized. This leads to the formation of haploid embryos that contain only the chromosomes from the mother plant.

» **Applications**: Ovule culture is particularly useful in species where male gametes (pollen) are difficult to manipulate or where other haploid induction techniques are less effective. It is often used in **fruit crops** and certain **vegetable species**.

Wide Crosses (Distant Hybridization)

Wide crosses, also known as **distant hybridization**, involve crossing two **genetically distant species** to produce **haploids**. This method relies on natural chromosome elimination processes that occur during the development of the hybrid embryo.

» **Mechanism**: In a wide cross, two distantly related species are crossed, producing hybrid embryos. However, during the early stages of embryo development, the chromosomes from one parent are selectively eliminated, resulting in haploid embryos that contain only the chromosomes from the other parent. For example, in **wheat × maize crosses**, the maize chromosomes are eliminated, and the resulting embryo contains only the wheat chromosomes.

» **Embryo Rescue**: In some cases, the hybrid embryo may not develop fully on its own, so researchers use a technique called **embryo rescue**, where the embryo is removed from the seed and cultured in a nutrient-rich medium to encourage its growth.

» **Applications**: Wide crosses are particularly effective in crops like **wheat**, where this method is commonly used to produce **haploid wheat embryos**. The resulting haploids can then undergo chromosome doubling to produce fully homozygous doubled haploids.

Step 2: Chromosome Doubling

Once **haploid cells** or **embryos** have been successfully induced, the next critical step in the development of **Doubled Haploids (DHs)** is the process of **chromosome doubling**. Haploid cells, which contain only a single set of chromosomes (**n**), are not viable as mature plants or organisms because they lack the full set of genetic material needed for normal growth, development, and reproduction. To overcome this limitation, **chromosome doubling** is performed, converting haploid cells into **diploid cells (2n)**, which contain two sets of chromosomes. This step is essential for producing viable **homozygous** individuals, as the resulting plants will have identical alleles for each gene, making them completely homozygous at all loci.

Mechanism of Chromosome Doubling

Chromosome doubling can be achieved using various techniques, most of which involve the application of chemicals or physical treatments to induce the doubling of chromosomes during cell division. Below are the common methods used to restore diploidy in haploid cells.

1. Colchicine Treatment

The most widely used method for inducing **chromosome doubling** is through the application of **colchicine**, a chemical compound derived from the autumn crocus plant (**Colchicum autumnale**). Colchicine interferes with **microtubule formation** during cell division, preventing the chromosomes from segregating properly, which leads to the duplication of chromosomes within the cell.

» **Mechanism**: Colchicine works by binding to **tubulin**, the protein that forms microtubules, which are essential for the formation of the **spindle apparatus** during mitosis. The spindle apparatus is responsible for separating the chromosomes into daughter cells. When colchicine disrupts the formation of microtubules, the spindle fibers cannot form, and the chromosomes are unable to separate during **anaphase**. As a result, the chromosomes remain within the same cell, effectively **doubling the chromosome number**.

» **Application**: Colchicine is typically applied to the **shoot apical meristem** of young haploid embryos or plants, which is the region where active cell division occurs. It can be applied as a solution by soaking seeds, seedlings, or tissues in colchicine, or it can be applied directly to the growing tips of the plants. Depending on the species, the concentration and duration of colchicine treatment can vary, but it is generally a short-term treatment, lasting anywhere from a few hours to a few days.

» **Outcome**: Once the treatment is completed, the haploid cells undergo mitosis, but the **chromosome number is doubled**, resulting in diploid cells. These diploid cells are fully **homozygous**, as both sets of chromosomes are identical copies of the original haploid genome. This homozygosity is crucial for breeding programs and genetic studies, as it ensures that all alleles are fixed and that the resulting plant or organism is genetically stable.

» **Advantages**: Colchicine is highly effective for inducing chromosome doubling and is widely used across many plant species. It has become a standard method in **plant breeding** due to its efficiency in producing homozygous diploid lines in a single generation.

» **Challenges**: Despite its widespread use, colchicine can be **toxic** to both the plants and the people handling it, so proper safety precautions must be followed. Additionally, colchicine treatment does not always result in 100% success, and treated plants must be carefully monitored for signs of successful doubling.

2. Other Doubling Agents

In addition to colchicine, other chemical agents can be used to induce **chromosome doubling**. These alternatives are sometimes preferred when colchicine is less effective or when specific species respond better to other compounds. Some of the commonly used alternatives include:

» **Oryzalin**: Oryzalin is a herbicide that, like colchicine, disrupts microtubule formation during cell division. It works in a similar manner by binding to tubulin and preventing the assembly of the **spindle apparatus**, leading to chromosome doubling. Oryzalin is often preferred in certain plant species, such as **grapes** and **ornamental plants**, where it is more effective than colchicine. It is less toxic to humans compared to colchicine, making it a safer option in some cases.

» **Trifluralin**: Trifluralin is another herbicide that interferes with microtubule formation, leading to chromosome doubling. It is commonly used in plant species that are difficult to treat with colchicine or oryzalin. Like oryzalin, trifluralin is also considered less toxic than colchicine, but its effectiveness can vary depending on the plant species being treated.

» **Mechanism**: Both **oryzalin** and **trifluralin** work by targeting the same molecular machinery as colchicine (i.e., tubulin in microtubules), but their specificity and effectiveness can differ depending on the species being treated. The choice of doubling agent often depends on species-specific responses and the success rates observed in earlier trials.

» **Application**: Similar to colchicine, these doubling agents are typically applied to the **shoot apical meristems** of haploid embryos or young plants. The treated plants are then grown to maturity and screened for signs of chromosome doubling.

» **Advantages**: Oryzalin and trifluralin provide viable alternatives to colchicine, especially for species where colchicine has low success rates. They also tend to have **lower toxicity** to humans, making them safer to handle in a laboratory or breeding environment.

3. Spontaneous Chromosome Doubling

In some species, **spontaneous chromosome doubling** can occur without the need for chemical treatments. Although the frequency of spontaneous doubling is generally much lower than induced doubling, it has been observed in several plant species.

» **Mechanism**: In certain species, cells naturally undergo chromosome doubling during their normal cell cycle without any external intervention. This is often due to **errors in mitosis**, where the chromosomes fail to segregate properly, resulting in diploid cells. In rare cases, these spontaneous events can lead to the formation of **doubled haploids** without any external chemical or physical treatments.

» **Advantages**: The primary advantage of spontaneous chromosome doubling is that it eliminates the need for **chemical treatments**, which can be both toxic and expensive. Spontaneous doubling is a **natural process**, meaning that no external agents are required, and the plants do not suffer from any chemical-induced stress.

» **Challenges**: The main limitation of relying on spontaneous doubling is that it occurs at a **low frequency**. This makes it an unreliable method for large-scale breeding programs or research, where consistent and predictable results are needed. As a result, breeders and researchers generally prefer induced methods, such as colchicine treatment, to ensure the success of chromosome doubling.

Step 3: Regeneration and Selection of Doubled Haploids

Once **chromosome doubling** has been successfully achieved, the next step in the creation of **Doubled Haploids (DHs)** is the **regeneration** of haploid embryos or cells into fully grown plants. These plants are genetically **homozygous** and **stable**, meaning they can be reliably used in breeding programs or genetic research. The regeneration process involves using **tissue culture techniques** to grow the haploid embryos into mature plants, which are then evaluated and selected based on desired traits or genetic markers.

Plant Regeneration

In most plant species, the process of regenerating **doubled haploid embryos** into full plants is achieved through **tissue culture** techniques. Tissue culture allows for the controlled development of haploid embryos under sterile, optimized laboratory conditions, ensuring that the plants grow into healthy, viable individuals.

» **Tissue Culture Techniques**: Tissue culture provides a controlled environment for the growth and development of **haploid embryos** that have undergone chromosome doubling. This method ensures that the plants are provided with the necessary nutrients, hormones, and environmental conditions (such as light and temperature) to promote growth.

- **Medium Composition**: The culture medium used for regenerating DH plants is carefully designed to include essential nutrients (e.g., macro- and micronutrients), vitamins, growth regulators (like **cytokinins** and **auxins**), and sometimes sugars to support embryo development. The composition of the medium varies depending on the plant species and the specific requirements of the tissue being cultured.
- **Sterile Conditions**: Tissue culture is carried out under sterile conditions to prevent contamination by bacteria or fungi, which could inhibit the growth of the plant embryos. The explants (plant tissues) are kept in closed, sterile containers or culture plates to ensure that the environment remains free of pathogens.
- **Hormonal Induction of Growth**: Specific plant growth regulators are added to the culture medium to stimulate cell division and differentiation, helping the doubled haploid embryos develop into full plants. **Cytokinins** promote shoot formation, while **auxins** encourage root development, ensuring the complete regeneration of the plant.

» **Transfer to Soil or Greenhouse**: Once the embryos have grown into small plants within the tissue culture environment, they are gradually transferred to more natural growing conditions, such as soil in a greenhouse or directly in the field. This step involves acclimatizing the plants to the external environment, allowing them to continue developing under standard growth conditions.

- **Acclimatization**: Acclimatizing the plants from the sterile tissue culture environment to natural growing conditions is an important step in the regeneration process. This often involves slowly adjusting the humidity, light intensity, and temperature to mimic natural conditions, allowing the plants to develop robust root systems and leaves.

» **Evaluation in Field or Greenhouse**: After regeneration, the DH plants are grown to maturity in the **field** or **greenhouse** to assess their growth, yield, and other phenotypic traits. This step is critical for identifying individuals that express the desired traits, such as **disease resistance**, **drought tolerance**, or **high yield**.

Selection of DH Plants

After the DH plants have been regenerated and transferred to field or greenhouse conditions, the next step is the **selection** of individuals that exhibit the desired traits. The complete **homozygosity** of DH plants allows breeders to make accurate selections based on either **phenotypic traits** or **molecular markers**, depending on the objectives of the breeding program.

» **Phenotypic Selection**: In some breeding programs, the DH plants are selected based on their **observable characteristics**, or phenotypes. Traits such as **plant height**, **flowering time**, **disease resistance**, and **yield** are evaluated in the field or greenhouse, and individuals that exhibit the best performance are selected for further breeding.

- **Uniformity**: Because DH plants are **completely homozygous**, they exhibit **genetic uniformity**, meaning that the plants are genetically identical at all loci. This uniformity ensures that phenotypic traits are consistent and reproducible in subsequent generations, making selection more straightforward.
- **Trait Assessment**: Breeders assess a range of agronomic traits during the selection process, such as **pest resistance**, **abiotic stress tolerance** (e.g., heat, drought), and **nutritional quality**. Since DH plants have no heterozygosity, the expression of these traits is stable, allowing for more reliable selection decisions.

» **Marker-Assisted Selection (MAS)**: In modern breeding programs, **marker-assisted selection (MAS)** is often used to complement phenotypic selection. **Molecular markers** linked to important traits (such as specific genes or QTLs) are used to identify individuals that carry the desired alleles. This approach is particularly useful for traits that are difficult to measure phenotypically or that are controlled by **multiple genes**.

- **Genetic Markers**: Molecular markers such as **SNPs** (Single Nucleotide Polymorphisms), **SSRs** (Simple Sequence Repeats), or **AFLPs** (Amplified Fragment Length Polymorphisms) are used to screen the DH population for specific alleles associated with desirable traits. MAS enables the rapid identification of plants carrying the target alleles, reducing the time and labor needed for phenotypic evaluation.
- **Efficiency**: The use of molecular markers significantly enhances the efficiency of the selection process by allowing breeders to **screen large populations** of DH plants quickly and accurately. By identifying plants that carry the desired genetic markers, breeders can focus their attention on individuals that have the greatest potential for success in the breeding program.

» **Trait Fixation and Breeding Stability**: The homozygosity of DH plants ensures that all selected traits are **fixed**, meaning they will be passed down to subsequent generations without segregation. This is particularly important in breeding programs aimed at developing new **pure lines** or hybrids, as it guarantees that the desired traits will be consistently expressed in offspring.

Regeneration and Selection: Critical for Genetic Studies and Breeding

The process of **regenerating doubled haploids** and **selecting individuals** based on phenotypic or molecular traits is a critical component of both **plant breeding** and **genetic research**. The ability to rapidly produce homozygous individuals in a single generation offers breeders and researchers the opportunity to study **trait inheritance**, develop **pure lines**, and accelerate the creation of new varieties.

» **Breeding Programs**: DH plants are frequently used in **hybrid breeding programs**, where the objective is to combine desirable traits from different parent lines into a single, stable variety. By selecting DH plants that carry the desired combination of traits, breeders can develop **elite lines** that serve as the parents of commercial hybrids.

» **Genetic Research**: In addition to their use in breeding, DH populations are invaluable for **QTL mapping** and **genetic studies**. Because DH plants are completely homozygous, they allow researchers to precisely study the relationship between **genotype** and **phenotype**, providing insights into the genetic architecture of complex traits.

Merits of Doubled Haploids (DHs)

1. **Rapid Homozygosity**

   - **Explanation**: The most significant advantage of DH populations is their ability to achieve **complete homozygosity** in just one generation. Traditional breeding methods require multiple generations of **selfing** to reach a homozygous state, which can take several years, depending on the species. In contrast, DH techniques allow for the rapid production of **genetically stable** lines in a single generation.
   - **Benefit**: This rapid production of homozygous lines makes DH populations highly efficient for **plant breeding**, **QTL mapping**, and **genetic studies**. It allows breeders to develop **pure lines** much faster, accelerating the breeding process and reducing the time to market for new varieties.
   - **Application**: DH populations are widely used in the development of **inbred lines** for crops such as maize, wheat, barley, and rice. These inbred lines can then be used as parent lines in **hybrid breeding** programs or for **genetic research**.

2. **Efficient for Mapping**

   - **Explanation**: DH populations are ideal for **QTL mapping** and other genetic studies because they are **completely homozygous** at all loci. This eliminates the complications of heterozygosity that are present in segregating populations, allowing for a more accurate assessment of **genotype-to-phenotype relationships**.
   - **Benefit**: The complete homozygosity of DHs ensures that any observed **phenotypic variation** is due to genetic differences between individuals rather than differences in heterozygous alleles. This precision is particularly important in **marker-assisted selection** (MAS), where breeders aim to identify markers linked to specific traits.
   - **Application**: DH populations are frequently used in **QTL mapping** studies to identify regions of the genome associated with important traits, such as disease resistance, yield, or abiotic stress tolerance.

3. **Reduced Time and Cost**
   - **Explanation:** The ability to produce homozygous lines in a single generation significantly reduces the **time and cost** associated with traditional breeding programs. Breeders can bypass several generations of selfing, which is required to produce inbred lines in conventional breeding.
   - **Benefit**: By reducing the time to homozygosity, DH populations allow for faster **development of new cultivars**, reducing both the **financial and temporal costs** of breeding programs. This is particularly advantageous for species with long generation times.
   - **Application**: DH technology has been extensively applied in the breeding of cereals, vegetables, and other crops, where time-to-market is critical for introducing new varieties with improved traits.

Demerits of Doubled Haploids (DHs)

1. **Limited Recombination**
   - **Explanation**: One limitation of DH populations is that they capture only **one recombination event per chromosome** because they are derived from **gametes** of an **F1 generation.** This limited recombination reduces the genetic variation within the DH population compared to other mapping populations like **recombinant inbred lines (RILs)**, which accumulate more recombination events over successive generations.
   - **Drawback**: The limited recombination makes it difficult to achieve high-resolution mapping for closely linked markers. In contrast, populations like RILs, which undergo multiple generations of recombination, offer greater precision for **fine-scale genetic mapping**.
   - **Limitation**: For example, when performing **QTL mapping**, DH populations may not provide sufficient recombination to break the linkage between closely associated markers, limiting their usefulness in **fine mapping**.
2. **Expensive Technology**
   - **Explanation**: The generation of DH populations requires **specialized techniques**, such as **anther culture**, **microspore culture**, or **ovule culture**, which can be both **technically demanding** and **costly**. These techniques require highly controlled environmental conditions and skilled personnel, making DH technology less accessible in resource-limited breeding programs or research institutions.
   - **Drawback**: The high cost of producing DH populations limits their widespread use, especially in small-scale breeding programs or in species where **DH production protocols** are not well established.
   - **Limitation**: In crops or species where **efficient DH production methods** have not been developed, or where resources are limited, researchers may rely on other population types, such as RILs or backcross populations, which do not require expensive tissue culture techniques.

Summary of Merits and Demerits

Merits	Description
Rapid Homozygosity	DH populations achieve complete homozygosity in one generation, allowing for faster production of stable lines.
Efficient for Mapping	The complete homozygosity of DHs enables more accurate genotype-to-phenotype mapping and reduces the effects of heterozygosity.
Reduced Time and Cost	Producing homozygous lines in a single generation significantly shortens the time and cost of breeding programs.

Demerits	Description
Limited Recombination	DH populations capture only one recombination event per chromosome, limiting their precision for fine-scale mapping.
Expensive Technology	The specialized techniques required for DH production are costly and technically demanding, limiting accessibility in certain contexts.

4. Backcross (BC) Populations

Backcross (BC) populations are widely used in both **plant and animal breeding**, particularly when breeders aim to introduce a **specific trait** from one parent into the genetic background of another, typically more desirable or commercially viable, parent. The creation of a backcross population involves crossing an **F1 individual** (resulting from the initial cross between two genetically distinct parental lines) with one of its parents, usually the **recurrent parent**. This process introduces recombination between the two parental genomes, enabling the introgression of specific **genes** or **genomic regions** from the **donor parent** into the recurrent parent's genetic background.

Repeated backcrossing over several generations is often employed to maintain most of the recurrent parent's genome while introducing and fixing the desired traits from the donor parent. **Backcrossing** is especially useful for **marker-assisted breeding, trait introgression**, and **genetic mapping**.

Mechanism of Backcrossing

The mechanism of **backcross population formation** relies on repeated crosses between the **F1 hybrid** and one of the original parents (usually the recurrent parent), followed by **selection** for the desired trait. Below is a detailed overview of the key steps involved in backcrossing:

Step 1: Initial Cross to Produce F1 Hybrid

The process of **backcrossing** begins with the creation of an **F1 hybrid**, which is a cross between two genetically distinct parent lines. The F1 hybrid serves as the intermediate progeny in the backcrossing process, combining genetic material from both the **donor parent** and the **recurrent parent**. Each parent brings unique traits to the cross, and the F1 generation serves as the starting point for further backcrossing to the recurrent parent in order to selectively introgress specific desirable traits from the donor parent.

Donor Parent

The **donor parent** is chosen for its possession of a **specific trait** that breeders wish to introduce into the recurrent parent's genetic background. This trait could be a **quantitative trait locus (QTL)** or a specific **gene** that imparts a desirable attribute such as **disease resistance, drought tolerance, pest resistance,** or **nutritional improvement.**

» **Trait of Interest:** The trait contributed by the donor parent is often a trait that is either **absent** or **underrepresented** in the recurrent parent. For example, the donor parent might carry resistance genes against pathogens like **rust, blight,** or **mildew,** or it might harbor genes that enhance **stress tolerance,** such as resistance to heat, salinity, or drought.

» **Exotic Germplasm:** In some cases, the donor parent comes from **wild relatives** or **exotic germplasm,** which may carry **unique genetic diversity** or alleles that are not present in modern cultivated varieties. These traits are valuable for breeding programs aiming to broaden the genetic base of commercial cultivars or to introduce **novel traits** that enhance adaptability to changing environmental conditions.

» **Application:** In modern breeding, the donor parent is often selected based on **molecular marker** data that identify specific regions of the genome associated with the desired trait. This enables precision in the selection process, ensuring that the trait is effectively introgressed into the recurrent parent during the backcrossing process.

Recurrent Parent

The **recurrent parent** is typically a **commercially valuable line,** often possessing **high yield, desirable agronomic traits,** or **marketable quality** characteristics. However, this parent may lack certain specific traits that are necessary for the breeding program, such as resistance to a particular **disease, insect pests,** or **environmental stresses.**

» **Elite Line:** The recurrent parent is usually an **elite cultivar** or breeding line that is already well-adapted to the target environment or market but needs improvement in specific areas. For instance, the recurrent parent may be a high-yielding rice variety that lacks resistance to **blast disease,** or a drought-tolerant maize variety that is susceptible to **corn borers.**

» **Desirable Agronomic Traits:** The recurrent parent is selected for its **overall genetic background,** which is advantageous for agronomic performance. These traits may include **high biomass, early flowering, good seed set,** or **superior grain quality.** However, to make the variety suitable for specific environments or markets, it may require the addition of certain traits from the donor parent.

» **Application:** The goal in using the recurrent parent is to retain the **majority of its genetic background** while selectively introducing the trait(s) from the donor parent through the backcrossing process. After several generations of backcrossing, the genome of the recurrent parent will be nearly restored, with the addition of the desired trait from the donor parent.

F1 Hybrid

The cross between the **donor parent** and the **recurrent parent** produces the **F1 hybrid**, which contains **50%** of its genetic material from each parent. The F1 hybrid is typically **heterozygous** at many loci, meaning that it carries one allele from the donor parent and one allele from the recurrent parent at each gene.

- **Genetic Recombination**: During the formation of the F1 hybrid, **recombination** occurs between the chromosomes of the donor and recurrent parents. This recombination shuffles the genetic material, resulting in offspring that have a mix of alleles from both parents. Recombination is essential for the **exchange of genetic information** and allows breeders to introgress specific regions of the donor parent's genome into the recurrent parent over successive backcross generations.
- **Heterozygosity**: The F1 hybrid is heterozygous at many loci because it inherits one allele from each parent. This heterozygosity is advantageous for introducing new alleles from the donor parent into the recurrent parent's genetic background. However, as backcrossing continues, the proportion of the donor genome decreases, and the recurrent parent's genome is restored with the targeted introgression of the donor trait.
- **Trait Segregation**: In the F1 generation, the **trait of interest** from the donor parent is often still present, but it may segregate in later backcross generations. This necessitates careful **phenotypic or genotypic selection** to ensure that the desired donor trait is retained while eliminating unwanted genetic material from the donor genome.

Importance of F1 Hybrid in Backcrossing

The F1 hybrid plays a critical role in the backcrossing process because it serves as the vehicle for introducing the desired donor trait into the recurrent parent's genetic background. By crossing the F1 hybrid back to the recurrent parent, breeders can progressively increase the proportion of the recurrent parent's genome while retaining the targeted genetic region from the donor parent.

- **First Backcross (BC1 Generation)**: In the first backcross generation (BC1), the F1 hybrid is crossed back to the recurrent parent. The resulting BC1 population will contain approximately 75% of its genetic material from the recurrent parent and 25% from the donor parent. Selection for the desired trait begins at this stage to ensure that the trait of interest is maintained in subsequent generations.
- **Subsequent Backcrosses**: Over successive backcross generations, the genome of the recurrent parent is progressively restored, with the donor genome being reduced to the **specific region** carrying the desired trait. After several generations, the backcross population will be nearly identical to the recurrent parent, except for the introgressed region from the donor parent.

Step 2: First Backcross (BC1 Generation)

The second step in the **backcrossing process** involves crossing the **F1 hybrid** (created from the initial cross between the donor and recurrent parents) back to the **recurrent parent**. This stage is critical for starting the process of **introgressing the desired trait** from the **donor parent** into the **genetic background** of the recurrent parent. The goal of the first backcross is to retain the **desired trait** contributed by the donor parent while gradually recovering the genetic background of the recurrent parent. This is the first of several rounds of backcrossing that will be conducted to achieve the final objective.

Backcrossing to Recurrent Parent

In the **first backcross (BC1)**, the **F1 hybrid** is crossed with the **recurrent parent**. The recurrent parent is typically chosen for its **desirable agronomic traits**, such as **high yield**, **stress tolerance**, or **market value**, but lacks the specific **trait of interest** found in the donor parent. By crossing the F1 hybrid with the recurrent parent, the goal is to recover a larger proportion of the recurrent parent's genome while still retaining the donor parent's **desired gene or genomic region.**

» **Genetic Makeup of BC1 Generation**: After the first backcross, the resulting **BC1 generation** will contain approximately **75% of its genetic material** from the **recurrent parent** and **25% from the donor parent.** This distribution is because the F1 hybrid contains equal genetic contributions (50%) from both parents, and backcrossing it to the recurrent parent increases the proportion of the recurrent parent's genome. However, recombination during meiosis introduces some **genetic variation** in the BC1 generation.

» **Recombination Events**: During gamete formation in the F1 hybrid, **recombination** occurs between the chromosomes inherited from the donor and recurrent parents. This recombination helps **shuffle the genetic material**, resulting in the transfer of **alleles** from the donor parent to the recurrent parent's genetic background. While the overall goal is to increase the proportion of the recurrent parent's genome, **recombination events** also help introduce genetic material from the donor parent, including the **desired trait.**

» **Importance of Backcrossing**: The first backcross is essential for **retaining the desired trait** while recovering the recurrent parent's genetic background. Over successive generations of backcrossing, breeders aim to progressively increase the genetic contribution of the recurrent parent while keeping the **target region** from the donor parent.

Selection for Desired Trait

After the BC1 generation is produced, the next crucial step is **selection** to ensure that the desired donor trait (e.g., **disease resistance, pest tolerance, quality attributes**) is retained in the offspring. Selection can be based on **phenotypic traits** or **molecular markers**, depending on the breeding goals and the availability of marker data.

» **Phenotypic Selection**: In some cases, breeders use **phenotypic selection** to evaluate the **visible traits** expressed by the BC1 individuals. For example, if the donor parent contributed a trait such as **resistance to a fungal disease**, breeders would screen the BC1 population by exposing the plants to the pathogen and selecting only those individuals that show **resistance**.

   - **Efficiency of Phenotypic Selection**: While phenotypic selection can be effective for traits that are easy to observe (e.g., disease resistance, plant height, yield), it may not be as efficient for **complex traits** or traits influenced by **environmental factors**. Moreover, **quantitative traits** that are controlled by multiple genes may not be reliably selected based solely on phenotype in early generations.

» **Marker-Assisted Selection (MAS)**: A more precise approach is to use **molecular markers** to select BC1 individuals that carry the desired **donor allele(s)**. **Marker-assisted selection (MAS)** is particularly useful when the desired trait is controlled by a specific gene or **quantitative trait locus (QTL)** that can be tracked using **genetic markers**. MAS allows breeders to identify individuals that carry the donor gene even before the trait is expressed phenotypically.

- **Use of Molecular Markers**: Molecular markers, such as **SNPs** (Single Nucleotide Polymorphisms), **SSRs** (Simple Sequence Repeats), or **AFLPs** (Amplified Fragment Length Polymorphisms), can be used to track the introgression of the **desired gene** from the donor parent into the recurrent parent's background. By genotyping the BC1 population, breeders can select individuals that carry the donor gene without the need for extensive phenotypic testing.
- **Efficiency of MAS**: **Marker-assisted selection** is highly efficient for traits that are difficult to observe phenotypically or for **early-stage selection** in the backcrossing process. MAS reduces the time and labor required for selection, especially when dealing with traits that are only expressed under specific environmental conditions or in later developmental stages.

Advantage of BC1 Generation

The BC1 generation offers several advantages in the **introgression process**, as it combines a significant proportion of the recurrent parent's genome with a smaller but important contribution from the donor parent. The **recurrent parent's genome** is progressively restored while the **desired trait** from the donor parent is retained through careful selection. Over successive generations, this process becomes more refined, allowing breeders to achieve **stable lines** that possess both the donor trait and the recurrent parent's overall genetic makeup.

» **Recurrent Parent Recovery**: The key advantage of the BC1 generation is that it brings breeders closer to recovering the recurrent parent's **desired agronomic background**, while maintaining the specific **genomic region** or **allele** from the donor parent. With each successive backcross, the **recurrent parent's genome** will be progressively restored, and the donor contribution will be reduced except for the target region carrying the desired trait.

» **Selection Refinement**: Selection in the BC1 generation allows breeders to refine their choices, ensuring that only those individuals that carry the **desired trait** are chosen for further backcrossing. This helps to **narrow the genetic focus**, ensuring that only the targeted alleles are introgressed while eliminating unwanted genetic material from the donor parent.

Challenges of BC1 Generation

While the **BC1 generation** represents an important step toward trait introgression, it also presents some challenges, particularly about **selection efficiency** and **recombination**.

» **Recombination and Linkage Drag**: During the formation of the BC1 population, **recombination** between the donor and recurrent parent genomes can lead to the **linkage drag** of undesirable alleles from the donor parent. **Linkage drag** refers to the unintended introgression of **linked alleles** that are physically close to the desired gene on the same chromosome. This can result in the retention of unwanted traits or genetic material from the donor parent, which may affect the overall performance of the recurrent parent's background.

- **Mitigating Linkage Drag:** To minimize linkage drag, breeders can use **fine mapping** techniques to **narrow down the target region** of the donor genome and select for individuals where recombination events have successfully broken the linkage between the desired trait and the unwanted alleles. This can be achieved through **genomic selection** or **marker-assisted backcrossing.**

» **Selection Intensity:** Another challenge in the BC1 generation is the need for **intensive selection** to ensure that only individuals carrying the **desired trait** are chosen for further backcrossing. If the desired trait is difficult to select for in early generations (e.g., if it is a **quantitative trait** or environmentally dependent), additional rounds of backcrossing and selection may be required to fix the trait.

Step 3: Repeated Backcrossing (BC2, BC3, etc.)

Once the **BC1 generation** has been created and selected for the desired donor trait, the next phase in the **backcrossing process** involves **repeated backcrossing** to the recurrent parent over multiple generations. This process allows breeders to progressively increase the proportion of the recurrent parent's genome in the population while retaining the **specific gene** or **genomic region** from the donor parent. Each successive backcross generation brings the genetic makeup of the population closer to the recurrent parent, with a smaller proportion of the donor parent's genome being retained.

Backcross to Recurrent Parent Again

Following the selection of individuals from the **BC1 generation** that exhibit or carry the **desired donor trait**, these selected individuals are backcrossed again to the **recurrent parent** to create the **BC2 generation.** The purpose of this backcross is to further reduce the genetic contribution of the donor parent while retaining the **targeted trait.**

» **Genetic Makeup of BC2 Generation:** The **BC2 generation** will contain approximately **87.5%** of its genetic material from the **recurrent parent** and only **12.5%** from the donor parent. This occurs because each backcross with the recurrent parent increases the proportion of its genome in the offspring. With each subsequent backcross, the genome of the recurrent parent is progressively restored.

» **Recombination and Selection:** During the formation of the BC2 generation, **recombination** occurs between the donor and recurrent genomes, further shuffling the genetic material. The goal is to select BC2 individuals that still carry the **desired trait** from the donor parent while further reducing unwanted genetic material from the donor parent.

» **Selection Process:** As in the BC1 generation, breeders use **phenotypic selection** or **marker-assisted selection (MAS)** to identify BC2 individuals that carry the desired trait. Selection ensures that only individuals with the **introgressed donor trait** are retained for further backcrossing to the recurrent parent.

Repeated Cycles of Backcrossing

The cycle of **backcrossing** and **selection** is repeated over several generations (e.g., BC3, BC4, etc.) to progressively increase the proportion of the recurrent parent's genome and reduce the donor parent's contribution. With each backcross, the offspring become genetically more similar to the recurrent parent, except for the specific **genomic region** or **gene** that was introgressed from the donor parent.

» **BC3 Generation**: In the **BC3 generation**, approximately **93.75%** of the genome is derived from the recurrent parent, while only **6.25%** comes from the donor parent. This is achieved by backcrossing selected BC2 individuals with the recurrent parent again and applying **selection** to ensure the donor trait is retained.

» **BC4 Generation**: In the **BC4 generation**, more than **96%** of the genome is from the recurrent parent. At this stage, the donor parent's contribution has been significantly reduced, except for the targeted genomic region.

» **BC5 Generation and Beyond**: By the **BC5 generation**, more than **98%** of the genome is derived from the recurrent parent, with only a small percentage of the donor parent's genome retained. At this point, the genome closely resembles that of the recurrent parent, but the **desired trait** or **gene** from the donor parent has been successfully introgressed.

Fixation of Desired Trait

Throughout the process of repeated backcrossing, **selection** is consistently applied to ensure that the **desired trait** from the donor parent is retained. By the final backcross generation, the targeted gene or genomic region from the donor parent will be **fixed** in a **homozygous state**, meaning that the trait will be stably expressed in all offspring.

» **Homozygosity**: As backcrossing continues, the desired donor allele is progressively fixed in the **homozygous** state. This means that all individuals in the final backcross generation will have two identical copies of the **desired gene** or **genomic region**, ensuring that the trait is expressed consistently in subsequent generations.

» **Molecular Markers for Selection**: In cases where **marker-assisted selection (MAS)** is used, breeders can track the **introgressed region** from the donor parent using molecular markers. This ensures that the desired gene is retained throughout the backcrossing process, even as the rest of the genome becomes increasingly similar to the recurrent parent.

» **Minimizing Linkage Drag**: During repeated backcrossing, **recombination events** help to break **linkage drag**, which occurs when unwanted genetic material from the donor parent is introgressed along with the desired gene. By selecting for individuals where recombination has separated the desired trait from unwanted alleles, breeders can reduce the negative effects of linkage drag.

» **Final Product**: After several generations of backcrossing and selection, the final backcross generation will closely resemble the **recurrent parent** in terms of overall genetic composition, but it will carry the **desired gene** or **trait** from the donor parent. This allows breeders to develop new lines that possess the **desired trait** while maintaining the **elite agronomic characteristics** of the recurrent parent.

Advantages of Repeated Backcrossing

1. **Restoration of Recurrent Parent's Genome**
   - Repeated backcrossing allows breeders to progressively restore the **genetic background** of the recurrent parent while introgressing a **specific trait** from the donor parent. This ensures that the final product retains the overall **agronomic quality** of the recurrent parent while gaining the beneficial trait from the donor.
2. **High Precision in Trait Introgression**
   - Through **phenotypic selection** or **MAS**, breeders can achieve **high precision** in retaining the donor trait while eliminating unwanted genetic material from the donor parent. This precision allows breeders to create new lines that are genetically similar to the recurrent parent but possess the desired trait.
3. **Reduced Linkage Drag**
   - With each round of backcrossing, **recombination** helps to reduce **linkage drag** by breaking the physical linkage between the desired gene and nearby unwanted alleles. This improves the performance of the final product by minimizing the negative effects associated with linked donor alleles.
4. **Stable Expression of Desired Trait**
   - The fixation of the donor gene in a **homozygous state** ensures that the **desired trait** will be stably expressed in all subsequent generations. This stability is particularly valuable in **commercial breeding programs**, where consistency and reliability are key.

Challenges of Repeated Backcrossing

1. **Time-Consuming Process**
   - Repeated backcrossing can be a **time-intensive** process, especially in crops or species with long life cycles. Each backcross generation requires a full cycle of growth, pollination, and selection, which can take years to complete, especially if multiple generations are needed to achieve the desired level of recurrent parent recovery.
2. **Risk of Linkage Drag**
   - Despite efforts to minimize it, **linkage drag** can still occur, especially if recombination events are limited in the genomic region surrounding the desired trait. This may result in the introgression of unwanted donor alleles that negatively affect the performance of the recurrent parent.
3. **Need for Effective Selection Tools**
   - The success of the backcrossing process depends on the availability of effective **selection tools**, such as molecular markers for **marker-assisted selection (MAS)**. Without these tools, breeders may struggle to efficiently track the introgressed donor gene and ensure that the trait is retained throughout the backcrossing process.

Step 4: Selfing or Sibling Mating to Create Homozygous Lines

After completing the process of **repeated backcrossing** (BC2, BC3, etc.) and ensuring that the **desired trait** from the **donor parent** has been successfully introgressed into the **recurrent parent's genetic background**, the next critical step is to create **homozygous lines**. This step is necessary to ensure that the **desired donor allele** or **genomic region** is **fixed** in a **homozygous state**, ensuring that the trait will be stably expressed in all subsequent generations. Achieving **homozygosity** for the donor trait is done either through **selfing** or **sibling mating** among the selected **backcross (BC) plants**.

Selfing to Create Homozygous Lines

One of the most effective methods to create **homozygous lines** is through **selfing**. In this process, the **BC plants** that carry the desired donor allele are allowed to **self-pollinate** to produce offspring that will be homozygous for the trait of interest.

» **Self-Pollination (Selfing)**: **Selfing** involves using the **pollen** from a plant to fertilize its own **ovules**, resulting in progeny that inherit both sets of chromosomes from the same parent. In plants, this is a commonly used method to fix traits and achieve **homozygosity**. Through self-pollination, the **heterozygous** state for the donor allele (if present) in the BC population can be transformed into a **homozygous state**.

- **Mechanism of Selfing**: Selfing leads to **segregation** of alleles during **meiosis**, and in the resulting progeny, **homozygous individuals** (with two identical copies of the desired donor allele) will be produced along with heterozygous and possibly recessive individuals. By selfing plants over successive generations, the desired allele can be **fixed** in the population.
- **Fixing the Donor Allele**: In the case of the backcrossing process, selfing allows breeders to **fix** the donor allele or genomic region in a **homozygous state**, ensuring that the trait is **uniformly expressed** across all offspring. This means that once homozygosity is achieved, the progeny will consistently express the desired trait, regardless of environmental or genetic factors.

» **Advantages of Selfing**: Selfing is a reliable way to achieve **homozygosity** without introducing additional genetic variation. It is commonly used in crops like **rice**, **wheat**, and **maize**, where self-pollination is feasible and helps to stabilize the trait of interest in the breeding population. Additionally, selfing speeds up the process of **trait fixation**, making it easier to establish pure lines.

Sibling Mating to Achieve Homozygosity

In cases where selfing is not feasible or desirable, breeders can use **sibling mating** (crossing closely related individuals from the same BC population) to achieve **homozygosity**. This process involves selecting **BC plants** that carry the **desired trait** and **mating them with their siblings** to produce progeny with a higher likelihood of being homozygous for the donor allele.

» **Sibling Crosses: Sibling mating** involves crossing individuals that are closely related, typically from the same **BC generation**. For instance, if both plants in a sibling pair carry one copy of the **donor allele**, crossing them can increase the chances of producing **homozygous**

offspring. Like selfing, sibling mating promotes **allelic segregation**, allowing breeders to screen for individuals that are **homozygous** for the desired trait.

» **Application of Sibling Mating**: Sibling mating is especially useful in cases where self-pollination is not common or where genetic diversity within the BC population needs to be preserved for a few more generations. For instance, in **outcrossing species**, where selfing is not a common reproductive strategy, **sibling mating** provides an alternative for achieving **homozygosity**.

» **Advantages**: While sibling mating may not achieve homozygosity as quickly as selfing, it helps to preserve some **genetic variation** within the population while still allowing for the **fixation of desired traits**. This method can also help mitigate the effects of **inbreeding depression** in species sensitive to self-pollination.

Selection for Homozygous Lines

After the process of selfing or sibling mating, the resulting progeny are **screened** to identify individuals that are **homozygous** for the desired donor allele. The goal is to ensure that the trait of interest has been **fixed** in the **genetic background** of the recurrent parent. Selection can be based on **phenotypic traits** or **molecular markers**.

» **Phenotypic Selection**: **Phenotypic selection** involves screening the progeny for the **expression of the desired trait**. For example, if the trait is **disease resistance**, the progeny may be exposed to the pathogen, and only those plants that exhibit resistance are selected. This ensures that only individuals that express the donor trait are carried forward in the breeding program.

- **Challenges of Phenotypic Selection**: While phenotypic selection can be effective, it may be difficult to select for quantitative traits or traits that are influenced by environmen**tal factors**. Additionally, phenotypic selection may not always provide an accurate representation of the **underlying genetic state**, especially in early generations where heterozygosity is still present.

» **Molecular Markers for Selection**: The use of **molecular markers** provides a more precise approach for identifying **homozygous individuals. Marker-assisted selection (MAS)** allows breeders to genotype the progeny and directly identify those that carry **two copies of the donor allele** for the desired trait.

- **SNPs and SSRs**: Molecular markers such as **SNPs (Single Nucleotide Polymorphisms)** or **SSRs (Simple Sequence Repeats)** can be used to track the **introgressed region** from the donor parent. By screening the progeny using these markers, breeders can determine which individuals are **homozygous** for the desired trait and ensure that the **gene** or **genomic region** has been fixed in the population.

- **Efficiency of MAS**: The use of molecular markers significantly improves the **efficiency of selection**, allowing breeders to identify **homozygous lines** without the need for extensive phenotypic testing. This is especially useful for traits that are difficult to observe or that are controlled by multiple genes.

Final Product of Backcrossing Process

The end result of the backcrossing process is the development of **homozygous lines** that carry the **desired donor trait** within the genetic background of the recurrent parent. These lines are **genetically stable**, meaning that the **trait of interest** will be consistently expressed in all subsequent generations.

» **Fixed Trait in Homozygous Lines**: Once homozygosity is achieved, the trait will be **fixed** in the population, meaning that all offspring will express the trait in a **uniform** manner. This is particularly important in **commercial breeding programs**, where consistent performance is critical for the development of new cultivars.

» **Genetically Uniform Lines**: The resulting homozygous lines are **genetically uniform**, meaning that they will express the desired trait in a stable and predictable manner. These lines can be used as the foundation for the development of **new cultivars, hybrids**, or **pure lines** that are suitable for **commercial production**.

Merits of Backcross Populations

1. **Focus on Specific Traits**
    - **Explanation**: One of the main advantages of backcross populations is their ability to **focus on specific traits** from the donor parent, such as **disease resistance, pest tolerance**, or **quality improvements**, while maintaining the desirable genetic background of the recurrent parent.
    - **Benefit**: This approach is especially useful in **marker-assisted breeding**, where breeders aim to introduce a single gene or genomic region from the donor parent into a high-yielding or commercially valuable variety. Backcrossing enables breeders to selectively **introgress desired traits** while keeping the majority of the recurrent parent's genome intact.
    - **Application**: Backcrossing is commonly used in crops like **maize, rice, wheat**, and **soybean**, where breeders aim to transfer specific traits (e.g., herbicide resistance, drought tolerance) into elite cultivars that are already adapted to local growing conditions.
2. **Useful for Introgression**
    - **Explanation**: Backcross populations are particularly effective for **introgressing desired genes** or genomic regions from **wild relatives** or **exotic germplasm** into elite cultivars. This technique allows breeders to access valuable traits that are not present in cultivated varieties but are found in **wild species** or **landraces**.
    - **Benefit**: Backcrossing is often used in **pre-breeding programs**, where traits such as **disease resistance, environmental tolerance**, or **nutritional improvements** are introgressed from **wild relatives** into commercial lines. The use of molecular markers helps to track the **introgressed region** and ensures that only the desired gene is retained.
    - **Application**: This method is frequently used in crops like **tomatoes,** where genes for **disease resistance** are transferred from wild relatives into commercial varieties to improve their resilience against pathogens.

Demerits of Backcross Populations

1. **Limited Genetic Diversity**
   - **Explanation:** One of the main limitations of backcross populations is the **loss of genetic diversity** over successive generations of backcrossing. As the population becomes more similar to the recurrent parent, the genetic diversity contributed by the donor parent is gradually lost.
   - **Drawback:** This reduced genetic diversity limits the **potential for discovering new traits** or studying the effects of complex genetic interactions. In populations where **genetic diversity** is important for the expression of certain traits, backcross populations may not be the best choice.
   - **Limitation:** For example, in studies of **complex traits** such as yield or abiotic stress tolerance, which are influenced by multiple genes, backcross populations may not capture the full range of genetic variation required to study these traits effectively.
2. **Long Development Time**
   - **Explanation:** The process of backcrossing can be **time-consuming**, especially when multiple generations of backcrossing are required to fully introgress the desired trait while maintaining the recurrent parent's genetic background.
   - **Drawback:** Each generation of backcrossing takes time, particularly in crops or species with long life cycles. If the desired trait is difficult to select for, or if molecular markers are not available, the process can become even more **labor-intensive** and **time-intensive**.
   - **Limitation:** For example, in crops with long generation times, such as **fruit trees** or **perennial plants**, backcrossing may take many years to complete. This limits the speed at which breeders can develop new cultivars, especially when compared to techniques like **doubled haploids**, which produce homozygous lines in a single generation.

Summary of Merits and Demerits

Mapping Population	Merits	Demerits
Recombinant Inbred Lines (RILs)	- High mapping resolution - Stable, long-term resource - Effective for fine-scale mapping	- Time-consuming to develop - Limited genetic diversity
Near-Isogenic Lines (NILs)	- Precise mapping of specific traits - Minimal genetic noise - Useful for dissecting QTL effects	- Time and labor-intensive - Limited genome coverage
Doubled Haploids (DHs)	- Rapid production of homozygous lines - Ideal for QTL mapping - Reduced time and cost for breeding	- Limited recombination - Expensive and technically demanding
Backcross (BC)	- Focused introgression of specific traits - Useful in marker-assisted breeding - Efficient for trait transfer	- Limited genetic diversity - Long development time

B. Marker-Assisted Selection (MAS)

Marker-assisted selection (MAS) is a powerful tool in modern plant and animal breeding programs that integrates **molecular markers** with traditional breeding techniques. The primary goal of MAS is to enhance the efficiency and precision of the selection process by allowing breeders to track specific **genes** or **quantitative trait loci (QTLs)** associated with desirable traits through the use of genetic markers. Instead of relying solely on **phenotypic traits**, which can be influenced by environmental factors and are often complex to measure, MAS provides a direct way to assess the **genetic makeup** of breeding individuals.

Introduction to Marker-Assisted Selection

In traditional breeding, selection is largely based on **observable characteristics** or **phenotypes**, such as **yield**, **disease resistance**, **fruit quality**, or **flowering time**. However, many traits of interest are controlled by multiple genes and are influenced by environmental factors, making phenotypic selection both **time-consuming** and **less precise**. This is particularly true for **quantitative traits**, such as **yield** or **drought tolerance**, where the trait is governed by the cumulative effect of many small gene variants (QTLs).

MAS addresses these challenges by using **DNA markers** to identify and select individuals carrying specific **genetic loci** associated with the trait of interest. The selection process becomes more **efficient** and **targeted** because breeders can directly evaluate the genetic potential of the breeding material based on the presence or absence of the desired **genomic region** or **alleles**.

How Marker-Assisted Selection Works

The success of **marker-assisted selection** relies on identifying **molecular markers** that are closely associated with the **genes** or **QTLs** controlling the trait of interest. These markers can then be used to screen the breeding population for individuals carrying the desired **allele(s)**, even before the trait is expressed phenotypically.

Here is a step-by-step outline of how MAS works in a typical breeding program:

1. Identification of Trait-Associated Markers

The **identification of molecular markers** that are closely linked to **genes** or **quantitative trait loci (QTLs)** controlling a trait of interest is the critical first step in **marker-assisted selection (MAS)**. The goal is to find **genetic markers** that are inherited along with the desired trait, which allows breeders to efficiently select individuals carrying the trait without needing to observe the trait phenotypically. This step is essential for ensuring the success of MAS because it enables breeders to work with **genetic information** rather than solely relying on the **phenotype**, which can be affected by environmental conditions and complex gene interactions.

Molecular Markers in MAS

The process of identifying **trait-associated markers** involves various techniques, including **QTL mapping**, **genome-wide association studies (GWAS)**, and **linkage analysis**. The choice of method depends on the complexity of the trait, the genetic diversity within the breeding population, and the availability of genomic resources for the species being studied.

- **QTL Mapping: Quantitative trait loci (QTL) mapping** is one of the most widely used techniques to identify genomic regions associated with **complex traits.** QTL mapping requires a segregating population (such as F2 or recombinant inbred lines) and the use of **molecular markers** to link variations in the trait with specific genomic regions. By analyzing the **co-segregation** of markers and phenotypic traits in the population, breeders can locate **QTLs** that contribute to the variation in the trait of interest.
- **Genome-Wide Association Studies (GWAS): GWAS** is another approach used to identify markers linked to traits. It involves scanning the entire genome for associations between genetic variants (typically **SNPs**) and specific traits in a large, diverse population. **GWAS** is particularly useful for identifying markers linked to **polygenic traits**, where many small genetic contributions influence the phenotype.
- **Linkage Analysis: Linkage analysis** is used to establish a relationship between genetic markers and the physical location of genes on chromosomes. By analyzing the frequency of **recombination events** during **meiosis,** breeders can estimate the distance between markers and traits, leading to the identification of **markers** closely associated with the desired trait.

Types of Molecular Markers Used in MAS

Several types of **molecular markers** are commonly used in MAS, each offering distinct advantages for different applications. The markers are used to **track specific regions** of the genome that carry **genes** or **QTLs** associated with the target trait.

- **Single Nucleotide Polymorphisms (SNPs)**: SNPs are **single base-pair differences** in the DNA sequence between individuals. These are the most abundant markers in the genome and provide **dense coverage** across the entire genome, making them ideal for **high-resolution genetic mapping** and **GWAS**. SNP markers are highly accurate and amenable to **high-throughput genotyping** technologies, making them widely used in **crop** and **livestock breeding**.
    - **Application in MAS**: SNPs are particularly valuable for identifying markers linked to **disease resistance, yield improvement,** or **abiotic stress tolerance** in plants. For instance, a breeder may identify an SNP marker tightly linked to a **disease-resistance gene** that confers resistance to a fungal pathogen. This marker can then be used to screen the breeding population, ensuring that individuals carrying the resistance allele are selected.
- **Simple Sequence Repeats (SSRs)**: Also known as **microsatellites**, SSRs consist of short, repeated sequences of DNA (2–6 base pairs). These markers are highly **polymorphic** due to variations in the number of repeat units between individuals, making them useful for detecting **genetic diversity** and **population structure**.
    - **Application in MAS**: SSR markers are widely used in MAS for traits such as **fruit quality, flowering time**, and **resistance to pests or diseases**. For example, a breeder working on grapevine improvement might use SSR markers linked to **powdery mildew resistance** to select resistant individuals for further breeding.

» **Restriction Fragment Length Polymorphisms (RFLPs): RFLPs** are based on differences in DNA fragment lengths generated by **restriction enzyme digestion.** This early type of molecular marker is reliable and **co-dominant**, meaning it can distinguish between **homozygous** and **heterozygous** individuals, but it is more labor-intensive and less commonly used today due to advances in more efficient marker systems.

- **Application in MAS**: In species where **RFLP markers** are still in use, they are valuable for tracking specific genes, such as those conferring **disease resistance** or **crop quality traits**. However, due to their labor-intensive nature, RFLPs have largely been replaced by more modern techniques like SNP genotyping.

» **Amplified Fragment Length Polymorphisms (AFLPs): AFLPs** are created by amplifying random DNA fragments after digestion with restriction enzymes. This technique provides a large number of polymorphisms in a single assay, making it useful for **genetic mapping** and **QTL identification.**

- **Application in MAS**: AFLPs are used for **marker-assisted breeding** in species where genomic resources are limited. For example, in some **wild relatives of crops** that lack detailed genomic information, AFLP markers can help breeders identify **genomic regions** associated with traits such as **drought tolerance** or **disease resistance.**

Example of Marker Identification in MAS

In a typical **crop breeding program**, identifying markers linked to traits such as **disease resistance** or **yield enhancement** is critical. For instance, suppose a breeder is working on improving disease resistance in **wheat.** The breeder would begin by identifying a **SNP marker** that is tightly linked to a **disease resistance gene**, such as the **Lr34 gene**, which provides resistance to **leaf rust**, a significant fungal disease in wheat.

1. **Initial Cross**: The breeder would first cross a **resistant donor line** (carrying the Lr34 gene) with a **susceptible elite variety** that has high yield potential but lacks the disease resistance gene.
2. **QTL Mapping**: The progeny from this cross would be evaluated for **disease resistance** and **genotyped** using SNP markers across the genome. Through **QTL mapping**, the breeder identifies a **SNP marker** on chromosome 7D that is tightly linked to the Lr34 gene.
3. **Marker Validation**: After identifying the marker, the breeder would **validate** its accuracy by testing it in multiple populations and environments to ensure that the SNP consistently predicts resistance to leaf rust.
4. **MAS Implementation**: Once the marker is validated, the breeder would use **marker-assisted selection** to screen new breeding populations. Individuals carrying the **favorable SNP allele** linked to Lr34 would be selected for further breeding, while those lacking the allele would be discarded.

By using **SNP markers** in this way, the breeder can select for **disease resistance** early in the breeding cycle, without needing to wait for the plants to be exposed to the pathogen in field trials. This saves time and resources while increasing the precision of the selection process.

2. Genotyping of Breeding Material

Once the **molecular markers** associated with the desired trait have been identified, the next critical step in **marker-assisted selection (MAS)** is to **genotype** the breeding population. This process involves determining the **genetic composition** of each individual in the population to assess which individuals carry the **favorable allele(s)** linked to the target trait. By genotyping the breeding material, breeders can make **informed selection decisions** early in the breeding process, significantly improving the **efficiency** and **precision** of their programs.

Genotyping Process

The process of **genotyping** involves several steps, beginning with the extraction of **DNA** from the individuals in the breeding population and ending with the analysis of their **genetic profiles**. The genotyping data allows breeders to distinguish between individuals that carry the desired **marker allele** and those that do not, enabling targeted selection for further breeding.

DNA Extraction

The first step in genotyping is to extract **DNA** from the breeding individuals. This step is crucial because the quality and quantity of DNA extracted directly affect the accuracy of the genotyping process. DNA extraction is typically performed on **leaf tissue** in plants or **blood samples** in animals, but the specific source of DNA can vary depending on the organism and the nature of the breeding program.

» **Plant DNA Extraction**: In plant breeding, DNA is often extracted from **young leaves** using chemical reagents that break down the cell walls, membranes, and proteins, leaving behind purified **genomic DNA**. The extracted DNA is then quantified to ensure that it is of sufficient quality for downstream genotyping.

» **Animal DNA Extraction**: In animal breeding, DNA is typically extracted from **blood, hair**, or **tissue samples** using similar chemical processes. The DNA extraction method must be optimized for each species to ensure that high-quality genetic material is available for analysis.

Molecular Techniques for Genotyping

After DNA extraction, the next step is to determine whether each individual carries the **favorable allele** associated with the trait of interest. This is achieved through various **molecular techniques**, each of which offers different levels of **accuracy, throughput**, and **cost-efficiency**. The most common methods include **PCR (Polymerase Chain Reaction), SNP arrays**, and **gel electrophoresis**, among others.

1. **Polymerase Chain Reaction (PCR)**
   - **PCR** is one of the most widely used techniques for **amplifying specific regions of DNA** that contain the **marker allele** of interest. This method uses **primers**, which are short DNA sequences complementary to the target region, to initiate DNA synthesis. Once amplified, the presence or absence of the **marker allele** can be determined.
   - **Process of PCR**: In the context of MAS, PCR is used to amplify the **region of the genome** where the desired marker is located. For example, if a breeder is selecting for **disease resistance** using an **SNP marker,** PCR primers specific to the **SNP region** would

be designed, and the DNA sample would be subjected to several cycles of **denaturation**, **annealing**, and **extension**. Each cycle results in the exponential amplification of the target region.

- **Genotype Detection**: After PCR amplification, the **amplified DNA** is analyzed to determine the presence of the **marker allele**. This can be done through **gel electrophoresis**, **SNP-specific probes**, or other detection systems. PCR is widely used in **low- to medium-throughput** breeding programs due to its flexibility and low cost.

2. **SNP Arrays**

   - **SNP arrays** are high-throughput genotyping platforms used to detect **single nucleotide polymorphisms (SNPs)** across the genome. These arrays contain thousands to millions of **probes** designed to detect specific **SNP variations** in the DNA sample. SNP arrays are particularly useful for **large-scale genotyping** efforts, where breeders need to analyze a vast number of individuals for multiple traits simultaneously.
   - **Process of SNP Array Genotyping**: DNA samples from the breeding population are hybridized to the SNP array, which contains **probes** complementary to the **SNP alleles** of interest. When the DNA binds to the probes, a **fluorescent signal** is generated, indicating whether the individual is **homozygous** or **heterozygous** for the SNP.
   - **Advantages of SNP Arrays**: SNP arrays are highly effective in **genotyping large populations** due to their ability to simultaneously screen for thousands of **SNP markers** across the genome. This makes them ideal for **genome-wide association studies (GWAS)**, **genomic selection**, and **marker-assisted selection** in large breeding programs.

3. **Gel Electrophoresis**

   - **Gel electrophoresis** is used to separate **DNA fragments** based on their size and charge, allowing breeders to visualize the **genotypic variation** at the marker locus. After PCR amplification, the DNA fragments are loaded into a **gel matrix** (typically agarose) and subjected to an **electric field**. **Shorter fragments** migrate faster through the gel than **longer fragments**, creating distinct **bands** that represent different alleles.
   - **Application in MAS**: Gel electrophoresis is commonly used in **SSR (Simple Sequence Repeat)** or **RFLP (Restriction Fragment Length Polymorphism)** genotyping, where variations in **DNA fragment lengths** are associated with the target trait. For example, a breeder using **SSR markers** to select for drought tolerance might observe distinct banding patterns corresponding to different alleles in the breeding population.

Determining Genotypes: Homozygous vs. Heterozygous

Once the genotyping results are obtained, breeders can determine whether each individual in the population is **homozygous** or **heterozygous** for the **marker allele**. This is crucial for making selection decisions, especially in **marker-assisted backcrossing (MABC)** or **introgression programs**, where breeders aim to **fix** the desired allele in a **homozygous state**.

» **Homozygous**: If an individual is **homozygous** for the desired marker allele, it means they have inherited **two identical copies** of the allele (one from each parent). In the case of

dominant or **recessive traits**, this is particularly important for ensuring the **consistent expression** of the trait in future generations.

» **Heterozygous:** If an individual is **heterozygous**, it means they carry **one copy of the desired allele** and **one copy of a different allele.** Heterozygous individuals can still express the trait, depending on the **inheritance pattern**, but breeders may need to cross these individuals further to produce **homozygous offspring**.

For example, if an SNP marker is used to track a **disease resistance gene**, the breeder can determine whether each individual in the population is **homozygous** (RR or rr) or **heterozygous** (Rr) for the **SNP allele** associated with the resistance gene. The presence of the **homozygous RR genotype** would indicate that the individual is fully resistant, while **heterozygous Rr** individuals might still express partial resistance, depending on the gene's mode of inheritance.

Advantages of Genotyping in MAS

1. **Early Selection**
   - **Genotyping** allows breeders to make **selection decisions early** in the breeding cycle, often at the **seedling stage**, before the trait is phenotypically expressed. This saves time and resources, especially for traits like **disease resistance**, which may only become apparent after exposure to pathogens.
2. **Precision**
   - By using **molecular markers**, genotyping enables more **precise selection** of individuals carrying the desired trait. This reduces the reliance on phenotypic evaluations, which can be influenced by environmental factors and may not fully reflect the individual's genetic potential.
3. **Genomic Selection**
   - Genotyping provides the foundation for **genomic selection (GS),** where breeders select individuals based on their **genomic breeding values** (GBVs) rather than just a single marker or trait. This is particularly valuable for improving **quantitative traits** like **yield** or **quality**, where many genes contribute to the phenotype.

3. Genotyping of Breeding Material

After the identification of molecular markers associated with the trait of interest, the next step in **marker-assisted selection (MAS)** is **genotyping the breeding population.** This step is crucial for determining which individuals within the population carry the **favorable allele** linked to the desired trait, allowing breeders to make informed decisions in advancing only the most genetically promising individuals.

DNA Extraction

The genotyping process begins with the extraction of **genomic DNA** from each individual in the breeding population. The source of the DNA can vary depending on the organism, but it is typically extracted from **leaf tissue** in plants or **blood samples** in animals. The quality of the extracted DNA is essential for successful genotyping, as degraded or contaminated DNA may lead to inaccurate results. The key steps involved in DNA extraction include:

1. **Tissue Collection**: In plants, leaves or other green tissue are harvested for DNA extraction, while in animals, blood, hair, or tissue samples may be used. The sample should be handled carefully to avoid degradation or contamination.
2. **Cell Lysis**: The harvested tissue is subjected to chemical treatment or mechanical disruption to break open the cell membranes, releasing the cellular contents, including DNA.
3. **DNA Purification**: After cell lysis, the DNA is separated from proteins, lipids, and other cellular debris. This is typically done using a series of chemical washes that selectively precipitate DNA. The purified DNA is then ready for use in downstream molecular techniques.

Once the DNA is extracted, it is quantified and checked for purity. High-quality DNA is essential for accurate and reproducible genotyping results.

Genotyping Techniques

Various molecular techniques are employed to determine the presence or absence of the **marker allele** in each individual. These techniques provide the foundation for selecting individuals with the desired genetic traits in the breeding population. The most commonly used genotyping methods include:

1. **Polymerase Chain Reaction (PCR)**
    - **PCR** is a widely used technique for amplifying specific regions of DNA associated with the marker allele. PCR involves using **primers** that are complementary to sequences flanking the target region of interest. Through multiple cycles of denaturation, annealing, and extension, the target region is amplified, making it detectable and analyzable.
    - **Application in MAS**: In marker-assisted selection, PCR is used to amplify the region containing the **marker allele**, which could be an **SNP**, **SSR**, or another marker type. After amplification, the DNA fragments are analyzed to determine whether the individual carries the **favorable allele**. For example, a breeder selecting for **disease resistance** might use PCR to amplify a segment of DNA containing an **SNP marker** linked to the resistance gene.
    - **Detection**: After amplification, the PCR products can be analyzed through **gel electrophoresis** or other detection methods to determine the **genotype** of each individual. The banding patterns or presence of fluorescent signals indicate whether an individual is **homozygous** (two copies of the favorable allele) or **heterozygous** (one copy of the favorable allele and one copy of the alternative allele).
2. **SNP Arrays**
    - **SNP arrays** are high-throughput genotyping platforms used to detect **single nucleotide polymorphisms (SNPs)** across the genome. These arrays contain thousands or even millions of probes designed to detect specific **SNPs** in the DNA sample. SNP arrays allow breeders to genotype large populations efficiently, making them ideal for **large-scale breeding programs** and **genomic selection**.

- **Application in MAS**: SNP arrays are widely used in **genomic selection** and **QTL mapping**, where multiple **SNP markers** associated with different traits are simultaneously screened. For example, in **wheat breeding**, a breeder may use an SNP array to screen for multiple traits, such as **disease resistance**, **yield potential**, and **stress tolerance**. Each SNP marker provides information about the presence or absence of the favorable alleles in the breeding material.
- **Detection:** The SNP array works by hybridizing the DNA sample to a chip containing **SNP-specific probes**. When the DNA binds to the probe, a **fluorescent signal** is generated, indicating the presence of the SNP allele. This technology enables breeders to identify whether an individual is **homozygous** or **heterozygous** for each SNP, providing valuable information for selection decisions.

3. **Gel Electrophoresis**
    - **Gel electrophoresis** is a traditional method used to separate **DNA fragments** based on their size and charge. After amplifying the DNA region of interest through PCR, the resulting DNA fragments are loaded into a gel matrix (usually agarose) and subjected to an **electric current. Shorter DNA fragments** migrate more quickly through the gel, while **longer fragments** move more slowly, resulting in a distinct **banding pattern**.
    - **Application in MAS**: Gel electrophoresis is often used in **SSR genotyping** or for analyzing PCR products in **RFLP assays**. For example, in a plant breeding program targeting **flowering time**, breeders may use **SSR markers** linked to early or late flowering. The banding pattern obtained through gel electrophoresis will reveal whether the individual carries the allele for early or late flowering.
    - **Detection**: The DNA bands are visualized under ultraviolet (UV) light after staining the gel with a **DNA-binding dye** such as **ethidium bromide**. The pattern of bands is compared between individuals to determine their **genotype** at the marker locus.

Example of SNP-Based Genotyping

Consider a breeder working on developing **disease-resistant** varieties of **rice**. In this case, an **SNP marker** associated with resistance to the **bacterial blight** pathogen (caused by *Xanthomonas oryzae*) has been identified. The breeder can use this SNP marker to genotype the breeding population and determine which individuals carry the resistance allele.

» **DNA Extraction**: The breeder begins by extracting **DNA** from all the rice plants in the breeding population.

» **SNP Genotyping**: The DNA is then genotyped using an **SNP array** or **PCR-based method**. If an SNP array is used, the DNA is hybridized to a chip containing probes for the **blight resistance SNP**.

» **Genotyping Results**: After genotyping, the breeder analyzes the data to determine which plants are **homozygous** for the resistance allele (RR), **heterozygous** (Rr), or **homozygous for susceptibility** (rr). Only those plants with the **RR** or **Rr** genotype are selected for further breeding.

Advantages of Genotyping in MAS

1. **Early Selection**
   - **Genotyping** allows breeders to make **early selections** before the trait of interest is fully expressed. For example, disease resistance or drought tolerance may only manifest later in the plant's life cycle, but genotyping allows breeders to select resistant individuals at the seedling stage, significantly shortening the breeding cycle.
2. **Increased Precision**
   - The use of **molecular markers** allows for more precise selection than relying solely on **phenotypic traits**, which can be influenced by environmental factors or measurement errors. By directly targeting the **genetic basis** of the trait, genotyping provides greater accuracy in selecting individuals that will express the desired trait.
3. **High-Throughput Capacity**
   - Techniques like **SNP arrays** offer **high-throughput genotyping**, enabling breeders to screen thousands of individuals for multiple traits simultaneously. This capability is particularly valuable in **large breeding programs** where efficiency and speed are critical for success.
4. **Improved Breeding Efficiency**
   - **Marker-assisted selection** combined with **genotyping** reduces the need for extensive field trials and repetitive phenotypic evaluations. By identifying the **genetic potential** of individuals early, breeders can focus their resources on the most promising plants or animals, accelerating the development of new varieties.

4. Selection of Individuals in Marker-Assisted Selection (MAS)

The **selection of individuals** based on **genotyping results** is one of the final and most critical steps in the **marker-assisted selection (MAS)** process. After identifying and genotyping the breeding population for the **desired marker alleles**, breeders can make **informed decisions** about which individuals should be advanced for further breeding. By focusing on the **genetic potential** of the breeding material, MAS allows for **early and precise selection**, improving the overall efficiency of breeding programs.

Selection Based on Genotyping Data

The **primary goal** of this step is to identify and select individuals that carry the **favorable marker allele** associated with the desired trait, such as **disease resistance, drought tolerance**, or **improved yield**. The individuals that do not carry the favorable allele are either discarded or **not advanced** in the breeding program. The advantage of using **genotyping data** is that it allows breeders to make **selection decisions** at an **early stage**, such as during the **seedling stage**, without waiting for the trait to be phenotypically expressed.

This step of **selecting individuals** is highly dependent on the **inheritance pattern** of the trait and the **breeding objective**. For example:

- **Homozygous Selection**: If the breeder's goal is to create a population where all individuals are **homozygous** for the desired trait, only individuals carrying **two copies of the favorable allele** will be selected. In the case of **recessive traits**, homozygous individuals (RR) are critical for ensuring that the trait will be expressed in future generations.
- **Heterozygous Selection**: In some cases, individuals that are **heterozygous** (Rr) for the marker allele may also be selected, particularly if the trait is **dominant** or if the breeding strategy involves **backcrossing** or **hybrid breeding**. These heterozygous individuals can then be crossed to further advance the desired trait.

  For example, in a **wheat breeding program** targeting **disease resistance**, individuals that are either **homozygous** (RR) or **heterozygous** (Rr) for an **SNP marker** linked to a **disease-resistance gene** can be selected. These individuals are then advanced to the next stage of the breeding cycle, while those lacking the marker (rr) are discarded.

Selection at the Early Stage

One of the major advantages of MAS is that **selection** can be done at an **early stage** in the breeding cycle, such as at the **seedling stage**. This ability to **select early** in the plant's or animal's life cycle is particularly beneficial because it **saves time** and **resources** compared to waiting for the trait to be expressed **phenotypically**.

- **Efficiency of Early Selection**: Many traits, such as **disease resistance** or **yield potential**, may not be observable until the plant reaches maturity or until the animal is exposed to certain environmental conditions. By using **genotyping** to select for the **favorable allele** early, breeders can avoid the time-consuming process of waiting for the full expression of these traits.
- **Reduction of Field Trials**: Early selection based on molecular markers also reduces the need for large-scale **field trials** or **phenotypic evaluations**, which are often labor-intensive and costly. Instead, breeders can focus their efforts on the most promising individuals, significantly **reducing costs** and **speeding up** the breeding process.

  For example, if a breeder is developing a **rice variety** that is resistant to **bacterial blight**, selecting individuals at the seedling stage that carry the **SNP marker** associated with resistance can prevent the need for inoculating large populations with the pathogen later in the breeding cycle. Only the individuals that are likely to express the **resistant phenotype** will be advanced.

Precision in Selection

The **precision** of selection in MAS is greatly improved compared to traditional breeding methods, which rely on **phenotypic observations** that can be influenced by **environmental factors**. By using **molecular markers** linked to the trait of interest, breeders can target the **genetic foundation** of the trait, leading to more accurate and consistent selection outcomes.

- **Avoiding Environmental Influence**: One of the challenges of traditional phenotypic selection is that environmental factors, such as **climate**, **soil conditions**, or **pathogen load**, can affect how a trait is expressed. For example, a plant may show **disease resistance** in one environment but be susceptible in another. MAS avoids this uncertainty by focusing on the **genetic markers** directly linked to the resistance gene, providing more reliable selection across different environments.

» **Improved Breeding Precision**: By focusing on the **genotype** rather than the **phenotype**, MAS allows breeders to **select more precisely** for **quantitative traits** that are controlled by multiple genes, such as **yield**, **quality**, or **stress tolerance**. This is particularly important in crops and animals where small genetic improvements can lead to significant gains in performance.

  For instance, in **maize breeding**, a breeder may use **SNP markers** linked to several **QTLs** associated with **drought tolerance**. By selecting individuals that carry the favorable alleles at these QTLs, the breeder can significantly improve the drought tolerance of the population without relying solely on field performance under variable conditions.

Marker-Assisted Backcrossing (MABC)

In many breeding programs, particularly when introducing a **specific trait** from a donor parent into an **elite line**, breeders use **marker-assisted backcrossing (MABC)**. This process involves **backcrossing** the selected individuals to the recurrent parent while using molecular markers to **track the introgressed allele**. MAS allows breeders to ensure that only individuals carrying the **favorable allele** from the donor parent are advanced, while the rest of the genome is progressively restored to the recurrent parent's genetic background.

» **Fixation of Desired Alleles**: MAS helps breeders **fix the desired allele** in the population through successive **backcross generations**. Individuals that carry the **target allele** but also contain the recurrent parent's genetic background are selected, ensuring that the final product expresses the desired trait while maintaining the elite line's overall genetic composition.

Example of Early Selection in Disease-Resistant Breeding

Consider a breeding program aimed at developing a **tomato variety** that is resistant to **Tomato Yellow Leaf Curl Virus (TYLCV)**. The breeder has identified an **SSR marker** linked to the **Ty-1 gene**, which confers resistance to the virus. After **genotyping** the breeding population, the breeder can **select early** for individuals that carry the **SSR marker** associated with the resistance allele.

» **Seedling Selection**: Instead of waiting for the plants to be exposed to the virus in the field (which would take several months), the breeder selects the **resistant individuals** at the seedling stage, based on their genotype. This approach saves time and resources while ensuring that only **resistant plants** are advanced in the breeding process.

» **Breeding Efficiency**: By selecting for the **SSR marker** early in the cycle, the breeder can quickly **eliminate susceptible plants** from the breeding population, focusing resources on developing high-yielding, virus-resistant varieties.

Challenges in Selection

While MAS offers significant advantages, there are challenges associated with the **selection of individuals** based on marker data:

1. **Recombination Events**: In some cases, **recombination** can occur between the marker and the **gene of interest**, leading to **false positives** or **false negatives**. This means that some individuals carrying the marker allele may not express the trait, or individuals without the marker may still possess the trait due to **recombination**.
2. **Incomplete Marker-Trait Association**: Not all markers are perfectly linked to the trait. In some cases, the marker may not account for the full **genetic variation** underlying the trait, requiring additional phenotypic validation to ensure that the selected individuals express the desired trait.

5. Breeding and Backcrossing in Marker-Assisted Selection (MAS)

Once individuals carrying the **desired marker allele(s)** have been identified through genotyping, the next critical step in the breeding process is to **cross** or **backcross** these selected individuals to **fix the trait** in the population. **Marker-assisted backcrossing (MABC)** is one of the most efficient methods used in modern plant and animal breeding programs to achieve this goal. It is particularly useful for **introgressing specific genes or quantitative trait loci (QTLs)** from a **donor parent** into the genetic background of an **elite recurrent parent** while minimizing the transfer of undesired genetic material from the donor parent.

Breeding and Backcrossing: Overview

Breeding involves the mating of selected individuals to create offspring that possess desirable traits. In **marker-assisted selection (MAS)**, the use of molecular markers allows breeders to target and track specific **genes** or **QTLs** associated with traits of interest, such as **disease resistance**, **stress tolerance**, or **yield improvement**.

In the context of **backcrossing**, the goal is to introduce a specific **gene or QTL** from a **donor parent** into the genetic background of an **elite line** (the **recurrent parent**) while gradually eliminating the unwanted genetic material from the donor. The breeding strategy involves multiple rounds of crossing between the **hybrid offspring** and the **recurrent parent** until the majority of the genome resembles that of the recurrent parent, with only the **target gene** or **QTL** from the donor parent retained.

Process of Marker-Assisted Backcrossing (MABC)

Marker-assisted backcrossing (MABC) allows breeders to use molecular markers to track the **desired genomic regions** and selectively breed individuals that carry the **favorable allele(s)**. The precision offered by **MABC** ensures that only the **targeted gene or QTL** is retained, reducing the **genetic drag** (the transfer of unwanted traits) often associated with traditional backcrossing.

The backcrossing process involves several steps:

1. **Initial Cross to Produce F1 Hybrid**
   - The first step is to create an **F1 hybrid** by crossing the **donor parent**, which carries the **target gene or QTL**, with the **recurrent parent**, which has desirable agronomic traits but lacks the specific trait from the donor parent.

- The resulting F1 individuals carry **50% of their genetic material** from each parent. They are typically heterozygous for the **target gene**, meaning they carry one copy of the allele from the donor and one copy from the recurrent parent.

2. **First Backcross (BC1 Generation)**
   - The **F1 hybrid** is backcrossed to the **recurrent parent** to produce the **BC1 generation.** This backcross introduces further recombination, but now the offspring carry approximately **75%** of their genome from the recurrent parent and **25%** from the donor parent.
   - **Marker-assisted selection** is used at this stage to ensure that only individuals carrying the **desired marker allele** from the donor parent are selected for further backcrossing. Individuals without the target allele are discarded or not advanced in the breeding program.

3. **Repeated Backcrossing (BC2, BC3, etc.)**
   - The process of backcrossing is repeated over multiple generations, with each successive generation (e.g., BC2, BC3) being crossed again with the recurrent parent. This progressively restores more of the **recurrent parent's genome** while retaining the **desired genomic region** from the donor parent.
   - By the time the **BC4** or **BC5** generation is reached, over **95% of the genome** will be derived from the recurrent parent, with only a small **introgressed region** from the donor parent remaining.
   - After each backcross generation, **genotyping** is performed to ensure that individuals carrying the **desired marker allele** are selected. This genotyping can be done using **molecular markers** such as **SNPs, SSRs,** or **RFLPs** that are linked to the target gene or QTL.

4. **Fixing the Desired Trait**
   - After the final backcross generation, individuals that carry the **desired allele** from the donor parent can be **selfed** (self-pollinated) or crossed with siblings to produce offspring that are **homozygous** for the target allele. This ensures that the trait will be consistently expressed in future generations.
   - The goal is to **fix** the desired gene or QTL in the population, meaning that the favorable allele becomes **homozygous** and is passed on to all offspring.

Advantages of Marker-Assisted Backcrossing (MABC)

MABC offers several advantages over traditional backcrossing methods, making it a preferred technique in many modern breeding programs:

1. **Precision in Selection**
   - The use of molecular markers allows breeders to **track the introgressed gene** or QTL throughout the backcrossing process with high precision. This reduces the risk of **genetic drag,** which occurs when unwanted donor traits are inadvertently transferred along with the target gene.

2. **Time Efficiency**

   - Traditional backcrossing often requires extensive **phenotypic evaluations** to ensure that the target trait is being retained. However, in **MABC**, selection can be made early and with greater accuracy using **molecular markers**, eliminating the need to wait for the trait to be expressed phenotypically.
   - This can significantly reduce the time needed to develop new varieties, especially for traits that are difficult or time-consuming to measure, such as **yield** or **stress tolerance**.

3. **Enhanced Breeding Efficiency**

   - MABC enhances breeding efficiency by enabling breeders to focus on **specific regions of the genome**. For example, if a breeder is attempting to introgress **disease resistance** from a wild relative into an elite cultivar, **MABC** ensures that only the **disease resistance gene** is transferred, while the elite background of the cultivar is maintained.
   - The ability to **fix the desired trait** quickly and effectively ensures that breeders can meet specific breeding objectives in a shorter timeframe.

Example of MABC: Introgressing Disease Resistance in Rice

A practical example of **marker-assisted backcrossing** can be seen in the development of **bacterial blight-resistant rice varieties**. In this case, a **donor parent** carrying a **resistance gene (Xa21)** for bacterial blight is crossed with an elite, high-yielding **recurrent parent** that lacks resistance.

» **Initial Cross**: The donor parent (resistant) is crossed with the recurrent parent (susceptible) to produce **F1 hybrids**. The F1 plants are heterozygous for the **Xa21 gene**.

» **First Backcross (BC1)**: The F1 hybrids are backcrossed with the recurrent parent to produce the **BC1 generation**. Individuals carrying the **Xa21 gene** are identified through **genotyping** and selected for further backcrossing.

» **Repeated Backcrossing**: The process is repeated for **several generations** (BC2, BC3, etc.), with each generation becoming more genetically similar to the recurrent parent but retaining the **Xa21 gene** from the donor parent.

» **Fixing the Trait**: After the final backcross generation, individuals are **selfed** to produce **homozygous** plants that consistently express **bacterial blight resistance**. The resulting plants are now **elite rice varieties** that combine **high yield** with **disease resistance**.

Challenges in Marker-Assisted Backcrossing

Despite its advantages, MABC does present some challenges:

1. **Recombination Events**

   - In some cases, recombination between the **marker** and the **gene of interest** can occur, leading to the loss of the desired allele even though the marker is present. Careful **marker validation** and the use of multiple markers flanking the gene can help reduce this risk.

2. **Limited by Available Markers**
    - The success of **MABC** relies on the availability of **closely linked markers** to the target gene or QTL. For some species or traits, the necessary markers may not be well-characterized, making MABC more challenging.
3. **Genomic Complexity**
    - For complex traits controlled by multiple genes, backcrossing becomes more difficult, as **multiple QTLs** need to be introgressed simultaneously. This requires a more sophisticated **marker system** and more extensive genotyping.

6. Validation of Trait Expression

Although **marker-assisted selection (MAS)** significantly improves the efficiency and precision of the breeding process, it is critical to **validate the expression** of the desired trait in the final product through **phenotypic evaluation.** While genotyping identifies individuals carrying the **favorable alleles** associated with the target traits, **genetic markers** may not always perfectly predict the expression of these traits in every environment. Thus, it is necessary to confirm that individuals selected based on **genetic markers** express the targeted trait under real-world conditions.

Phenotypic Evaluation for Trait Validation

The process of **validating trait expression** involves observing and measuring the **phenotypic characteristics** of selected individuals in either **field trials** or **controlled environments**. This step ensures that the **genotypic selection** through MAS corresponds to the **phenotypic performance** of the breeding material, especially for complex traits such as **yield, disease resistance,** or **stress tolerance.**

Steps in Trait Validation:

1. **Field Trials:**
    - Field trials involve planting and evaluating the selected individuals under **natural field conditions** to determine whether they express the desired trait. The performance of these individuals is assessed for **agronomic traits** such as **disease resistance, yield, flowering time, growth rate,** and **quality** parameters.
    - For example, in a **wheat breeding program** aimed at developing **rust-resistant varieties,** individuals selected based on MAS markers for **rust resistance genes** are planted in rust-prone environments to validate their resistance. The plants are monitored for their **infection levels, yield performance,** and other agronomic traits to ensure that the selected markers accurately predict rust resistance under field conditions.
2. **Controlled Environment Trials:**
    - Controlled environment trials involve growing the selected individuals in a **greenhouse, growth chamber,** or **laboratory setting** under specific and controlled conditions. This allows breeders to evaluate **traits** like **drought tolerance, disease resistance,** or **nutrient use efficiency** under conditions that simulate stress or disease pressures.
    - For instance, **disease resistance** may be validated by inoculating plants with the **pathogen**

in a controlled environment to observe their response. Similarly, **drought tolerance** can be assessed by subjecting plants to **water-limited conditions** in a growth chamber to evaluate their **physiological response** and **biomass accumulation**.

3. **Multi-Environment Testing:**
   - Some traits, such as **yield** and **stress tolerance**, are **quantitative** and influenced by **multiple genes** and **environmental interactions**. For these complex traits, **multi-environment testing** is essential to validate whether the selected individuals perform consistently across different **locations, climates**, and **growing seasons**.
   - In such trials, the performance of the breeding material is assessed under varying **soil conditions, temperature fluctuations**, and **water availability** to determine if the **trait expression** is stable and consistent. This step helps confirm that the **genetic potential** identified through MAS is fully realized in practical, real-world agricultural settings.

Importance of Validation in Marker-Assisted Selection

1. **Marker-Trait Linkage:**
   - While **molecular markers** are highly effective in tracking specific genes or **QTLs** linked to desirable traits, there can be cases where the **marker-trait linkage** is not perfect. For example, **recombination events** during meiosis can cause the marker to segregate away from the target gene, resulting in **false positives** or **false negatives** in the selection process.
   - Validating the **phenotypic expression** of the trait ensures that the individuals carrying the marker also express the desired trait. If discrepancies are observed, breeders can adjust their marker strategy or identify additional markers to improve selection accuracy.
2. **Environmental Influence:**
   - Many traits, especially **quantitative traits**, are influenced by **environmental factors** such as temperature, moisture, soil conditions, and pathogen load. Even if an individual carries the **favorable allele** for a specific trait, its expression may vary depending on the environment.
   - For example, **drought tolerance** in plants may depend not only on specific genes but also on the **interaction** between the genotype and the environment. **Field trials** in various environments help breeders determine how well the trait is expressed under different **growing conditions** and ensure the selection of **stable** and **robust** individuals.
3. **Trait Penetrance and Expression:**
   - In some cases, the **penetrance** (the proportion of individuals with a particular genotype that actually express the associated phenotype) of the trait may vary. **Incomplete penetrance** can occur due to complex gene interactions, **epistasis**, or **modifier genes**.
   - Validating trait expression helps breeders assess whether the **genotypic markers** used for selection reliably lead to **trait expression** in the breeding population. If **trait penetrance** is low, additional phenotypic selection or marker refinement may be necessary to achieve the desired results.

Examples of Trait Validation in Different Breeding Programs

1. **Disease Resistance in Soybean:**

    - In a soybean breeding program focused on **Phytophthora root rot resistance**, MAS may be used to select individuals carrying **QTLs** associated with resistance. However, to validate that these individuals are truly resistant, breeders conduct **field inoculation trials**, where the selected individuals are exposed to **Phytophthora** in a disease-prone environment. Plants are evaluated for **root rot severity, yield loss**, and overall performance to ensure that the marker-based selections align with field performance.

2. **Yield Improvement in Maize:**

    - Yield is a **complex quantitative trait** controlled by many genes. In a **maize breeding program** aiming to improve **grain yield**, MAS is used to select for **QTLs** associated with yield potential. However, validation through **multi-environment yield trials** across different regions and climatic conditions is critical to confirm that the selected individuals consistently deliver higher yields in the field.

3. **Drought Tolerance in Rice:**

    - **Drought tolerance** is another complex trait that is highly dependent on the **genotype-environment interaction.** In a rice breeding program, individuals selected for **drought tolerance QTLs** through MAS may be tested in **drought-prone** regions to assess their **physiological response** to water stress, **biomass production**, and **grain yield.** The performance of these individuals under water-limited conditions will confirm whether the drought-tolerant QTLs identified through markers lead to significant improvements in the field.

Advantages of Trait Validation

1. **Ensures Genetic and Phenotypic Correlation:**

    - Validating the **phenotypic expression** of the trait confirms the **correlation** between the **genotypic selection** and **trait expression** in real-world conditions. This helps breeders ensure that their **marker-assisted selections** are truly effective.

2. **Increases Confidence in Breeding Decisions:**

    - By confirming that the individuals selected based on markers actually express the target trait, breeders can advance those individuals with **greater confidence**, knowing that they will perform well under various conditions.

3. **Improves Breeding Accuracy:**

    - Validation helps identify **gaps** in the selection process, such as **recombination issues** or **environmental influences**, that may need to be addressed through additional **marker development** or **phenotypic selection**.

Challenges in Trait Validation

1. **Time and Resource Intensive:**
    - Conducting field trials and controlled environment testing can be **time-consuming** and **resource-intensive**, particularly for complex traits or those that require multi-environment testing. This adds an additional layer of effort to the MAS process.
2. **Trait Complexity:**
    - For **polygenic traits** or traits with **high genotype-environment interaction**, validation becomes more challenging, as it is difficult to predict how the trait will behave across different environments. This often requires extensive **field trials** and **multi-year testing**.

Types of Molecular Markers Used in MAS

Several types of **molecular markers** can be used in MAS, each with different advantages and applications. The choice of marker depends on the **trait of interest**, the available **resources**, and the **genomic knowledge** of the species being studied.

1. **Single Nucleotide Polymorphisms (SNPs)**
    - SNPs are **single-base pair variations** in the DNA sequence that are abundant throughout the genome. Due to their high density, SNPs are widely used in **high-throughput genotyping** and **association mapping**.
    - SNP markers are particularly valuable in **large-scale breeding programs** because they allow breeders to assess thousands to millions of genetic loci simultaneously using platforms like **SNP arrays** or **genotyping-by-sequencing (GBS)**.
2. **Simple Sequence Repeats (SSRs)**
    - SSRs, also known as **microsatellites**, are short, repetitive DNA sequences found throughout the genome. These markers are highly polymorphic, making them useful for **genetic diversity studies**, **population genetics**, and **trait mapping**.
    - SSRs are co-dominant markers, meaning they can distinguish between **homozygous** and **heterozygous** individuals, providing detailed genetic information for **marker-assisted selection**.
3. **Restriction Fragment Length Polymorphisms (RFLPs)**
    - RFLPs are one of the earliest types of molecular markers and are based on differences in the lengths of DNA fragments generated by restriction enzyme digestion. These markers are reliable and co-dominant, but they are labor-intensive compared to newer technologies like SNPs.
4. **Amplified Fragment Length Polymorphisms (AFLPs)**
    - AFLPs are generated by amplifying random DNA fragments using **PCR**. These markers are dominant and provide a large number of polymorphisms in a single assay, making them useful for **genetic mapping** and **QTL identification**.

Applications of Marker-Assisted Selection

MAS has broad applications in both **plant** and **animal breeding.** It is particularly useful for the **rapid introgression** of desirable traits into elite breeding lines and for improving the precision of **trait selection**.

1. **Disease Resistance**
   - MAS is commonly used to introduce **disease resistance genes** into commercial crop varieties. By using molecular markers linked to resistance genes, breeders can quickly screen for individuals that carry the resistance allele without having to expose the plants to the pathogen.
   - For example, in wheat breeding, markers linked to genes conferring resistance to **rust** and **powdery mildew** are used to develop resistant varieties.
2. **Abiotic Stress Tolerance**
   - MAS is also applied to develop crops that can tolerate **abiotic stresses** such as **drought, salinity, heat**, and **cold**. Markers linked to **QTLs** associated with stress tolerance are used to select individuals that can thrive under suboptimal environmental conditions.
   - In rice, markers linked to **drought tolerance genes** are used to develop varieties that perform better in regions prone to water scarcity.
3. **Yield Improvement**
   - Complex traits like **yield** are often governed by multiple genes, making MAS particularly useful in identifying the specific **QTLs** responsible for yield improvements. By selecting individuals with favorable alleles at these QTLs, breeders can develop high-yielding varieties more efficiently.
4. **Nutritional Quality**
   - MAS is also used to improve the **nutritional content** of crops. For example, markers linked to genes controlling **protein content**, **vitamin levels**, or **antioxidants** can be used to develop varieties with enhanced nutritional profiles.

Advantages of Marker-Assisted Selection

1. **Increased Efficiency**
   - MAS accelerates the breeding process by allowing breeders to select for desirable traits early in the breeding cycle, often at the **seedling stage**, before the trait is expressed phenotypically.
2. **Improved Precision**
   - By using molecular markers, breeders can **target specific genes** or genomic regions associated with the trait of interest. This precision reduces the chances of introgressing unwanted traits from the donor parent and minimizes **genetic drag**.

3. **Cost and Time Savings**
   - MAS reduces the number of field trials and phenotypic evaluations required to develop new varieties, saving both **time** and **resources**. This is particularly valuable for traits that are difficult to measure phenotypically or require several growing seasons to evaluate.
4. **Selection for Complex Traits**
   - MAS allows breeders to **dissect complex traits** controlled by multiple genes and select for favorable combinations of alleles at different loci. This makes it easier to improve traits like **yield**, **quality**, and **stress tolerance**, which are influenced by many genes.

Limitations of Marker-Assisted Selection

1. **Requires Prior Knowledge of Markers**
   - MAS depends on the availability of **well-characterized markers** that are closely linked to the trait of interest. For species with limited genomic resources, this can be a challenge.
2. **Marker-Trait Linkage**
   - The success of MAS relies on the strength of the **linkage** between the marker and the target gene. If recombination occurs between the marker and the gene, the marker may no longer accurately predict the presence of the desired allele.
3. **Cost of Genotyping**
   - Although the cost of genotyping has decreased significantly, high-throughput genotyping technologies like **SNP arrays** or **next-generation sequencing (NGS)** can still be expensive, especially for small-scale breeding programs.

C. Linkage Disequilibrium (LD) and Marker-Trait Association

Linkage disequilibrium (LD) is a fundamental concept in genetics, referring to the non-random association of alleles at different loci. In other words, LD occurs when certain combinations of alleles or genetic markers are found together more often than would be expected by chance. Understanding LD is crucial for identifying **marker-trait associations** because it allows researchers to determine whether specific genetic markers are linked to traits of interest, such as disease resistance, yield improvement, or drought tolerance in plants and animals.

The Concept of Linkage Disequilibrium (LD)

Linkage disequilibrium (LD) is a key concept in genetics that describes the non-random association of alleles at two or more loci. In a **randomly mating population**, genetic recombination during meiosis tends to shuffle alleles between loci that are not physically close to each other on a chromosome, leading to a state of equilibrium. However, when alleles at two loci are **inherited together more frequently** than would be expected by chance, the loci are said to be in **linkage disequilibrium**.

LD is essential in **genetic studies** because it helps researchers understand how genes and genetic markers are inherited together and how specific alleles influence traits, including diseases or agronomic features in crops and animals.

Factors Influencing Linkage Disequilibrium

The strength and persistence of LD between two loci are influenced by several genetic and population-based factors:

1. **Genetic Distance:**
    - **Genetic distance** between loci plays a fundamental role in LD. The closer two loci are on a chromosome, the less likely they are to be separated by **recombination** during meiosis. This physical proximity leads to strong LD between closely linked markers because the chances of recombination breaking up the association between the alleles at these loci are low.
    - For example, if two genes are 10 centimorgans apart, the likelihood of recombination occurring between them is lower than for genes that are 50 centimorgans apart. As a result, **tightly linked markers** on the same chromosome tend to display **strong LD,** whereas markers that are far apart on the chromosome (or on different chromosomes) are typically in **linkage equilibrium,** where recombination breaks up allele combinations more readily.
2. **Recombination Frequency**:
    - **Recombination** during meiosis reshuffles alleles between loci. **High recombination rates** lead to a faster breakdown of LD between loci over successive generations. In regions with **low recombination rates,** LD tends to persist for longer periods because recombination events are less frequent, maintaining the association between alleles.
    - For instance, regions near **centromeres** or **telomeres** often exhibit lower recombination frequencies, leading to extended LD in those areas. In contrast, regions of the genome with high recombination frequencies, such as those near **hotspots,** experience a rapid decay in LD.
3. **Selection Pressure**:
    - **Natural or artificial selection** can maintain LD between loci if certain combinations of alleles provide a selective advantage. For example, if a particular **allele combination** at two loci confers **disease resistance,** those alleles may be inherited together across generations due to **positive selection.**
    - In breeding programs, **marker-assisted selection (MAS)** can result in **artificial selection pressure,** where breeders select individuals with favorable alleles at specific loci. This selection pressure maintains LD between **markers** and **genes** associated with desirable traits, such as **yield** or **stress tolerance** in crops.
4. **Population Structure**:
    - LD patterns can also be influenced by the **structure of a population,** including **population size, migration, admixture,** and **genetic drift.** Smaller populations are more likely to exhibit higher levels of LD due to limited genetic diversity and fewer recombination events. In contrast, **large populations** with **random mating** and **frequent migration** tend to break down LD more quickly due to higher genetic diversity and recombination.

- **Founder effects** or **bottlenecks** in a population can also increase LD, as a smaller subset of alleles is inherited from a limited number of individuals. For example, in a crop breeding population derived from a small number of founders, LD may be higher compared to a population with a more diverse gene pool.

Measuring Linkage Disequilibrium

LD is typically quantified using two key statistical measures: **D'** and **r^2**. These measures provide insight into the degree of association between alleles at different loci.

1. **D' (D prime):**
   - **D'** measures the **normalized difference** between the observed frequency of an allele combination and the expected frequency under random association. A value of **D' = 1** indicates **complete linkage disequilibrium**, meaning that the alleles at the two loci are always inherited together. A **D'** value of **0** indicates that the loci are in **linkage equilibrium**, meaning that the alleles are inherited independently.
   - **D'** is particularly useful when assessing the historical relationship between loci, as it is less affected by the **allele frequencies** of individual markers.
2. **r^2**:
   - **r^2** measures the **correlation** between allele frequencies at two loci and is often used in **genome-wide association studies (GWAS)**. **r^2** quantifies the extent to which the genotype at one locus can predict the genotype at another locus. An **r^2** value of **1** indicates **complete correlation** between the loci, while an **r^2** value close to **0** suggests no correlation.
   - In **GWAS**, **r^2** is particularly useful for assessing **marker-trait associations** because it directly measures the predictability of an allele at one locus based on the allele at another locus. For example, an **r^2** value of **0.8** or higher is often used as a threshold for strong LD between loci.

Implications of Linkage Disequilibrium for Marker-Trait Association

The study of **linkage disequilibrium** is fundamental for identifying **marker-trait associations**, particularly in the context of **QTL mapping** and **GWAS**. Understanding LD allows researchers and breeders to use **molecular markers** to identify genetic regions associated with traits of interest and to track these markers across generations.

» **Marker-Assisted Selection (MAS)**: In MAS, LD between a **marker** and a **trait-associated gene** is exploited to select individuals with desirable traits based on their genetic markers. If a marker is in strong LD with a QTL for disease resistance, breeders can use the marker to identify resistant individuals early in the breeding cycle, even if the trait is not yet phenotypically expressed.

» **Genome-Wide Association Studies (GWAS)**: **GWAS** relies heavily on LD to identify **genomic regions** associated with traits. In a GWAS, researchers analyze thousands of markers across the genome to find associations between specific **SNPs** and traits such as yield, disease resistance, or height in plants and animals. LD enables the identification of **QTL regions** by tracking the inheritance of **linked markers**.

Case Studies of Linkage Disequilibrium and Marker-Trait Association

1. **Wheat Rust Resistance:**
   - In wheat, **leaf rust resistance genes** like **Lr34** have been identified through **QTL mapping** and GWAS. Markers linked to **Lr34** exhibit **high LD** with the gene, allowing breeders to use these markers to track rust resistance in breeding populations.
2. **Maize Drought Tolerance**:
   - **Maize GWAS studies** have revealed markers in **linkage disequilibrium** with **drought tolerance QTLs.** These markers are used in **MAS** to select for maize lines that perform better under water-limited conditions. Strong LD between these markers and the QTLs helps ensure accurate selection.
3. **Grape Berry Quality**:
   - In grape breeding, **SSR markers** linked to traits like **soluble solids content** and **acidity** are in **LD** with key **QTLs** controlling these traits. Breeders use these markers to improve **berry quality** in new grapevine varieties.

Marker-Trait Association

Marker-trait association refers to the relationship between a **genetic marker** and a specific **trait.** If a genetic marker is in **linkage disequilibrium** with a gene or quantitative trait locus (QTL) controlling a trait, the presence of the marker can be used to **predict the presence** of the trait in an individual. This forms the basis of **marker-assisted selection (MAS),** where breeders use molecular markers to identify individuals carrying desirable alleles linked to the trait of interest.

Marker-trait associations are commonly identified through approaches like **QTL mapping** or **GWAS.** These techniques rely on detecting **statistical associations** between the presence of specific markers and the expression of traits within a population.

Case Studies of Marker-Trait Association in LD

Case Study 1: Linkage Disequilibrium in Maize and Drought Tolerance

In maize (**Zea mays**), drought tolerance is a critical trait for improving crop yields, especially in regions prone to water scarcity. **GWAS** studies in maize have been used to identify markers linked to drought tolerance through LD analysis.

» **Population and Methodology**: Researchers studied a diverse population of maize inbred lines and hybrids using **SNP markers** spread across the genome. These markers were analyzed to identify regions of the genome in **linkage disequilibrium** with drought tolerance traits, such as **leaf senescence** (premature leaf aging) and **yield stability** under water-limited conditions.

» **Findings**: Through **GWAS,** several regions of the maize genome were identified that were in significant **linkage disequilibrium** with drought tolerance traits. In particular, **SNPs** on chromosomes 1 and 8 were strongly associated with genes involved in **water uptake,**

stomatal regulation, and **root development**. These SNPs were linked to **QTLs** that had been previously identified for drought tolerance in maize.

» **Application**: The markers identified through this study are now used in **marker-assisted selection (MAS)** programs to develop maize varieties with improved drought tolerance. The use of LD in identifying marker-trait associations allows breeders to target specific **QTL regions** associated with enhanced water-use efficiency and yield stability.

Case Study 2: Linkage Disequilibrium in Wheat for Disease Resistance

In **wheat (Triticum aestivum)** breeding, resistance to **leaf rust** (caused by *Puccinia triticina*) is an important trait due to the widespread impact of the disease on global wheat production. LD-based marker-trait associations have helped identify markers linked to **leaf rust resistance genes**.

» **Population and Methodology**: Researchers used a combination of **QTL mapping** and **GWAS** to study a population of **wheat breeding lines** segregating for leaf rust resistance. The study focused on identifying **SNP markers** in **linkage disequilibrium** with known **leaf rust resistance genes**, such as **Lr34** and **Lr46**.

» **Findings**: The analysis revealed several **SNP markers** on chromosome 7D that were in strong **linkage disequilibrium** with the **Lr34** gene, a well-known **adult plant resistance gene** that confers durable resistance to leaf rust. These SNPs were found to be tightly linked to the gene, providing a reliable marker for selecting **resistant individuals** in breeding programs.

» **Application**: Wheat breeders now routinely use these **SNP markers** in **marker-assisted selection** to incorporate the **Lr34** gene into new wheat varieties. The use of LD and marker-trait associations has accelerated the breeding of **rust-resistant wheat**, helping to mitigate the impact of this devastating disease.

Case Study 3: Linkage Disequilibrium in Grapes for Berry Quality

Grapevines (Vitis vinifera) are widely grown for fruit and wine production, and traits related to **berry quality**—such as **sugar content, acid levels**, and **color intensity**—are crucial for market value. Marker-trait associations through LD analysis have been instrumental in identifying **QTLs** related to these quality traits.

» **Population and Methodology**: Researchers conducted a **GWAS** on a population of grape varieties and breeding lines using **SSR** and **SNP markers**. LD analysis was performed to identify markers associated with **soluble solids content (SSC)**, which is a key determinant of grape sweetness and overall berry quality.

» **Findings**: The study found several **SNPs** in **linkage disequilibrium** with QTLs on chromosomes 2 and 18, which were strongly associated with high **soluble solids content**. These SNP markers were linked to genes involved in **sucrose metabolism** and **sugar transport**, critical for determining the sweetness of the grapes.

» **Application**: Breeders now use these **SNP markers** in their **selection programs** to develop grape varieties with improved **sugar content** and **berry quality**. The ability to predict these traits through marker-trait associations reduces the reliance on long, costly field trials.

Case Study 4: Linkage Disequilibrium in Soybean for Yield Improvement

Soybean (Glycine max) is a globally important crop, and **yield improvement** is a primary goal of soybean breeding programs. **Marker-trait associations** have been used to identify genetic regions linked to **high yield** through LD analysis.

» **Population and Methodology**: Researchers studied a population of elite soybean cultivars using a high-density **SNP array** to assess LD and identify **QTLs** linked to yield-related traits, such as **seed size, pod number**, and **biomass accumulation**.

» **Findings**: Several **QTL regions** on chromosomes 5, 6, and 20 were identified in **strong LD** with SNP markers linked to high **pod number** and **seed size.** These regions were associated with genes involved in **photosynthesis efficiency, hormone regulation**, and **seed development**.

» **Application**: These markers are now employed in **genomic selection** and **MAS** programs to breed soybeans with **increased yield potential**. By leveraging LD-based marker-trait associations, breeders can efficiently select for individuals with higher genetic potential for yield improvement.

Advantages of Using Linkage Disequilibrium in Marker-Trait Association

1. **Increased Mapping Resolution:**

   - **Linkage disequilibrium (LD)** improves the precision of **quantitative trait loci (QTL) mapping** and **genome-wide association studies (GWAS)** by identifying genomic regions that are strongly associated with a trait. This allows researchers to pinpoint **closely linked genetic regions** responsible for the trait. By focusing on genomic regions with strong LD, researchers can narrow down the number of candidate genes that may control a particular trait, thus improving **mapping resolution**. In species with complex genomes, such as wheat or maize, LD helps in locating traits more effectively within the vast number of genetic markers.

2. **Enhanced Marker-Assisted Selection (MAS):**

   - LD-based marker-trait associations are highly beneficial in **marker-assisted selection (MAS).** Breeders can use **molecular markers** in strong LD with a desirable trait to identify and select individuals carrying the **favorable allele** early in the **breeding cycle.** This is especially useful for traits that are **difficult to phenotype**, such as **disease resistance** or **stress tolerance.** With LD, breeders can perform **early selection**, reducing the reliance on extensive field trials or waiting for traits to express phenotypically. For example, in disease resistance breeding, markers in LD with resistance genes can allow early and accurate selection of resistant lines without needing pathogen inoculation in the field.

3. **Efficient Utilization of Resources:**

   - **LD-based approaches** enable efficient screening of large populations, reducing the need for expensive and time-consuming **phenotypic evaluations**. By using markers in LD with a trait of interest, breeders can quickly identify individuals with the desired trait, significantly cutting down on the **number of plants** or animals that need to be grown, tested, or maintained. This leads to more **cost-effective breeding programs,** especially for traits that are labor-intensive to measure or that only manifest under certain environmental conditions, such as **drought tolerance** or **yield** under stress.

Challenges in LD-Based Marker-Trait Association

1. **Breakdown of Linkage Disequilibrium Over Time:**
    - As generations pass, **recombination events** occur during meiosis, gradually breaking down **linkage disequilibrium** between genetic markers and the target trait. This can lead to a **loss of association** between the marker and the trait, especially in species with **high recombination rates** or in populations with **long breeding histories**. In crops such as **maize** or **rice**, the effectiveness of markers linked to traits may diminish over successive breeding cycles, requiring **constant validation** and updates to the marker set.
2. **Population Structure Effects**:
    - **Population structure** and **admixture** can introduce **false positives** in **marker-trait associations**. When different subpopulations within a species have varying allele frequencies, the **non-random association** between alleles can lead to the appearance of false **marker-trait linkages**. This issue is particularly problematic in species with **high population stratification**, such as cattle or maize. To address this, researchers must account for **population structure** in their analyses using methods like **principal component analysis (PCA)** or **mixed linear models (MLM)** to correct for stratification.
3. **Environmental Influences**:
    - **Environmental factors** can significantly affect the expression of **quantitative traits**, making it harder to establish strong **marker-trait associations**. For instance, traits like **yield, growth rate**, and **stress tolerance** can vary dramatically depending on **climate, soil type**, or **water availability**. This complicates the interpretation of LD-based studies, as the **phenotypic expression** of a trait may not always align with the **genotypic marker data**. To ensure reliable marker-trait associations, breeders must conduct **multi-environment validation** to confirm that the identified markers consistently correlate with the trait across different environmental conditions.

D. Marker-Assisted Pre-Breeding Programs

Marker-assisted pre-breeding (MAB) refers to the use of molecular markers to facilitate the introduction of desirable traits from **wild relatives** or **unimproved genetic material** into elite breeding lines. Pre-breeding is a critical step in plant and animal breeding programs because it broadens the **genetic base** by incorporating **valuable genes** for traits such as **disease resistance, stress tolerance, yield improvement**, and **nutritional quality** into breeding populations. This process provides breeders with **genetic resources** that can be used in **advanced breeding programs** to develop improved varieties.

The use of **molecular markers** in pre-breeding programs allows for the **precise transfer** of **target genes** and **quantitative trait loci (QTLs)** from exotic or wild sources into elite germplasm, improving the efficiency and speed of traditional breeding methods.

Importance of Marker-Assisted Pre-Breeding

Pre-breeding is essential for enhancing the **genetic diversity** available to breeders. In many crops and livestock species, modern breeding has led to a **narrowing of the genetic base** due to intense selection for a few desirable traits. This genetic uniformity increases vulnerability to **pests, diseases,** and **environmental stresses**, which can have devastating consequences for food production.

Marker-assisted pre-breeding programs aim to address this issue by incorporating genes from **wild relatives** or **landraces** that have evolved in diverse environments and often carry beneficial alleles for traits like **drought tolerance, pest resistance**, or **heat tolerance**.

Process of Marker-Assisted Pre-Breeding

Marker-assisted pre-breeding is a critical process used in breeding programs to introduce valuable traits from wild relatives, landraces, or unimproved genetic resources into elite breeding lines. The process integrates molecular marker technologies to efficiently track the introgression of desired genes or quantitative trait loci (QTLs), thereby minimizing unwanted genetic drag and speeding up the breeding cycle. The goal of pre-breeding is to broaden the genetic base of crops and livestock, ensuring better resilience, quality, and productivity in commercial varieties. Below is a detailed breakdown of the steps involved in marker-assisted pre-breeding.

1. Identification of Donor Material

The first and most crucial step in marker-assisted pre-breeding is the identification of suitable **donor material** that carries the genes or QTLs for the trait of interest. This step typically involves searching for wild relatives, landraces, or unimproved varieties that possess the genetic diversity needed to improve target traits such as **disease resistance, abiotic stress tolerance**, or **nutritional enhancement**.

Sources of Donor Material:

» **Wild relatives**: These are non-domesticated species closely related to the crop species and are often more genetically diverse. They may carry beneficial alleles for traits like **pest resistance** or **drought tolerance**, which are absent or lost in modern varieties.

» **Landraces**: Traditional varieties developed over centuries by farmers are often adapted to local environmental conditions and may have traits like **heat tolerance** or **salinity resistance**.

» **Genebanks and Germplasm Collections**: These resources store a wide range of genetic material from different varieties and wild species, offering breeders access to a rich genetic pool for introducing new traits.

Screening for Target Traits:

Once the potential donor material is identified, the next step is to **screen** the candidates for the presence of target genes or QTLs using **molecular markers**. This screening ensures that the chosen material carries the **desired alleles** for traits such as:

» **Disease resistance** (e.g., resistance to **rusts** or **blights** in cereals)

» **Abiotic stress tolerance** (e.g., **drought, salinity, cold**, or **heat tolerance**)

» **Nutritional quality** (e.g., increased **iron** or **zinc** content for **biofortification**)

The use of **SNP markers, SSR markers**, or **AFLP markers** enables precise identification of **genetic regions** associated with these traits.

2. Crossing with Elite Germplasm

After identifying the donor material with the desired traits, the next step is to cross this material with an **elite breeding line** (also known as the recurrent parent). The elite parent is usually a high-yielding, well-performing line that is well-adapted to commercial production but lacks the specific trait carried by the donor material.

Creation of F1 Hybrids:

» The cross between the **donor parent** (carrying the desired trait) and the **recurrent parent** (elite germplasm) results in **F1 hybrids**. These hybrids combine the genetic material from both parents.

» However, **F1 individuals** often carry both **desired traits** (e.g., disease resistance from the donor) and **undesirable traits** from the wild donor material (e.g., poor yield or late flowering), requiring further selection.

3. Marker-Assisted Backcrossing (MABC)

The goal of **marker-assisted backcrossing** (MABC) is to selectively retain the **desired trait** from the donor parent while restoring the **genetic background** of the elite recurrent parent. This process is critical for reducing **genetic drag**—the transfer of undesirable traits from the donor parent—while maintaining the elite traits of the recurrent parent.

Steps in Marker-Assisted Backcrossing:

» **First Backcross (BC1)**: The **F1 hybrid** is crossed with the **recurrent parent** to create the **BC1 generation**, which contains about 75% of its genetic material from the recurrent parent and 25% from the donor. **Molecular markers** are used to identify BC1 individuals that carry the desired trait from the donor parent, such as **disease resistance**.

» **Subsequent Backcrosses (BC2, BC3, etc.)**: Selected BC1 individuals are backcrossed again to the recurrent parent, producing the **BC2 generation**, which has about 87.5% of its genome from the recurrent parent. The process of **backcrossing** and **selection** is repeated for several generations (usually BC3 to BC5), with each generation further restoring the genetic background of the elite recurrent parent while retaining the desired trait.

Marker-Assisted Selection (MAS) in MABC:

» Molecular markers tightly linked to the **target gene** or QTL are used at each backcross generation to ensure the desired trait is retained.

» **Marker-assisted selection (MAS)** reduces the number of plants that need to be screened, as breeders can focus on selecting individuals with the target alleles at each generation.

By the **final backcross generation** (BC5 or BC6), the genome is almost entirely derived from the recurrent parent (approximately 98% or more), except for the **introgressed region** carrying the desired gene or QTL from the donor parent.

4. Validation and Selection of Pre-Breeding Lines

After completing the backcrossing process, the next step is to create **homozygous lines** that carry the **desired trait**. This involves **selfing** (self-pollination) or **sibling mating** between selected BC individuals to produce lines that are homozygous for the introgressed region.

Steps in Validation:

» **Phenotypic Evaluation**: Pre-breeding lines are subjected to **field trials** or **greenhouse evaluations** to validate that the introgressed trait is **expressed phenotypically**. For example, lines with disease resistance are tested under pathogen pressure, while drought-tolerant lines are grown in water-limited environments.

» **Multi-Environment Testing**: The pre-breeding lines are tested across different **geographical locations** and **environmental conditions** to ensure **trait stability** and to rule out any unwanted effects from the donor genome, such as poor yield or late flowering.

By evaluating the pre-breeding lines in **real-world conditions**, breeders ensure that the introgressed traits not only function as expected but also that the agronomic qualities of the elite recurrent parent are maintained.

5. Development of Improved Breeding Lines

Once pre-breeding lines are validated, they become valuable resources for **advanced breeding programs** aimed at producing new commercial varieties. These lines serve as a **genetic bridge** between the wild donor material and elite breeding lines, facilitating the **transfer of new traits** into commercial varieties.

Steps in Developing Improved Breeding Lines:

» **Crossing with Elite Lines**: Pre-breeding lines are crossed with other elite lines to produce new hybrid varieties that carry the **desired traits** while maintaining high agronomic performance.

» **Hybrid Breeding Programs**: Pre-breeding lines may also be used in **hybrid breeding programs**, where they serve as parents to generate hybrid offspring with enhanced resistance, stress tolerance, or yield traits.

Applications:

» Pre-breeding lines developed through marker-assisted pre-breeding have been used to improve crops for traits such as **disease resistance** (e.g., **rust resistance** in wheat), **drought tolerance** (e.g., in maize), and **nutritional quality** (e.g., **biofortified rice** with increased iron and zinc content).

Applications of Marker-Assisted Pre-Breeding

1. **Disease Resistance in Wheat:**

    - **Rust diseases** (such as **stem rust** and **leaf rust**) are major threats to global wheat production. **Marker-assisted pre-breeding** has been instrumental in introducing rust resistance genes from wild relatives of wheat, such as **Aegilops tauschii**. Molecular markers linked to rust resistance genes like **Sr2, Lr34,** and **Yr18** have been used to introgress these genes into elite wheat varieties through **backcrossing.**

- The resulting pre-breeding lines carry strong rust resistance without compromising yield potential, making them valuable genetic resources for developing new rust-resistant wheat varieties.

2. **Drought Tolerance in Maize:**
   - **Wild relatives of maize,** such as **teosinte,** possess alleles for **drought tolerance** that are absent in modern maize varieties. Marker-assisted pre-breeding has facilitated the transfer of **drought tolerance QTLs** from teosinte into commercial maize lines.
   - Using SNP markers, researchers have identified QTLs associated with root arc**hitecture, stomatal regulation,** and **water-use efficiency,** which are crucial for improving drought tolerance in maize. These pre-breeding lines have been integrated into breeding programs to produce drought-tolerant maize hybrids.

3. **Biofortification in Rice:**
   - Pre-breeding programs aimed at **biofortifying** rice have focused on increasing the content of **micronutrients** like **iron** and **zinc.** Wild rice species and landraces with high micronutrient content have been identified as donor parents for these traits.
   - **Marker-assisted selection (MAS)** has been used to transfer QTLs linked to **iron** and **zinc accumulation** into high-yielding rice varieties. These pre-breeding lines are now being used in breeding programs to develop **nutrient-rich rice varieties** that help address micronutrient deficiencies in developing countries.

4. **Salt Tolerance in Tomatoes:**
   - Wild relatives of tomato, such as **Solanum pimpinellifolium,** carry **salt tolerance genes** that enable them to survive in saline soils. Pre-breeding programs have used molecular markers to transfer these genes into commercial tomato lines.
   - Through **marker-assisted backcrossing,** QTLs associated with **salt tolerance** have been introgressed into elite tomato lines. The resulting pre-breeding lines can thrive in saline environments, providing a valuable resource for developing tomatoes suited for cultivation in areas affected by soil salinity.

Advantages of Marker-Assisted Pre-Breeding Programs

1. **Enhanced Precision:**
   - The use of molecular markers in pre-breeding enables precise tracking of the **introgressed traits,** reducing the **genetic drag** from wild or unimproved donor parents. This increases the efficiency of transferring only the desired gene(s) while maintaining the agronomic qualities of the elite parent.

2. **Accelerated Breeding Process**:
   - Marker-assisted pre-breeding accelerates the process of trait introgression by eliminating the need for extensive **phenotypic selection** in the early generations. This saves time, allowing breeders to move quickly from initial crosses to the development of elite pre-breeding lines.
3. **Broadens Genetic Diversity**:
   - Pre-breeding programs increase the **genetic diversity** of breeding populations by incorporating valuable alleles from **wild relatives** or **exotic germplasm**. This diversity is critical for **improving resilience** to environmental stresses and developing crops that can thrive in diverse conditions.
4. **Efficient Use of Resources**:
   - By using molecular markers, breeders can focus their resources on the individuals most likely to carry the desired trait, reducing the need for large-scale **phenotypic evaluations** and increasing the overall efficiency of the breeding process.

Challenges in Marker-Assisted Pre-Breeding

1. **Limited Availability of Markers:**
   - For some species, particularly those without well-developed genomic resources, the availability of molecular markers linked to important traits may be limited. This can slow down the pre-breeding process and make trait introgression less efficient.
2. **Undesirable Traits from Wild Relatives**:
   - While wild relatives often carry useful traits, they may also introduce **undesirable agronomic traits**, such as poor yield or delayed flowering. It is challenging to balance the retention of desirable traits while eliminating the negative effects.
3. **Time and Cost**:
   - Although marker-assisted pre-breeding accelerates certain aspects of the breeding process, it still requires significant investment in **genomic resources, molecular marker development**, and **laboratory infrastructure** for genotyping and marker validation.

References

Acquaah, G. (2012). *Principles of Plant Genetics and Breeding* (2nd ed.). Wiley-Blackwell.

Adams, R. P., & Adams, J. E. (1992). *Conservation of Plant Genes: DNA Banking and In Vitro Technology*. Academic Press.

Allard, R. W. (1960). *Principles of Plant Breeding*. John Wiley & Sons.

Aravind, J., Kaur, V., Wankhede, D. P., & Nanjundan, J. (2020). EvaluateCore: Quality Evaluation of Core Collections. R package version 0.1.1. Available at: https://aravind-j.github.io/EvaluateCore/

Bahl, P. N., Salimath, P. M., & Patnaik, S. K. (1997). Genetics and breeding of pulse crops. *Plant Breeding Reviews*, 15, 181–227.

Berg, T. (2009). The importance of managing crop genetic resources. *Plant Breeding Reviews*, 25, 121–149.

Bhullar, N. K., Street, K., Mackay, M., Yahiaoui, N., & Keller, B. (2009). Unlocking wheat genetic resources for the molecular identification of previously undescribed functional alleles at the *Pm3* resistance locus. *Proceedings of the National Academy of Sciences of the United States of America, 106*(23), 9519–9524. https://doi.org/10.1073/pnas.0904156106

Bioversity International. (2007). Developing crop descriptor lists: Guidelines for developers. *Bioversity Technical Bulletin*. Bioversity International, Rome, Italy, pp. 71.

Brown, A. H. D. (1989). The case for core collections. In A. H. D. Brown, O. H. Frankel, D. R. Marshall, & J. T. Williams (Eds.), *The Use of Plant Genetic Resources* (pp. 136–165). Cambridge University Press, Cambridge.

Brown, A. H. D., Frankel, O. H., Marshall, D. R., & Williams, J. T. (1989). *The Use of Plant Genetic Resources*. Cambridge University Press.

Collard, B. C. Y., & Mackill, D. J. (2008). Marker-assisted selection: An approach for precision plant breeding in the twenty-first century. *Philosophical Transactions of the Royal Society B*, 363(1491), 557–572.

Duvick, D. N. (2001). Biotechnology in the 21st century: Impacts on plant breeding. *Crop Science*, 41, 641–651.

Dwivedi, S. L., Perotti, E., & Ortiz, R. (2008). Towards molecular breeding of reproductive traits in cereal crops. *Plant Biotechnology Journal*, 6, 529–559.

El-Bouhssini, M., Street, K., Amri, A., Mackay, M., Ogbonnaya, F. C., Omran, A., Abdalla, O., Baum, M., Dabbous, A., & Rihawi, F. (2010). Sources of resistance in bread wheat to Russian wheat aphid (*Diuraphis noxia*) in Syria identified using the focused identification of germplasm strategy (FIGS). *Plant Breeding, 130*(1), 96–97. https://doi.org/10.1111/j.1439-0523.2009.01759.x

Endresen, D. T. F., Street, K., Mackay, M., Bari, A., & De Pauw, E. (2011). Predictive association between biotic stress traits and eco-geographic data for wheat and barley landraces. *Crop Science, 51*(5), 2036–2055. https://doi.org/10.2135/cropsci2010.11.0656

FAO. (2011). *Second Global Plan of Action for Plant Genetic Resources for Food and Agriculture*. Rome, Italy.

Federer, T. W. (1956). Augmented Designs. *The Hawaiian Planters' Record, 4*, 191–208.

Forster, B. P., & Thomas, W. T. B. (2005). Doubled haploids in crop improvement. *Plant Breeding Reviews*, 25, 57–88.

Frankel, O. H. (1984). Genetic perspectives of germplasm conservation. In W. Arber, K. Limensee, W. J. Peacock, & P. Starlinger (Eds.), *Genetic Manipulation: Impact on Man and Society* (pp. 161–170). Cambridge University Press, Cambridge.

Frankel, O. H., & Brown, A. H. D. (1984). Plant genetic resources: An urgent need for conservation. *Nature*, 312(5981), 197–198.

Gepts, P. (2006). Plant genetic resources conservation and utilization. *Crop Science*, 46(5), 2278–2292.

Gouesnard, B., Bataillon, T. M., Decoux, G., Rozale, C., Schoen, D. J., & David, J. L. (2001). MSTRAT: An algorithm for building germplasm core collections by maximizing allelic or phenotypic richness. *Journal of Heredity, 92*(2), 93–94. https://doi.org/10.1093/jhered/92.2.93

Gupta, P. K. (2006). *Biotechnology and Genomics*. Rastogi Publications.

Hallauer, A. R., Carena, M. J., & Miranda, J. B. (2010). *Quantitative Genetics in Maize Breeding* (2nd ed.). Springer.

Hoisington, D., Khairallah, M., & Reeves, T. (1999). Plant genetic resources: What can they contribute toward increased crop productivity? *Proceedings of the National Academy of Sciences USA*, 96, 5937–5943.

Jain, H. K., & Kharkwal, M. C. (2004). *Plant Breeding: Mendelian to Molecular Approaches*. Narosa Publishing House.

Jansen, J., & van Hintum, T. (2007). Genetic distance sampling: A novel sampling method for obtaining core collections using genetic distances with an application to cultivated lettuce. *Theoretical and Applied Genetics, 114*(3), 421–428. https://doi.org/10.1007/s00122-006-0436-4

Jeong, S., Kim, J. Y., Jeong, S. C., Kang, S. T., Moon, J. K., & Kim, N. (2017). Geno Core: A simple and fast algorithm for core subset selection from large genotype datasets. *PLoS ONE, 12*(7), e0181420. https://doi.org/10.1371/journal.pone.0181420

Khazaei, H., Street, K., Bari, A., Mackay, M., & Stoddard, F. L. (2013). The FIGS (Focused Identification of Germplasm Strategy) approach identifies traits related to drought adaptation in *Vicia faba* genetic resources. *PLoS ONE, 8*(5), e63107. https://doi.org/10.1371/journal.pone.0063107

Khush, G. S. (2001). Green revolution: The way forward. *Nature Reviews Genetics*, 2(10), 815–822.

Kim, K. W., Chung, H. K., Cho, G. T., Ma, K. H., Chandrabalan, D., Gwag, J. G., Kim, T. S., Cho, E. G., & Park, Y. J. (2007). Power Core: A program applying the advanced M strategy with a heuristic search for establishing core sets. *Bioinformatics, 23*(16), 2155–2162. https://doi.org/10.1093/bioinformatics/btm313

Klug, W. S., & Cummings, M. R. (2003). *Concepts of Genetics* (7th ed.). Pearson Education.

Krishnan, R. R., Sumathy, R., Ramesh, S., Bindroo, B., & Naik, G. V. (2014). SimEli: Similarity elimination method for sampling distant entries in development of core collections. *Crop Science, 54*(3), 1070–1078. https://doi.org/10.2135/cropsci2013.07.0467

Lande, R., & Thompson, R. (1990). Efficiency of marker-assisted selection in the improvement of quantitative traits. *Genetics*, 124(3), 743–756.

Lewin, B. (2004). *Genes VIII.* Oxford University Press.

Mackay, M. C. (1995). One core collection or many? In T. Hodgkin, A. H. D. Brown, T. J. L. van Hintum, & A. A. V. Morales (Eds.), *Core Collections of Plant Genetic Resources* (pp. 199–210). Wiley, Chichester.

Mackay, M. C., & Street, K. (2004). Focused identification of germplasm strategy—FIGS. In C. K. Black, J. F. Panozzo, & G. J. Rebetzke (Eds.), *Proceedings of the 54th Australian Cereal Chemistry Conference and the 11th Wheat Breeders' Assembly* (pp. 138–141). Royal Australian Chemical Institute, Melbourne.

Mahajan, R. K., Sapra, R. L., Srivastava, U., Singh, M., & Sharma, G. D. (2000). *Minimal Descriptors for Characterization and Evaluation of Agri-horticultural Crops (Part I).* National Bureau of Plant Genetic Resources, New Delhi, pp. 230.

Marschner, H. (1995). *Mineral Nutrition of Higher Plants* (2nd ed.). Academic Press.

McCouch, S. R., Chen, X., & Panaud, O. (1997). Microsatellite marker development, mapping, and applications in rice genetics and breeding. *Plant Molecular Biology*, 35, 89–99.

Mohan, M., Nair, S., & Bhagwat, A. (1997). Genome mapping, molecular markers, and marker-assisted selection in crop plants. *Molecular Breeding*, 3(2), 87–103.

Mondal, T., & Peter, S. (2020). Evolutionary principles in plant breeding. *Genetics Journal*, 45(3), 401–415.

Odong, T. L., Jansen, J., van Eeuwijk, F. A., & van Hintum, T. J. L. (2013). Quality of core collections for effective utilization of genetic resources: Review, discussion, and interpretation. *Theoretical and Applied Genetics, 126*(2), 289–305. https://doi.org/10.1007/s00122-012-1971-y

Odong, T. L., van Heerwaarden, J., Jansen, J., van Hintum, T. J. L., & van Eeuwijk, F. A. (2011). Statistical techniques for defining reference sets of accessions and microsatellite markers. *Crop Science, 51*(6), 2401–2411. https://doi.org/10.2135/cropsci2011.02.0117

Parry, M. A. J., Madgwick, P. J., & Bayon, C. (2012). Improving resource use efficiency in wheat through genetic improvement. *Plant Breeding Reviews*, 36, 1–47.

Paterson, A. H., Bowers, J. E., & Bruggmann, R. (2009). The Sorghum bicolor genome and the diversification of grasses. *Nature*, 457, 551–556.

Poehlman, J. M., & Borthakur, D. (1995). *Breeding Asian Field Crops*. Oxford & IBH Publishing Co.

Rai, R. (2007). Genetics and Plant Breeding. In *Introduction to Plant Biotechnology*. National Research Centre on Plant Biotechnology.

Rani, S. J., & Usha, R. (2013). Transgenic plants: Types, benefits, public concerns, and future. *Journal of Pharmacy Research*, 6(8), 879–883.

Reddy, B. V. S., Ashok Kumar, A., & Ramesh, S. (2010). Genetic enhancement for drought tolerance in sorghum. *Plant Breeding Reviews*, 33, 297–309.

Rehman, S., & Abbas, M. (2012). Role of biotechnology in developing stress-tolerant crops. *Plant Breeding Reviews*, 18, 211–240.

Reif, J. C., Melchinger, A. E., & Frisch, M. (2005). Genomic selection in plant breeding: Methods and applications. *Trends in Plant Science*, 10(9), 421–428.

Sharma, J. R. (1994). *Principles and Practice of Plant Breeding*. Tata McGraw Hill Publishing.

Singh, B. D. (2006). *Plant Breeding: Principles and Methods*. Kalyani Publishers.

Singh, K., Gupta, K., Tyagi, V., & Rajkumar, S. (2020). Plant genetic resources in India: Management and utilization. *Vavilov Journal of Genetics and Breeding, 24*(3), 306–314. https://doi.org/10.18699/VJ20.661

Singh, R. P., Hodson, D. P., & Jin, Y. (2006). Current status, likely migration, and strategies to mitigate the threat to wheat production. *Euphytica*, 148(3), 271–284.

Singh, S. R., & Banik, B. C. (2011). Advanced methods in plant breeding. *Asian Journal of Horticulture*, 6, 151–154.

Smale, M., & Drucker, A. G. (2007). Managing crop diversity on farmers' fields. *Plant Genetic Resources: Characterization and Utilization*, 5(1), 3–15.

Snustad, D. P., & Simmons, M. J. (2006). *Principles of Genetics* (4th ed.). John Wiley & Sons.

Tanksley, S. D., & McCouch, S. R. (1997). Seed banks and molecular maps: Unlocking genetic potential from the wild. *Science*, 277(5329), 1063–1066.

Tester, M., & Langridge, P. (2010). Breeding technologies to increase crop production in a changing world. *Science*, 327(5967), 818–822.

Thachuk, C., Crossa, J., Franco, J., Dreisigacker, S., Warburton, M., & Davenport, G. F. (2009). Core Hunter: An algorithm for sampling genetic resources based on multiple genetic measures. *BMC Bioinformatics, 6*(1), 243. https://doi.org/10.1186/1471-2105-10-243

Tiwari, K. N. (2007). *Soil Fertility and Nutrient Management*. Indian Journal of Fertilizer.

Upadhayaya, H. D., & Peterson, A. H. (2022). Mini core germplasm collections for developing climate-resilient crop cultivars and achieving enhanced genetic gains in plant breeding programs. *Indian Journal of Plant Genetic Resources, 35*(1), 69–73.

Upadhyaya, H. D., Reddy, K. N., & Gowda, C. L. L. (2009). Managing and enhancing the use of germplasm. *Plant Genetic Resources: Characterization and Utilization*, 7(2), 95–105.

van Hintum, T. J. L., Brown, A. H. D., Spillane, C., & Hodgkin, T. (2000). Core collections of plant genetic resources. *Plant Breeding Reviews*, 19, 1–20.

Vavilov, N. I. (1992). Origin and geography of cultivated plants. *Cambridge University Press.*

Xu, Y., & Crouch, J. H. (2008). Marker-assisted selection in plant breeding: From theory to practice. *Crop Science*, 48, 391–407.

Zohary, D., & Hopf, M. (2000). *Domestication of Plants in the Old World* (3rd ed.). Oxford University Press.